Advances in Silicon Solar Cells

Shadia Ikhmayies
Editor

# Advances in Silicon Solar Cells

Springer

*Editor*
Shadia Ikhmayies
Department of Physics
Al Isra University
Amman, Jordan

ISBN 978-3-319-69702-4    ISBN 978-3-319-69703-1 (eBook)
https://doi.org/10.1007/978-3-319-69703-1

Library of Congress Control Number: 2017963208



Printed on acid-free paper

This Springer imprint is published by Springer Nature
The registered company is Springer International Publishing AG
The registered company address is: Gewerbestrasse 11, 6330 Cham, Switzerland

# Contents

**Effective Light Management in Thin Silicon Wafers** . . . . . . . . . . . . . . . 1
Zhi-Peng Ling

**Optoelectronic Characteristics of Passivated and Non-passivated Silicon Quantum Dot** . . . . . . . . . . . . . . . . . . . . . . . . . . . . . . 25
A. Laref

**Absorption by Particulate Silicon Layer: Theoretical Treatment to Enhance Efficiency of Solar Cells** . . . . . . . . . . . . . . . . . . . . . . . . . . . 53
Alexander A. Miskevich and Valery A. Loiko

**Modeling and Simulation of New Generation of Thin-Film Silicon Solar Cells Using Efficient Light-Trapping Structures** . . . . . . . . . . . . . . 109
R.S. Dubey and S. Saravanan

**Optical Anisotropy and Compositional Ratio of Conductive Polymer PEDOT:PSS and Their Effect on Photovoltaic Performance of Crystalline Silicon/Organic Heterojunction Solar Cells** . . . . . . . . . . . 137
Hajime Shirai, Qiming Liu, Tatsuya Ohki, Ryo Ishikawa, and Keiji Ueno

**Flexible, Stretchable, and Biodegradable Thin-Film Silicon Photovoltaics** . . . . . . . . . . . . . . . . . . . . . . . . . . . . . . . . . . . . . . . 161
Xing Sheng, Shuodao Wang, and Lan Yin

**Silicon Nanocrystal-Based Organic/Inorganic Hybrid Solar Cells** . . . . . . . . . . . . . . . . . . . . . . . . . . . . . . . . . . . . . . . . 177
Yi Ding and Tomohiro Nozaki

**Organic–Inorganic Hybrid Silicon Solar** . . . . . . . . . . . . . . . . . . . . . . . . 205
Yingfeng Li, Younan Luo, and Meicheng Li

**Recent Advances in the Use of Silicon-Based Photocathodes for Solar Fuel Production** . . . . . . . . . . 229
Ahmad M. Mohamed, Basamat S. Shaheen, Aya M. Mohamed, Ahmad W. Amer, and Nageh K. Allam

**Silicon Nanowire Solar Cells** . . . . . . . . . . 269
Guijun Li and Hoi-Sing Kwok

**Si Nanowire Solar Cells: Principles, Device Types, Future Aspects, and Challenges** . . . . . . . . . . 299
Mrinal Dutta, Lavanya Thirugnanam, and Naoki Fukata

**Index** . . . . . . . . . . 331

# Effective Light Management in Thin Silicon Wafers

**Zhi-Peng Ling**

**Abstract** Crystalline silicon (c-Si)-based photovoltaics have dominated the global market share over the past decade. To progress toward utility-scale adoption, cost reduction plans are necessary, in which one option is to reduce the silicon material used. This comes at a cost of lower photo-absorption and generation, particularly for near-infrared photons due to their much higher absorption depth as compared to short-wavelength photons. In this chapter, the different methods to enhance light trapping for near-infrared photons are mentioned, and in particular the methodology on the proper design of a one-dimensional conductive distributed Bragg reflector (DBR) scheme is introduced. Both experimental and simulation results in this chapter consistently demonstrate the feasibility of integrating a conductive DBR scheme at the rear of a heterojunction silicon wafer solar cell (a type of c-Si-based photovoltaics technology) for enhanced photo-generation at the target long-wavelength regions (i.e., 900 ± 200 nm). The methodology presented here can be easily extended to other target wavelengths of interest and also not limited to solar cells applications alone.

## 1 Introduction

Crystalline silicon-based photovoltaics [1] have dominated the global market share (~90%) over the past decade, arising from the reductions in module pricing and manufacturing costs. However, to be attractive enough for utility-scale adoption, plans for further cost reduction are necessary [2]. Reducing the silicon material used is an attractive proposition, as the crystalline silicon wafer was found to contribute about 20–25% of the total cost of crystalline panels. However, thinner silicon wafers present new challenges in terms of processing, and lower photo-absorption of incoming photons, due to a significantly higher absorption depth

Z.-P. Ling (✉)
Department of Electrical & Computer Engineering, National University of Singapore, Solar Energy Research Institute of Singapore, Singapore, Singapore
e-mail: gordon.ling@nus.edu.sg



S. Ikhmayies (ed.), *Advances in Silicon Solar Cells*,
https://doi.org/10.1007/978-3-319-69703-1_1

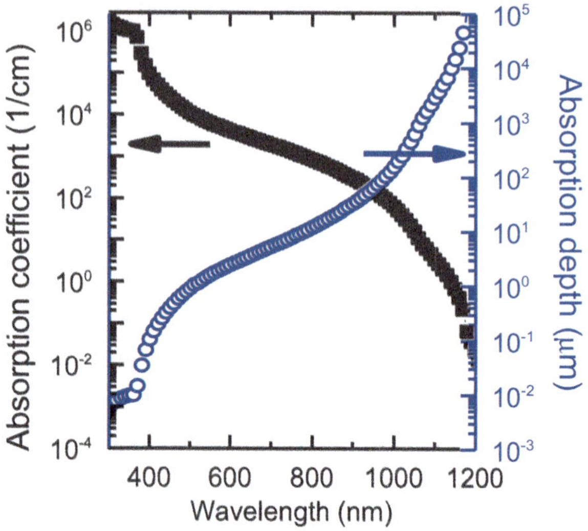


**Fig. 1** Wavelength-dependent absorption coefficient and corresponding absorption depth for a silicon wafer

required for long-wavelength photons [3] as compared to short-wavelength photons (see Fig. 1). From the wavelength-dependent absorption coefficient characteristics of Si wafers at 300 K, the absorption depth ($L$) which is inversely correlated to the absorption coefficient implies a larger absorption depth for long-wavelength photons as compared to short-wavelength photons. For illustration purposes, Fig. 1 shows that $L \approx 160$ μm at $\lambda = 1000$ nm and $L \approx 0.1$ μm at $\lambda = 400$ nm. Hence, it is clear that thinner silicon wafers are not able to absorb long-wavelength photons effectively, which leads to reduced photo-generation and a lower short-circuit current density ($J_{SC}$). Consequently, the key challenge with thinner wafers is to enhance the optical path length by trapping long-wavelength photons within the absorber for a duration as long as possible in order to increase the photo-generation rate.

## 2 Different Methods for Light Trapping

Light-trapping schemes can be implemented at both the front and rear surfaces of the solar cell. Several approaches had been adopted, which include randomly roughened surfaces [4], periodic gratings [5–8], nano-element arrays [9–11], plasmonic structures [12–16], antireflective coatings [17, 18], and distributed Bragg reflectors (DBR), among others. Among these different options, the most well-known and industrially adopted light-trapping scheme is the coupling of random pyramid textured front surfaces with a suitable single-layer or double-layer antireflective coating such as silicon nitride or silicon nitride/aluminum oxide stack. In contrast to a planar front surface, a textured front surface can significantly increase the optical path length of an incoming photon by up to 50 times when a silicon absorber is utilized based on theoretical studies [19]. This is a result of light being coupled obliquely into the silicon absorber, allowing the incoming photons to travel a much further distance than the actual device thickness. By further adopting a suitable antireflection coating material with the optimal refractive indices and thickness, the front-surface reflection can be further reduced, leading to enhanced photo-generation in the bulk silicon absorber. Light trapping for the rear surface can also be established by techniques such as one-dimensional and two-dimensional periodic gratings. Periodic gratings had been theoretically studied and experimentally demonstrated [20–25] to enhance the photo-generation due to a higher degree of scattering and reflection. However, a careful optimization of the periodicity and the height of the gratings is required, which increases process complexity and costs.

On the other hand, utilizing a simple one-dimensional distributed Bragg reflector for rear-surface light trapping appears to be an attractive option as well. Previous studies had included using at least one insulating or moderately conducting layer such as combinations of $Si/Si_3N_4$ or $Si/SiO_2$ stacks [26, 27], porous silicon [28, 29], $TiO_2$, [30–32] or conductive materials like ITO [33–35]. Considering the ease and merits of integrating a DBR scheme at the rear of a thin silicon wafer for light trapping, the subsequent sections focus on the proper design of a DBR-based light-trapping scheme.

## 3 Design Methodology of a Distributed Bragg Reflector

Similar to a semiconductor with a defined bandgap within which no electronic states can exist, an analogous situation is observed for photons in a distributed Bragg reflector (equivalently a one-dimensional photonic crystal [36]) as seen in Fig. 2. If films with a periodically varying dielectric function (or index of refraction) are stacked together, the reflections and refractions of light from the various interfaces can create photonic bandgaps, which will forbid photon propagation in certain directions for a certain (designed) photon wavelength $\lambda_0$ within a certain bandwidth $\Delta\lambda_0$.

To realize the DBR, periodic stacks of two layers (DBR unit blocks) with different refractive indices $n_1$ and $n_2$ are chosen, whereby $n_1 > n_2$. For the DBR unit block, the thickness of the individual layers $d_1$ and $d_2$ is typically chosen to satisfy a quarter-wavelength stack at the target peak reflectance wavelength $\lambda_0$ [36] as:

$$\boldsymbol{d} = \frac{\boldsymbol{\lambda}_0}{4\boldsymbol{n}}, \tag{1}$$

where $d$ is a function of the target peak reflectance wavelength $\lambda_0$ and the refractive index $n$ of that layer at that wavelength. The objective of using a quarter-

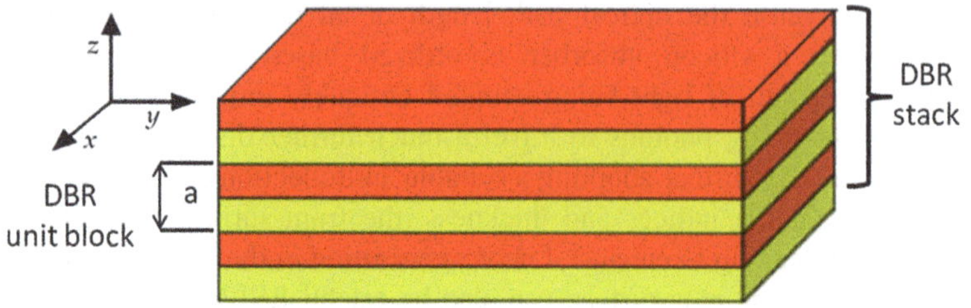


**Fig. 2** Schematic of a one-dimensional photonic crystal ("distributed Bragg reflector," DBR). The DBR consists of several DBR unit blocks. Each DBR unit block consists of alternating layers of two materials with different dielectric constants, with a period $a$ (From Ref. [37])

wavelength stack is to achieve quarter-wavelength optical thickness for the individual layers, so that for a DBR with multiple stacks of periodic DBR unit blocks, only wavelengths within the designed peak reflectivity regions interfere constructively, while the rest interfere destructively to give lower or zero reflectance. The overall optical reflectance at the target wavelength and its bandwidth is thus highly enhanced, so that the designed distributed Bragg reflector acts as a mirror.

It is also of relevance to find out the wavelength range (bandwidth) around the target peak reflectance wavelength for which high optical reflectance is expected. From experimental and theoretical analysis, the bandwidth can be approximated via the gap–midgap ratio $\Delta\lambda_0/\lambda_0$ [36]:

$$\frac{\Delta\lambda_0}{\lambda_0} = \frac{4}{\pi} \sin^{-1}\left(\frac{|n_1 - n_2|}{n_1 + n_2}\right), \tag{2}$$

where $\Delta\lambda_0$ denotes the bandwidth at the target peak reflectance wavelength $\lambda_0$. The bandwidth $\Delta\lambda_0$ defines the photonic gap between the first two bands of a quarter-wavelength stack, analogous to a bandgap in a semiconductor material. If we define a refractive index contrast $R$ such that $R = n_1/n_2$, the above equation can be rewritten as:

$$\frac{\Delta\lambda_0}{\lambda_0} = \frac{4}{\pi} \sin^{-1}\left(\frac{|R - 1|}{R + 1}\right), R = \frac{n_1}{n_2}. \tag{3}$$

With an increasing refractive index contrast, the gap–midgap ratio $\Delta\lambda_0/\lambda_0$ increases as evident in Fig. 3, indicating a wider reflection bandwidth $\Delta\lambda_0$ for the DBR unit block.

The optical reflectance of a DBR can be expressed as [38]:

$$r = \frac{n_0(n_2)^{2N} - n_s(n_1)^{2N}}{n_0(n_2)^{2N} + n_s(n_1)^{2N}}, \tag{4}$$

where $n_0$, $n_1$, $n_2$, and $n_s$ are the refractive indices of the ambient, of Bragg material 1, of Bragg material 2, and of the substrate, respectively, and N is the number of

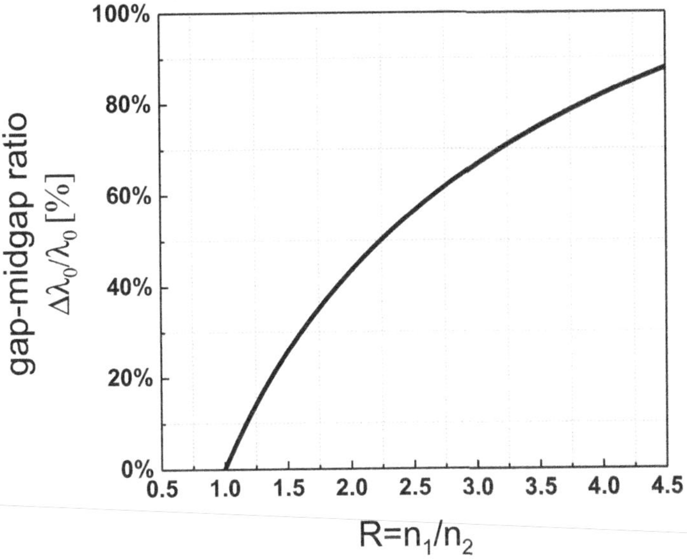


**Fig. 3** Gap–midgap ratio $\Delta\lambda_0/\lambda_0$ of a DBR unit block as a function of the refractive index contrast $R = n_1/n_2$

DBR unit blocks making up the DBR. From Fig. 4, we can observe two phenomena: (1) the optical reflectance at the peak reflectance wavelength increases with the number of DBR unit blocks, and (2) by using materials with higher refractive index contrast, it is possible to approach the maximum optical reflectance with a reduced number of DBR unit blocks. As an example, Fig. 4 shows that a *hypothetical* DBR with a refractive index contrast of 1.5 would require 8 DBR unit blocks to approach unity reflection, whereas a DBR with a refractive index contrast of 3.5 would only need three DBR unit blocks. Thus, when considering DBRs, it is logical to utilize materials with a high refractive index contrast in order to reduce the number of DBR unit blocks required, as well as to benefit from a wider reflection bandwidth.

It is to be noted that the above calculations have been based on the refractive index $n$ of the different materials and do not involve the extinction coefficients $k$. Generally, the calculations suffice for dielectric materials possessing a wide bandgap, as their extinction coefficients are either zero or very low at the targeted long-wavelength regions (~1000 nm). Some examples include silicon dioxide, silicon nitride, aluminum oxide, or titanium dioxide.

However, if conductive materials were utilized, the constituent films generally have a non-zero extinction coefficient or equivalently the presence of parasitic absorption in targeted wavelength regions. Hence, the design has to take into consideration (i) the wavelength dependence of the complex refractive indices as well as (ii) the absorption induced by the various layers. Accordingly, the gap–midgap ratio $\Delta\lambda_0/\lambda_0$ used to determine the reflection bandwidth in Eq. 2 should be more accurately stated as shown in Eq. 5 instead:

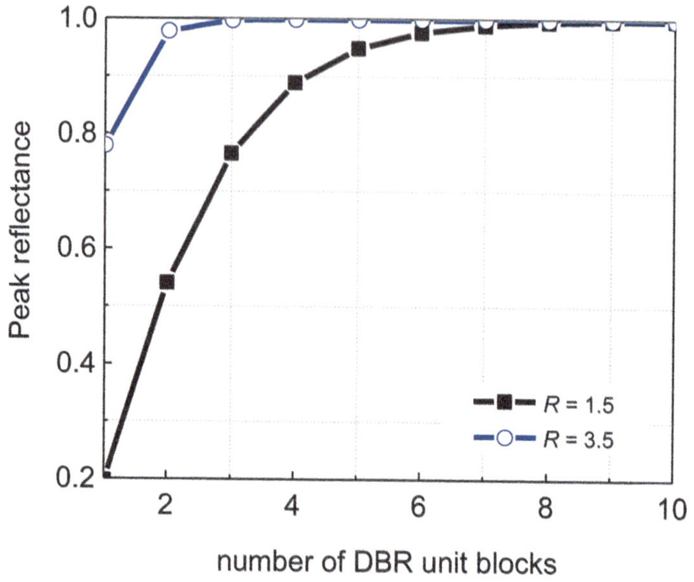


**Fig. 4** Calculated peak reflectance of a DBR at the peak reflectance wavelength $\lambda_0$ with respect to the number of DBR unit blocks used, assuming air as ambient ($n_0 = 1$) and glass as substrate ($n_s = 1.51$) and using $R = n_1/n_2$. Two hypothetical scenarios are shown: (i) DBR unit blocks with a refractive index contrast $R = 1.5$ and (ii) a refractive index contrast $R = 3.5$

$$\frac{\Delta\lambda_0}{\lambda_0} = \frac{4}{\pi}\sin^{-1}\left(\frac{|\tilde{n}_1 - \tilde{n}_2|}{\tilde{n}_1 + \tilde{n}_2}\right), \tag{5}$$

where $\tilde{n}_1$ and $\tilde{n}_2$ are the complex refractive indices for the first and second conductive layer and can be further expanded as $\tilde{n}_1 = n_1 + ik_1$ and $\tilde{n}_2 = n_2 + i k_2$, respectively. The refractive index and extinction coefficient of the constituent layer are indicated as $n$ and $k$, respectively. Multiplying both the top and bottom terms with the complex conjugate of the denominator and after some simplifications, Eq. 6 can be obtained. It can be easily verified that Eq. 6 reduces to Eq. 2 if the selected materials do not exhibit absorption, by setting $k_1$ and $k_2$ equal to zero.

$$\frac{\Delta\lambda_0}{\lambda_0} = \frac{4}{\pi}\sin^{-1}\left(\frac{n_1^2 - n_2^2 + k_1^2 - k_2^2 + 2i(n_2k_1 - n_1k_2)}{(n_1 + n_2)^2 + (k_1 + k_2)^2}\right). \tag{6}$$

Conveniently, the arcsine function can be expressed as a Taylor series expansion. For the condition $|x| \leq 1$, one obtains:

$$\sin^{-1}x = \sum_{n=0}^{\infty}\frac{(2n)!}{2^{2n}(n!)^2(2n+1)}x^{2n+1} = x + \frac{x^3}{6} + \frac{3x^5}{40} + \frac{5x^7}{112} + \dots. \tag{7}$$

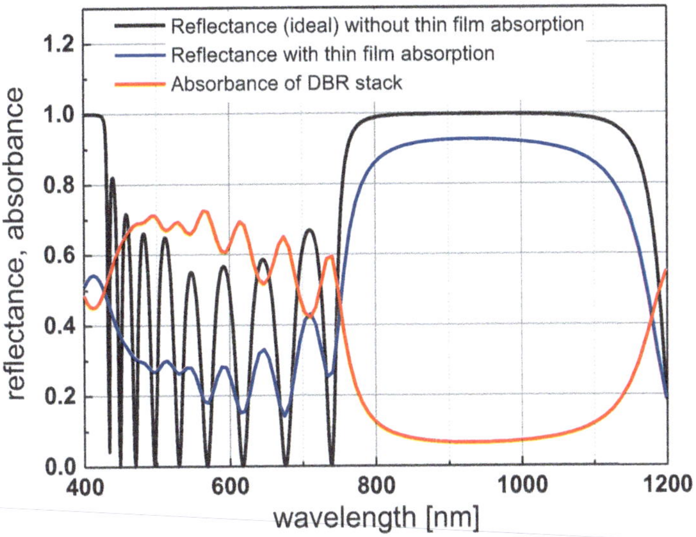


**Fig. 5** Comparison of the calculated reflectance and absorbance of an air/five DBR unit blocks/air structure for two scenarios: (i) no absorption from the conductive thin films (zero extinction coefficient) and (ii) with absorption. The harmonics of the reflectance spectra matches for both scenarios, while the lower peak reflectance at the target wavelength of 900 nm for the practical case can be attributed mainly to absorption in the conductive thin films

It is sufficient to use the summation of the first four terms, as the contributions from higher-order terms to the final evaluated results are minimal. In addition, since the evaluated function in Eq. 7 is still a complex number, it is more appropriate to express it using the Euler's formula:

$$\frac{\Delta\lambda_0}{\lambda_0} = (a + ib) = re^{i\theta}, \tag{8}$$

where $r = \sqrt{a^2 + b^2}$ and $\theta = \arctan(b/a)$.

To illustrate the influence of parasitic absorption on the optical reflectance at the target wavelength range, the transfer matrix calculation [39] can be utilized, using the optical constants from two different conductive materials (e.g., n-doped microcrystalline silicon and aluminum-doped zinc oxide as seen in Fig. 5 and elaborated in subsequent sections), and compared for two scenarios: (i) with absorption (non-zero extinction coefficients) and (ii) without absorption (assuming zero extinction coefficients). For the same targeted peak reflectivity wavelength range of 900 ± 200 nm, and utilizing an air/five DBR unit blocks/air configuration, the reflectance spectra harmonics and bandwidth match for both the ideal case (no absorption) and the practical case (with film absorption). However, for the practical case, the parasitic absorption by the conductive DBR stacks leads to a clear drop (~10%) in the calculated optical reflectance within the targeted wavelength range, which cannot be neglected.

Despite the inherent parasitic absorption losses exhibited by conductive thin-film materials, the key motivations for using a conductive DBR scheme include (1) a reduced contact resistance and (2) a direct metal contact as compared to the dielectric or hybrid (dielectric/conducting) DBR stacks which demand additional procedures (such as combination of photolithography, reactive ion etching, or laser ablation) to create metal contacts to the substrate. Careful optimization of these additional steps is necessary to avoid damaging the substrate surface, which would result in higher recombination rates and lower cell efficiency.

## 4 Experimental, Characterization, and Simulation Methods

For the selected choice of conductive materials, it is important to be able to extract important information pertaining to the selected material, such as the complex refractive indices, film thickness, and sheet resistance. In addition, prior to the integration of both conductive materials, the independent optimization of each conductive material is often necessary.

In this chapter, to realize the conductive DBR stacks, two different heavily doped semiconductor films: n-type hydrogenated microcrystalline silicon, μc-Si:H (n), and n-type (aluminum) zinc oxide, ZnO:Al, were utilized. The first is widely used in thin-film silicon solar cells, while the latter is a transparent conductive oxide (TCO) material with applications in the solar and display industries. For characterization purposes, individual films were deposited onto planar glass substrates. The μc-Si:H(n) films were deposited in a conventional RF (13.56 MHz) parallel-plate direct plasma-enhanced chemical vapor deposition reactor, while the ZnO:Al films were deposited by RF magnetron sputtering, using a ceramic ZnO:$Al_2O_3$ target with 2% $Al_2O_3$ by weight [37]. Table 1 summarizes the most important deposition parameters of the two investigated films. For initial characterization, the thickness of the n-type silicon film was chosen to vary from 20 to 80 nm, in 20 nm steps, whereas the ZnO:Al film was chosen to vary from 80 to 140 nm, in 20 nm steps.

The transmission and reflection characteristics of each sample (film/glass) were obtained using a UV-VIS-IR spectrophotometer (PerkinElmer, Lambda 950). The measurements were performed from the film side. The raw optical data were fitted using a commercial software (Coating designer, v3.44 [40]), giving the optical constants of each film. In addition, the thickness and sheet resistance of each film were measured by stylus profilometry (Bruker, Dektak 150) and a four-point probe (Napson Corporation, Cresbox), respectively.

Next, using the optical constants of the two conductive thin films, the wavelength-dependent gap–midgap ratio $\Delta\lambda_0/\lambda_0$ was calculated in order to select the optimum peak reflectance wavelength $\lambda_0$ with its bandwidth $\Delta\lambda_0$, with a focus on the near-IR photons (~1000 nm) for which light trapping needs to be optimized for c-Si solar cells. The theoretically proposed Bragg quarter-wavelength stacks were then realized on planar glass substrates, whereby various numbers (up to 10) of DBR unit blocks were deposited. The measured reflectance of the resulting

**Table 1** Overview of process parameters for the μc-Si:H (n) and the ZnO:Al thin-film layers

| | μc-Si:H(n) | ZnO:Al |
|---|---|---|
| Pressure (mTorr) | 1900 | 3 |
| Substrate temperature (°C) | 210 | 100 |
| $H_2$ flow (sccm) | 150 | – |
| $SiH_4$ flow (sccm) | 5 | – |
| $PH_3$ flow (2% in $H_2$) (sccm) | 2 | – |
| Ar flow (sccm) | – | 23 |
| RF power density (W) | 85 | 200 |

conductive DBRs was then compared with numerically calculated values, using FreeSnell [39], a commercial software which adopts the transfer matrix method [41]. Furthermore, all experiments were repeated on aluminum-coated glass substrates, in order to study the influence of a rear metal contact (which is typically present at the rear side of silicon solar cells).

Finally, the suitability of incorporating the developed conductive DBR at the rear of a heterojunction silicon wafer solar cell (a type of c-Si photovoltaic technology) was numerically evaluated. The influences of the enhanced internal interface reflectance and the enhanced conductance of the DBRs on the current–voltage characteristics of the solar cell are discussed.

## 5 Optical Constants of the Conductive Thin Films

The fitted optical constants of the 60 nm thick n-doped μc-Si:H and 80 nm thick ZnO:Al films are shown in Fig. 6. It is observed from Fig. 6 that the developed n-doped silicon film is indeed of a microcrystalline nature, with its optical constants lying between pure amorphous silicon and crystalline silicon [42]. The film structure of the n-doped silicon film is thickness dependent, whereby the film is mainly amorphous at low thickness (20 nm) and becomes more crystalline with increasing thickness, and the optical constants change accordingly. On the other hand, it is observed that the optical constants of the ZnO:Al films are relatively insensitive to the film thickness (in the 80–140 nm range).

## 6 Thickness Optimization for Conductive DBR Thin Films

Since the gap–midgap ratio of a conductive distributed Bragg reflector varies with the wavelength-dependent complex refractive indices, it is more appropriate to show the upper and lower peak reflectance limits, $\lambda_0 + \frac{\Delta\lambda_0}{2}$ and $\lambda_0 - \frac{\Delta\lambda_0}{2}$, as a function of the peak reflectance wavelength $\lambda_0$ (see Fig. 7).

To obtain Fig. 7, the complex refractive indices $(n, k)$ of the μc-Si:H(n) and ZnO:Al thin films were deduced from Fig. 6 (from 400 to 1150 nm, using 50-nm

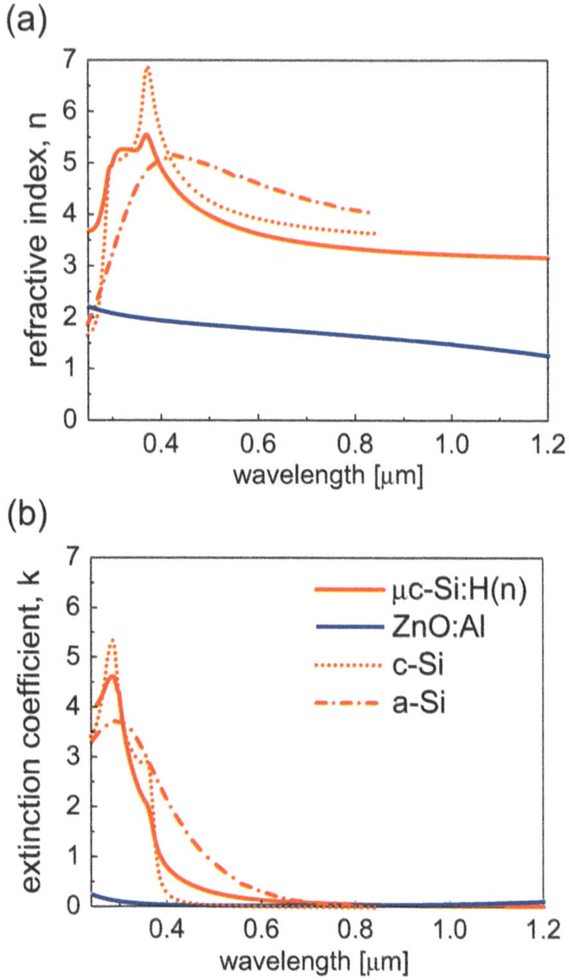


**Fig. 6** Measured (**a**) refractive index *n* and (**b**) extinction coefficient *k* of the investigated n-type hydrogenated microcrystalline silicon film and TCO film (ZnO:Al). Also shown, for comparison, are the corresponding values for high-quality monocrystalline silicon [43] and PECVD amorphous silicon [44]

intervals), and the upper/lower peak reflectance limits were calculated using Eq. 5. Considering the bandgap (~1.12 eV) of the crystalline silicon absorber, the upper peak reflectance should be chosen at ~1100 nm. This is shown as dashed line in Fig. 7. Thus, a central peak reflectance wavelength at ~900 nm is deduced, with a corresponding lower peak reflectance limit of ~700 nm. In other words, in order to enhance the internal reflection of long-wavelength photons up to 1100 nm within the silicon solar cell absorber, the thickness of the μc-Si:H(n) and ZnO:Al films should be chosen to satisfy a quarter-wavelength stack, aiming for a target peak reflectance wavelength at ~900 nm. According to this design rule, a high optical reflectance for the 900 ± 200 nm wavelength range is expected, and a corresponding gap–midgap ratio $\Delta\lambda_0/\lambda_0$ of ~0.45 is determined. Based on the above specifications, the optimized layer thickness for the μc-Si:H(n) and ZnO:Al

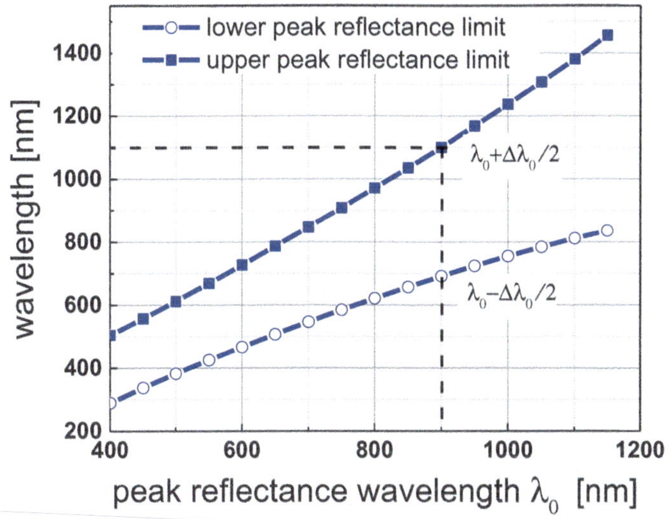


**Fig. 7** Calculated upper and lower peak reflectance limits as a function of the peak reflectance wavelength based on the wavelength-dependent refractive indices of μc-Si:H(n) and ZnO:Al

**Table 2** Calculated thickness for the μc-Si:H(n) and ZnO:Al films to achieve a peak reflectance at 900 nm

| | μc-Si:H(n) | ZnO:Al |
|---|---|---|
| Peak wavelength (nm) | 900 | 900 |
| Refractive index *n* | 3.28 | 1.58 |
| Calculated thickness required (nm) | 69 | 142 |

films to be used in a conductive DBR at the rear of a c-Si-based solar cell will be 69 nm and 142 nm, respectively (see Table 2).

## 7 Conductive DBR Performance on Different Substrates

Prior to device integration, the performance of the developed conductive DBRs can be evaluated on different substrates and compared with simulation results to provide more insights.

## *7.1 Conductive DBR on Glass Substrates*

Using the optimized layer thickness (Table 2), characterization was performed firstly on planar glass substrates using various numbers of DBR unit blocks. Figure 8 shows a cross-sectional scanning electron microscopy (SEM) image of one of the fabricated DBRs on glass, while Fig. 9 shows the measured reflectance of air/DBR/

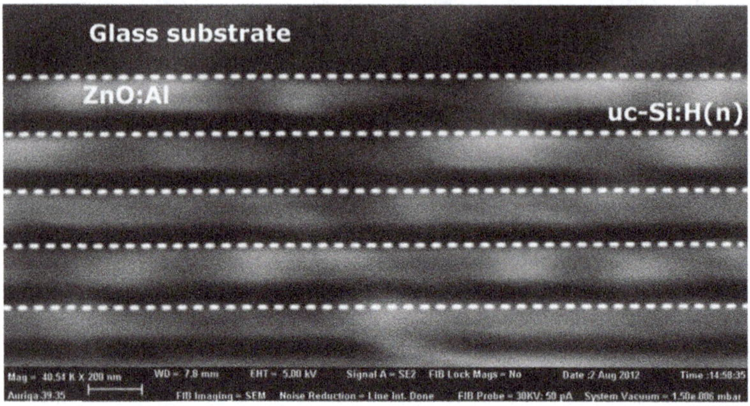


**Fig. 8** Cross-sectional SEM image of a fabricated distributed Bragg reflector consisting of 142 nm thick ZnO:Al and 69 nm thick μc-Si:H(n) films, deposited on a glass substrate. The white dashed lines define the individual DBR unit blocks

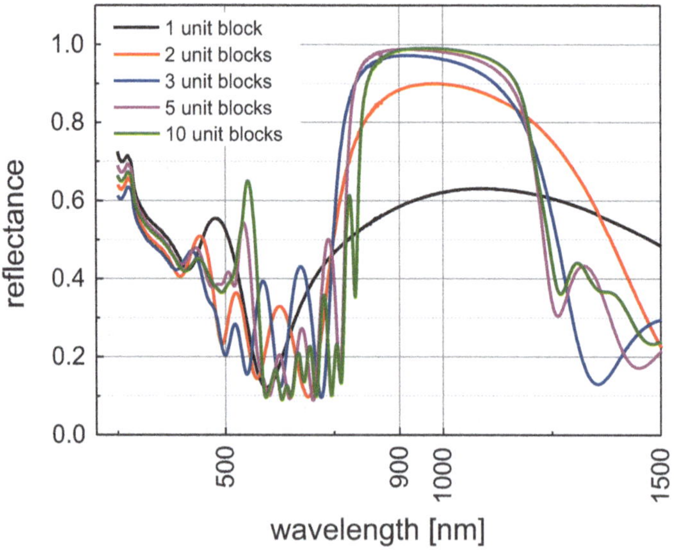


**Fig. 9** Measured optical reflectance as a function of wavelength for air/DBR/glass samples with five different numbers of DBR unit blocks. The samples were illuminated from the film side

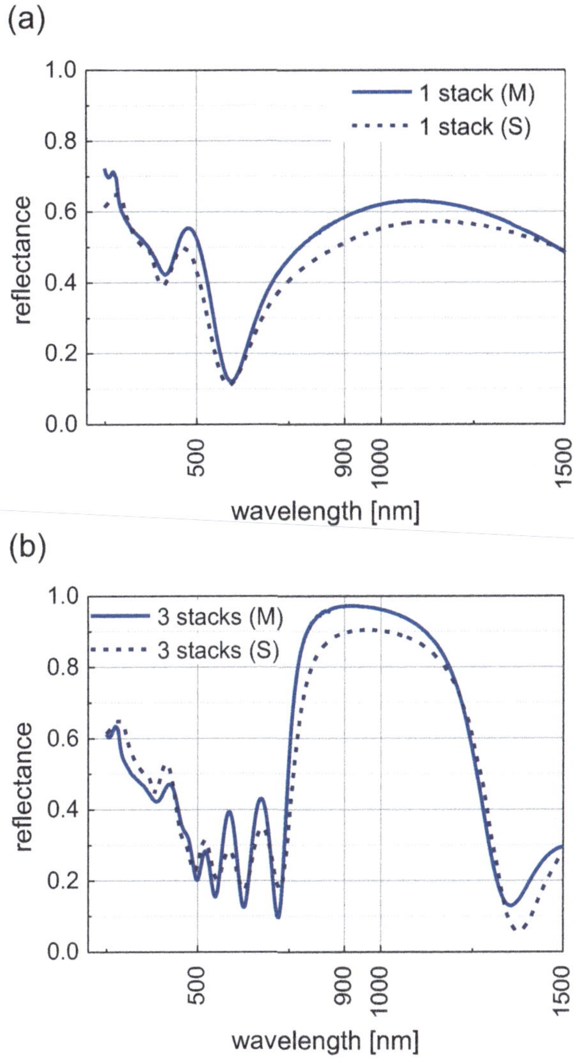


**Fig. 10** Comparison of measured (M) and simulated (S) results for (**a**) one DBR unit block on a glass substrate and (**b**) three DBR unit blocks on a glass substrate. Each DBR unit block consists of optimized thickness for μc-Si:H(n) and ZnO:Al

glass samples with five different numbers of DBR unit blocks, illuminated from the film side. It can be observed that the optical reflectance in the targeted wavelength regions increases with the number of DBR unit blocks. For three DBR unit blocks, the peak reflectance is about 97%. For ten DBR unit blocks, it increases further to 99% at the design wavelength. These very high numbers confirm the feasibility of using these two thin-film materials for forming a high-performance distributed Bragg reflector.

Figure 10 shows a comparison of the experimental and simulated reflectance results, for the case of one and three DBR unit blocks. In general the simulations

Fig. 11 Influence of the number of DBR unit blocks on the measured sheet resistance and peak reflectance at the target wavelength of 900 nm

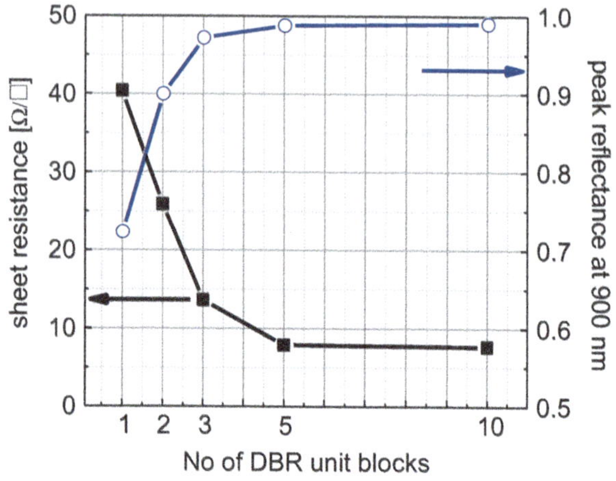


agree quite well with the measured data, although some deviations in the peak reflectance values were observed. These were very likely due to slight thickness variations in the fabricated samples.

Besides improving the optical reflectance at the target wavelength range with increasing number of DBR unit blocks, the sheet resistance also decreases, i.e., from 40 Ω/□ using only one DBR unit block to 8 Ω/□ using five DBR unit blocks (see Fig. 11). This is beneficial when applied as a rear contact system of a silicon-based solar cell. Five DBR unit blocks are already sufficient to achieve very high optical and electrical performance. These results compare favorably with the various conductive DBR schemes reported in Refs. [32–35, 45], where the peak reflectivity wavelength of their reported conductive DBR stacks ranges from 426 to 550 nm, with peak reflectivity ranging from 73 to 90% and sheet resistance in the range of 35 to 50 Ω/□ for a 4-stack to 8-stack configurations.

## 7.2 *Conductive DBR on Metal Substrates*

Given the presence of a rear metal contact in the typical silicon wafer solar cells, the measured and simulated optical reflectance of the developed conductive DBR on a metal-coated substrate was also evaluated. Simulation studies on air/DBR/Al samples in Fig. 12 predicted significantly higher optical reflectance using only one DBR unit block, if an Al-coated substrate is used instead of a glass substrate (i.e., ~93% optical reflectance in comparison to ~57% at the same target peak wavelength), benefiting from the inherent high reflectivity of the metal coating. Metal-based back surface reflectors had been reported; however, their high extinction coefficient in the long-wavelength region [46] makes them unsuitable for high-efficiency solar cell applications. In comparison, the extinction coefficients of our doped films are significantly lower in the target wavelength range, and hence a significant absorption of long-wavelength photons can be avoided. Increasing the number of DBR

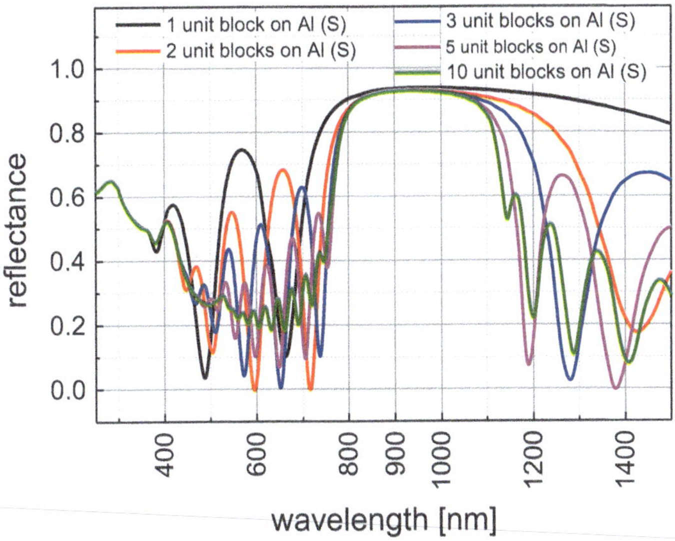


**Fig. 12** Simulated (S) reflectance of the optimized DBR on an Al-coated substrate with different numbers of DBR unit blocks

unit blocks does not increase the peak reflectance (see Fig. 12). It only leads to a reduced reflection bandwidth, which is undesired. Hence, it appears adequate to use only one DBR unit block if a full-area rear metal contact is adopted. It is further noted that the predicted lower reflectance in the short-wavelength region (< 700 nm) is of no relevance, as most of these photons would have been absorbed within the first 10 μm from the front silicon surface.

## *7.3 Conductive DBR Integration at Device Level*

Given the promising performance from independent measurements on glass and metal substrates, it is of interest to evaluate the final device performance when the developed conductive DBR is integrated at the device level. Using a calibrated three-dimensional solar cell simulation model running on Sentaurus TCAD, the influence of the conductive DBR on device performance is predicted when applied at the rear of a heterojunction silicon wafer solar cell (a type of c-Si-based photovoltaic technology). A detailed description of the calibrated simulation input parameters is presented in Refs. [47, 48].

Figure 13 shows a schematic of the investigated heterojunction silicon wafer solar cell structure with a planar rear surface using either (a) standard thickness for the a-Si:H(n) BSF (20 nm) and ZnO:Al thin films (80 nm) or (b) optimized DBR layer thickness for μc-Si:H(n) BSF (69 nm) and ZnO:Al (142 nm). The influence of a rear DBR on the optical as well as electrical solar cell performance was investigated and compared to the reference solar cell not using a DBR as sketched in Fig. 13a.

Fig. 13 Schematic of a modified heterojunction silicon wafer solar cell structure with a planar rear surface using either (a) standard thickness of 20 nm and 80 nm for the a-Si:H(n) BSF and ZnO:Al thin films or (b) optimized layer thickness of 69 nm and 142 nm for μc-Si:H (n) BSF and ZnO:Al in order to achieve high internal optical reflectance at a peak wavelength of 900 nm

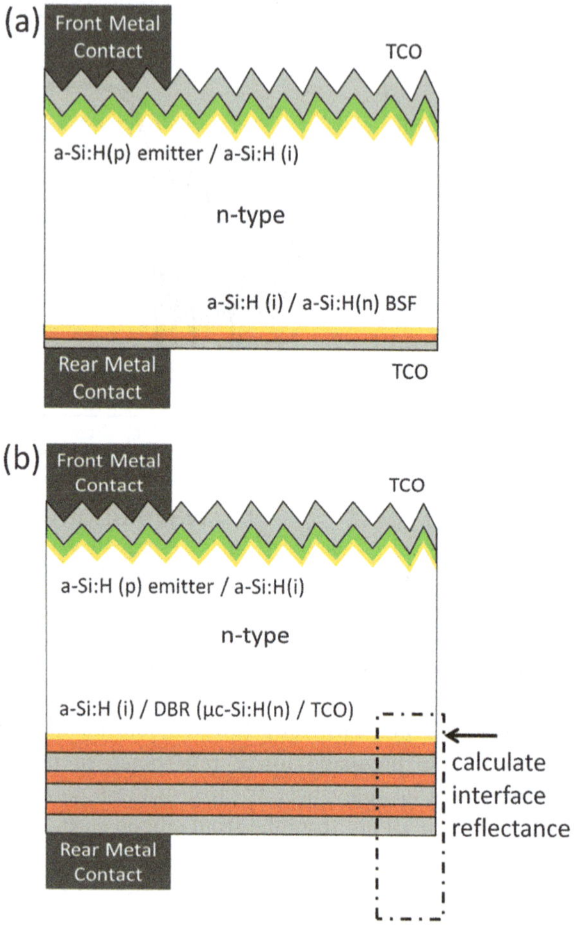


For this device structure, the rear interface reflectance can be calculated using a slightly modified version of the transfer matrix program from Burkhard et al. [49, 50]. Based on a c-Si(n)/a-Si:H(i)/DBR(μc-Si(n)/TCO)/air configuration (see dotted region in Fig. 13b), this program can calculate the internal optical reflectance at the c-Si rear interface, as well as the parasitic absorbance and total transmittance of the individual layers, for incident photons in the 300–1200 nm range.

### 7.3.1 Reflectance, Absorptance, and Transmittance

Figure 14 showed that by increasing the number of DBR unit blocks, a significant drop in transmitted light (i.e., unabsorbed photons) is observed, while the interface reflectance increased markedly and resembles that of our conductive DBR on glass and Al-coated substrates. Hence, higher optical generation rate near the rear of the

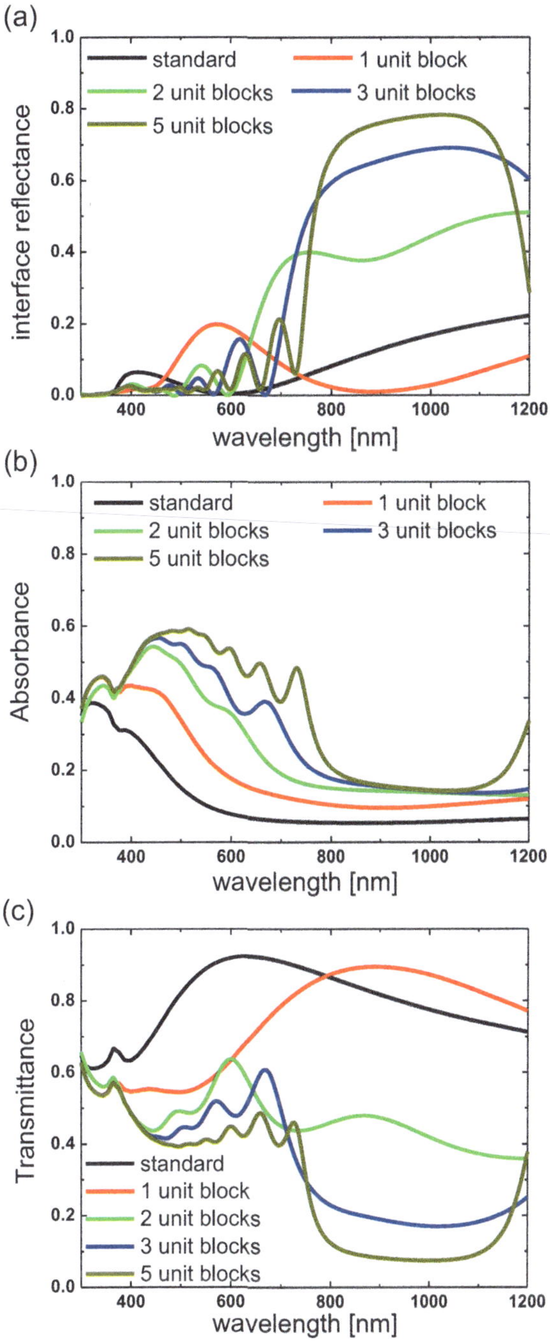


**Fig. 14** (**a**) Calculated internal interface reflectance, (**b**) combined absorbance by the conductive DBRs, and (**c**) transmittance through the conductive DBRs for a heterojunction silicon wafer solar cell sketched in Fig. 13b, i.e., a c-Si/a-Si:H(i)/DBR(μc-Si/TCO)/air configuration

c-Si(n) substrate arising from the improved interface reflectance is expected. On the contrary, the combined absorbance by the conductive DBR also increased with the number of DBR unit blocks, since the extinction coefficients of both conductive materials are non-zero. Hence, the resulting optical generation rate at the rear surface is a trade-off between increasing rear optical reflectance and increasing parasitic absorbance with more DBR unit blocks. It is worthy to note that a single DBR unit block alone cannot enhance the optical generation rate at the rear, as evident from the low interface reflectance in the target wavelength range (700–1100 nm) (see Fig. 14a). A minimum of two DBR unit blocks is needed. This is due to the fact that a Bragg reflector requires several periodic DBR unit blocks in order to achieve an overall high optical reflectance. Furthermore, the incident light will interact with the intrinsic a-Si:H buffer layer present at the interface. It is again noted that the parasitic absorbance for high-energy photons (i.e., with wavelength $<$ 800 nm) is irrelevant for the device performance, as most of these photons will not reach the rear of the solar cell.

### 7.3.2 Optical Generation/Photo-Generation Current Density

Finally, the complete heterojunction silicon wafer solar cell structure as sketched in Fig. 13b is simulated using Sentaurus TCAD for varying numbers (up to 5) of DBR unit blocks. Three-dimensional optical and electrical simulations have been performed. The resulting optical generation rates above the non-contacted regions of Fig. 13 already demonstrate the feasibility of using multiple DBR unit blocks at the rear of the solar cell (see Fig. 15).

The optical generation rate (or equivalently the collectable photocurrent) is increasing if multiple DBR unit blocks are used. As already discussed, using only one DBR unit block is insufficient. In that case the optical generation profile is lower than the standard configuration (i.e., using no DBR and applying the standard thickness of the μc-Si:H(n) and ZnO:Al layers). For the heterojunction silicon wafer solar cell architecture using five DBR unit blocks on a thick silicon wafer (220 μm), the improved internal rear reflectance leads to an enhanced optical generation rate, which corresponds to an enhanced integrated photo-generation current density of 42 mA $cm^{-2}$ (see Fig. 15b). A noticeable optical enhancement starts from 50 μm onward, which demonstrates the feasibility of this concept using ultrathin wafers as well. It is to be noted that the short-circuit current density of the solar cell is generally lower than its photo-generation current density due to recombination losses in the bulk or at surfaces or interfaces.

### 7.3.3 Current–Voltage Characteristics

Figure 16 and Table 3 sum up the results obtained from the 3D electrical simulations performed with Sentaurus TCAD. Some key observations are:

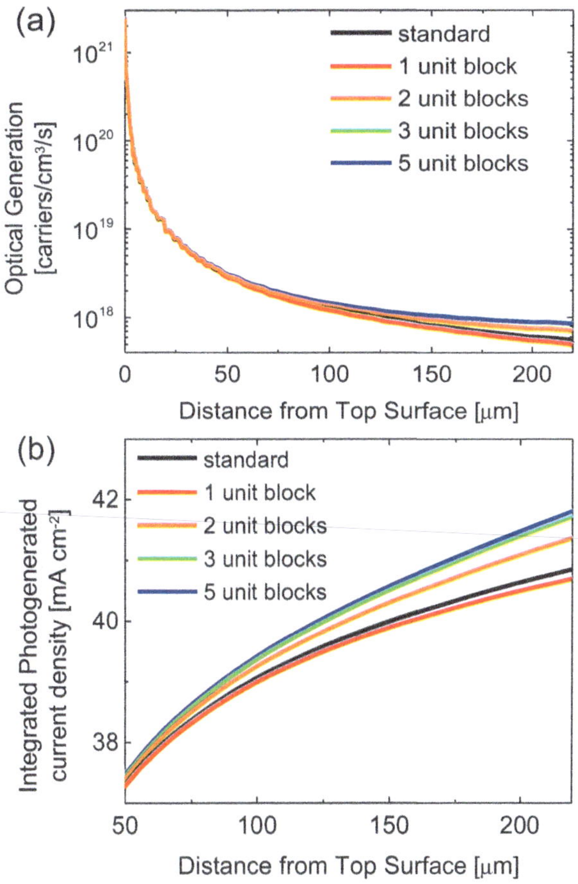


**Fig. 15** (**a**) Simulated optical generation profile for the non-contacted regions in the schematic of Fig. 13. (**b**) Corresponding integrated photo-generation current density as a function of the position within the wafer

(a) Using five DBR unit blocks, an efficiency improvement of 1.6% absolute can be achieved. The simulated device has a significantly higher short-circuit current density as compared to the standard configuration (i.e., a $J_{sc}$ increase of 1.7 mA/ $cm^2$). Furthermore, a significantly lower series resistance is observed in agreement with the findings outlined earlier. This leads to an overall one-sun solar cell efficiency approaching 23.5% (which is to be compared to the simulated standard cell efficiency of 21.9%).

(b) If only one DBR unit block is used, a lower short-circuit current density (36.7 mA/$cm^2$) as compared to the standard configuration (36.9 mA/$cm^2$) is predicted (as discussed above). However, due to the lower series resistance using thicker layers, a higher fill factor exceeding 81% is obtained, and the simulated cell efficiency (~ 22.2%) still exceeds the standard case.

(c) Generally, the open-circuit voltage using the conductive DBR is slightly lower compared to the standard configuration. This can be attributed to the increased absorption and higher recombination rates associated with higher carrier densities.

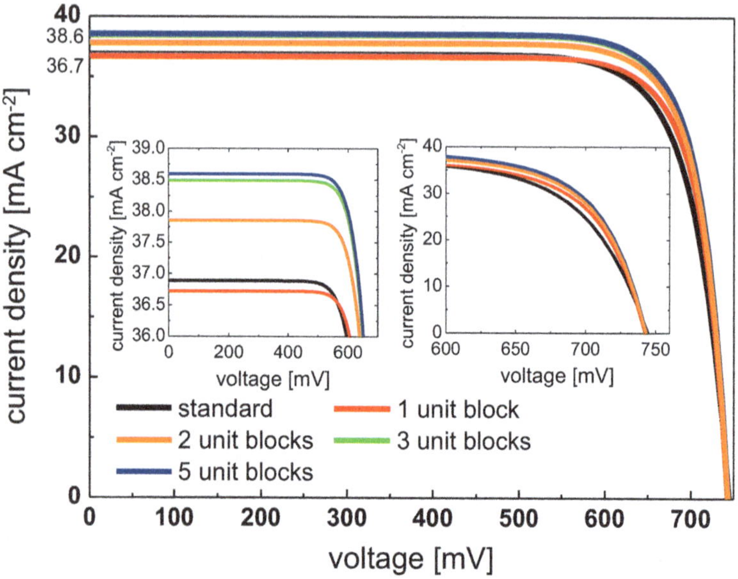


**Fig. 16** Simulated J–V characteristics of heterojunction silicon wafer solar cells as sketched in Fig. 13, using variable numbers of DBR unit blocks. The insets show a zoom in at the short-circuit current density and near the maximum power point regions

**Table 3** Simulated solar cell performance of the heterojunction silicon wafer solar cells as sketched in Fig. 13

| | Substrate thickness (μm) | Efficiency (%) | $J_{sc}$ (mA/cm$^2$) | $V_{oc}$ (mV) | FF (%) |
|---|---|---|---|---|---|
| Standard | 220 | 21.86 | 36.89 | 744.5 | 79.61 |
| 1 DBR unit block | 220 | 22.25 | 36.73 | 742.6 | 81.58 |
| 2 DBR unit block | 220 | 22.97 | 37.86 | 743.4 | 81.63 |
| 3 DBR unit block | 220 | 23.39 | 38.50 | 743.3 | 81.72 |
| 5 DBR unit block | 220 | 23.46 | 38.60 | 743.6 | 81.73 |
| Standard | 50 | 20.67 | 32.60 | 761.9 | 83.22 |
| 5 DBR unit block | 50 | 21.46 | 33.75 | 761.7 | 83.50 |

Given the potential cost savings if thinner wafers are used, we compare the device performance using either a thick substrate (220 μm) or a thin substrate (50 μm) and utilizing either the standard a-Si:H(n)/ZnO:Al thickness or applying five DBR unit blocks [μc-Si:H(n)/ZnO:Al] at the c-Si rear surface (see Table 3 and Fig. 17). In comparison, a lower short-circuit current density of ~33.8 mA/cm$^2$ is predicted for a thin substrate using five DBR unit blocks, which is nevertheless a relative improvement of ~3.5% as compared to the standard case (no rear DBR, standard a-Si:H(n)/ZnO:Al thickness). In addition, the thin substrate shows a higher open-circuit voltage than the thick substrate, in accordance with experimental

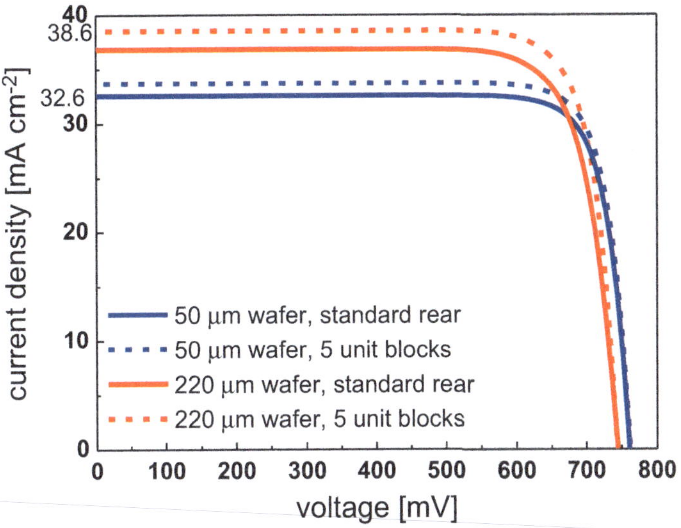


**Fig. 17** Comparison of the simulated J–V characteristics of the heterojunction Si solar cell in Fig. 13 with the standard rear a-Si:H(n)/ZnO:Al film thickness or the five DBR unit blocks [µc-Si:H(n)/ZnO:Al] for both thick (220 µm) and thin (50 µm) substrates

findings by Panasonic [51]. The increase in open-circuit voltage with decreasing wafer thickness can be attributed to a reduced bulk recombination given that the surface passivation is already so good that surface recombination is no longer a dominant recombination channel. Given that the photogenerated excess carriers have to travel a shorter distance to reach the contacts, the fill factor also improves. Correspondingly, the simulation predicts that it should be possible to attain a one-sun cell efficiency of ~21.5% for silicon substrates as thin as 50 µm (using the in-house processed heterojunction layers). Further improvement to the heterojunction layers can allow even higher cell efficiencies to be reached.

The simulation results clearly demonstrate the feasibility of using conductive DBR stacks of periodically stacked µc-Si:H(n) and ZnO:Al films at the rear of a heterojunction silicon wafer solar cell, thereby increasing the efficiency by 7.3% or 3.8% (relative), using thick or thin wafers, respectively.

## *7.4 Comparison with State-of-the-Art Concepts*

It is relevant to compare the predicted performance using the developed conductive DBR stacks to dielectric DBR stacks or hybrid (dielectric/conducting) DBR stacks which have been reported in literature. For example, Zeng et al. [26] have reported the application of a hybrid DBR stack consisting of $SiO_2$/Si films on a thick (675 µm) crystalline silicon substrate to provide a relative cell efficiency gain of

~8% as compared to the standard cell. This compares favorably with our simulated relative cell efficiency gain of 7.3% for a 220 μm thick substrate, despite the inherent parasitic absorption losses exhibited by the conductive thin-film materials. Pure dielectric stacks are expected to perform even better for light trapping in the near-IR wavelength regions since parasitic absorption losses are entirely absent; however additional procedures are required to create metal contacts to the substrate, which do increase process complexity. Using conductive DBR stacks instead presents a balance between reduced number of processing steps and delivering sufficiently high rear optical reflectance for light-trapping purposes.

## 8 Conclusions

In this chapter, the importance of light trapping for solar cells utilizing thin silicon wafers is emphasized, in which the suitable placement of a conductive distributed Bragg reflector at the rear of the silicon absorber presents itself to be an attractive solution. The design considerations and the feasibility of a conductive distributed Bragg reflector (DBR) for solar cell applications were further elaborated from both the perspective of numerical calculations and experimental results. When the designed conductive DBR was applied at the rear of a 220 μm thick heterojunction silicon wafer solar cell, an efficiency increases up to 7.3% (relative), i.e., from 21.9% to 23.5% (absolute) is predicted, using five DBR unit blocks. When a much thinner silicon wafer is considered (50 μm), a relative increase in cell efficiencies by 3.8%, reaching ~21.5% had been predicted. The improvement in cell performance stems from the enhanced internal optical reflectance of long-wavelength photons and the reduced contact resistance.

While the experimental results in this chapter had been based on suitably designed periodic stacks of μc-Si:H(n) and ZnO:Al films (DBR unit blocks) for near-infrared wavelength photons, the reader should benefit from the design methodology outlined in this chapter and be able to apply these approaches to any type of conductive films and for any target wavelength regions of interest.

## References

1. S.W. Glunz, R. Preu, D. Biro, in *Comprehensive Renewable Energy*, ed. by A. Sayigh (Elsevier, Oxford, 2012), pp. 353–387
2. D.M. Powell, M.T. Winkler, H.J. Choi, C.B. Simmons, D.B. Needleman, T. Buonassisi, Energ. Environ. Sci. **5**, 5874–5883 (2012)
3. M.A. Green, M.J. Keevers, Prog. Photovolt. Res. Appl. **3**, 189–192 (1995)
4. M.A. Green, *High Efficiency Silicon Solar Cells*, 1st edn. (Trans Tech Publications, Zürich, 1987), pp. 1–240
5. C. Heine, R.H. Morf, Appl. Optics **34**, 2476–2482 (1995)
6. C. Eisele, C.E. Nebel, M. Stutzmann, J. Appl. Phys. **89**, 7722–7726 (2001)

7. F. Llopis, I. Tobías, Sol. Energy Mater. Sol. Cells **87**, 481–492 (2005)
8. C.-H. Chang, J.A. Dominguez-Caballero, H.J. Choi, G. Barbastathis, Opt. Lett. **36**, 2354–2356 (2011)
9. W.J. Nam, L. Ji, V.V. Varadan, S.J. Fonash, J. Appl. Phys. **111**, 123103 (2012)
10. W.J. Nam, D. Fischer, Z. Gray, N. Nguyen, L. Ji, D. Neidich, S.J. Fonash, IEEE J. Photovoltaics **5**, 28–32 (2015)
11. W. J. Nam, S. J. Fonash, L. Ji, V. V. Varadan, in *2012 38th IEEE Photovoltaic Specialists Conference*, 3–8 June 2012
12. S.J. Fonash, *Light Trapping in Solar Cell and Photo-Detector Devices* (Academic Press, Boston, 2015), pp. 33–48
13. P. Mandal, S. Sharma, Renew. Sustain. Energy Rev. **65**, 537–552 (2016)
14. K.R. Catchpole, A. Polman, Opt. Express **16**, 21793–21800 (2008)
15. H. Heidarzadeh, A. Rostami, M. Dolatyari, G. Rostami, Appl. Optics **55**, 1779–1785 (2016)
16. Y.H. Jang, Y.J. Jang, S. Kim, L.N. Quan, K. Chung, D.H. Kim, Chem. Rev. **116**, 14982–15034 (2016)
17. M. Cid, N. Stem, C. Brunetti, A.F. Beloto, C.A.S. Ramos, Surf. Coat. Technol. **106**, 117–120 (1998)
18. N. Sahouane, A. Zerga, Energy Procedia **44**, 118–125 (2014)
19. E. Yablonovich, G.D. Cody, IEEE Trans. Elect. Dev. **ED-29**, 300–305 (1982)
20. F.J. Haug, T. Söderström, O. Cubero, V. Terrazzoni-Daudrix, X. Niquille, S. Perregeaux, C. Ballif, MRS Proc. 1101 (2011). https://www.cambridge.org/core/journals/mrs-online-proceedings-library-archive/article/periodic-textures-for-enhanced-current-in-thin-film-silicon-solar-cells/9AA982E474D51427AE0D8080C3C67680
21. J. Krc, M. Zeman, A. Campa, F. Smole, M. Topic, MRS Proc. 910 (2011). https://www.cambridge.org/core/journals/mrs-online-proceedings-library-archive/article/novel-approaches-of-light-management-in-thinfilm-silicon-solar-cells/8C578252B8FF5E52B07B9D6BA0F706E4
22. S. Solntsev, O. Isabella, D. Caratelli, M. Zeman, IEEE J. Photovoltaics **3**, 46–52 (2013)
23. O. Isabella, A. Campa, M. Heijna, W. J. Soppe, R. v. Erven, R. H. Franken, H. Borg, M. Zeman, in *23rd European Photovoltaic Solar Energy Conference and Exhibition*, Valencia, Spain
24. S. Solntsev, M. Zeman, Energy Procedia **10**, 308–312 (2011)
25. Z. Yu, A. Raman, S. Fan, Opt. Express **18**, A366–A380 (2010)
26. L. Zeng, Y. Yi, C. Hong, J. Liu, N. Feng, X. Duan, L.C. Kimerling, B.A. Alamariu, Appl. Phys. Lett. **89**, 111111–111113 (2006)
27. L. Zeng, P. Bermel, Y. Yi, B.A. Alamariu, K.A. Broderick, J. Liu, C. Hong, X. Duan, J. Joannopoulos, L.C. Kimerling, Appl. Phys. Lett. **93**, 221105–221103 (2008)
28. M.B.H. Breese, D. Mangaiyarkarasi, Opt. Express **15**, 5537–5542 (2007)
29. J. Diener, N. Kunzner, D. Kovalev, E. Gross, V.Y. Timoshenko, G. Polisski, F. Koch, Appl. Phys. Lett. **78**, 3887–3889 (2001)
30. K. Jiang, A. Zakutayev, J. Stowers, M.D. Anderson, J. Tate, D.H. McIntyre, D.C. Johnson, D.A. Keszler, Solid State Sci. **11**, 1692–1699 (2009)
31. S.-H. Park, D.-H. Lee, C.-S. Kim, H.-D. Jeong, Jpn. J. Appl. Phys 50 **4**, 012503–012503 (2011)
32. L. Cheng-Chung, L. Meng-Chi, C. Sheng-Hui, K. Chien-Cheng, Jpn. J. Appl. Phys. **51**, 052602 (2012)
33. W.C. Tien, A.K. Chu, Sol. Energy Mater. Sol. Cells **120**(A), 18–22 (2014)
34. M.F. Schubert, J.K. Kim, S. Chhajed, E.F. Schubert, in *Thin-Film Coatings for Optical Applications IV*, San Diego, 26 August 2007
35. M.F. Schubert, J.-Q. Xi, J.K. Kim, E.F. Schubert, Appl. Phys. Lett. **90**, 141115 (2007)
36. J.D. Joannopoulos, S.G. Johnson, J.N. Winn, R.D. Meade, *Photonic Crystals: Molding the Flow of Light*, 2nd edn. (Princeton University Press, New York, 2008), pp. 44–65
37. Z.P. Ling, T. Mueller, A.G. Aberle, R. Stangl, IEEE J. Photovoltaics **4**, 1320–1325 (2014)
38. C.J.R. Sheppard, Pure Appl. Opt. **4**, 665–669 (1995)

39. A. Jaffer. (2010). *The freesnell thin-film optical simulator*. Available: http://people.csail.mit.edu/jaffer/FreeSnell/
40. W. Theiss. (2008). *CODE*. Available: http://www.wtheiss.com/
41. M. Born, E. Wolf, *Principles of Optics: Electromagnetic Theory of Propagation, Interference and Diffraction of Light*, 7th edn. (Cambridge University Press, Cambridge, 1999), pp. 1–986
42. Z.P. Ling, J. Ge, R. Stangl, A. Aberle, T. Mueller, J. Mater. Sci.Chem. Eng. **1**, 1–14 (2013)
43. M. Zeman, D. Zhang, in *Physics and Technology of Amorphous-Crystalline Heterostructure Silicon Solar Cells*, ed. by W. G. J. H. M. v. Sark, L. Korte, F. Roca (Springer, Berlin/Heidelberg, 2012), pp. 13–43
44. W. Luft, Y.S. Tsuo, *Hydrogenated amorphous silicon alloy deposition processes*, 1st edn. (Marcel Dekker, Inc, Colorado, 1993), pp. 1–344
45. W.C. Tien, A.K. Chu, Opt. Express **22**, 3944–3949 (2014)
46. U. Manna, J. Yoo, S.K. Dhungel, M. Gowtham, U. Gangopadhyay, K. Kim, J. Yi, J. Korean Phys. Soc. **46**, 1378–1382 (2005)
47. Z.P. Ling, Design, fabrication and characterisation of thin-film materials for heterojunction silicon wafer solar cells, PhD Thesis, National University of Singapore, Singapore, 2014
48. Z.P. Ling, S. Duttagupta, F. Ma, T. Mueller, A.G. Aberle, R. Stangl, AIP Adv. **5**, 077124 (2015)
49. G.F. Burkhard, E.T. Hoke, Transfer Matrix Optical Modeling (2010). http://web.stanford.edu/group/mcgehee/transfermatrix/index.html publisher: McGehee Group
50. G.F. Burkhard, E.T. Hoke, M.D. McGehee, Adv. Mater. **22**, 3293–3297 (2010)
51. T. Mishima, M. Taguchi, H. Sakata, E. Maruyama, Sol. Energy Mater. Sol. Cells **95**, 18–21 (2011)

# Optoelectronic Characteristics of Passivated and Non-passivated Silicon Quantum Dot

A. Laref

**Abstract** In this chapter, our recent research about the impact of hydrogen passivations and size on the electronic and optical features of silicon quantum dots will be reviewed. A theoretical modeling will be presented for silicon quantum dot with spherical topologies and treating their corresponding physical properties. This recent study was conducted by means of first principle calculations to explore the energy band gap versus the radius of Si quantum dots for passivated and non-passivated surface. The optimization of the structures of quantum dots was performed for both passivated and unpassivated quantum dots with various sizes. The interesting features for the electronic characteristic, such as the energy band gaps are higher in the case of hydrogenated surface than the unpassivated case. Accordingly, both quantum confinement and surface passivation provide information concerning the electronic and optical characters of Si quantum dots. The passivation impact on the surface dangling bonds with hydrogen atoms as well as the contribution of surface states on the gap energy are also presented. The hydrogen passivation influence increases the energy gap than that of pure silicon quantum dots. The significant character of the confinement and surface passivation on the optical properties are reviewed. The previous experimental determinations have shown that the optical properties of these dots were significantly affected by the quantum confinement effects. Overall, the hydrogen saturation surface controls principally the ground-state geometry, the energy gap, and optical absorption of Si quantum dots with the change of radius size. It was inferred in our previous study that the insertion of hydrogen could lead to the alteration of the electronic structure of silicon quantum dots. The saturated surface by hydrogen atoms has also a main contribution on the spatial distribution of the highest occupied and lowest unoccupied molecular orbitals. The hydrogen effect on optical absorption spectra and the static dielectric constant are also reviewed. Exclusively, the absorption threshold

A. Laref (✉)
Department of Physics and Astronomy, Science Faculty, King Saud University, Riyadh, King Saudi Arabia, Saudi Arabia

Department of Physics, National Taiwan University, Taipei, Taiwan
e-mail: amel_la06@yahoo.fr



S. Ikhmayies (ed.), *Advances in Silicon Solar Cells*,
https://doi.org/10.1007/978-3-319-69703-1_2

relationship of Si nanoparticles on the radius and hydrogenation surmise a decrease in the quantum confinement effect. The absorption spectra illustrated that the absorption properties are intimately accompanied with the surface saturation as well the radius of the dots. This theoretical finding could assist the comprehension of the microscopic mechanism which is spectacular for the devices performance and the potential application in nanotechnologies. This could highlight the significant optical parameters of silicon quantum dots for the purpose to comprehend the optical properties in the photoluminescence process of finite-size dots. The recent work about the optical absorption showed that the nanostructured Si could possess a very high luminescence in the visible regime as reported in the experimental inspection.

## 1 Introduction

Solid-state semiconducting nanostructures represent a very active realm of research. Exclusively, quantum dots yield confinement of charge carriers (holes, electrons, and excitons) in the entire three spatial dimensions that arise in discrete energy, localized wave functions, and further fascinating physical and advanced device characteristics [1]. The challenge is to create functional nanomaterials possessing a confined size distribution and ultimately to elucidate the relationship between shape, size, geometry, properties, and interfaced with the substrate. The preparation, synthesis, and growth of nanomaterials on substrate with manageable dimensions, shapes, and orientation are nontrivial. The most crucial task is to contemplate a convenient and reproducible technique for sizeable scale synthesis, besides to a compatible and simple integration into traditional microelectronic processing techniques. Frequently, the physical aspects of nanomaterials are ruled via the manipulation of particle size, granting the design of the growth process with the prerequisites and appealing applications of the ultimate nanodevices. Interestingly, these nanomaterials own outstanding properties that are not present in their bulk counterparts. Intriguingly, semiconductor quantum dots (QDs) have aroused paramount interest due to their novel technological applications that are realizable in the physics of zero-dimensional structures. It is indispensable to examine the physical aspects of these prodigious nanomaterials, the so-called semiconductor QDs [2–4]. These promising systems possess extraordinary electronic and optical characteristics that are tunable artificially.

The immense attempts have been centralized to tailor the material dimension at nanometer scale and thereby to achieve desirable properties that can be produced by varying the shape. For instance, elemental silicon (Si) owns very distinct properties versus the change of size. Meanwhile silicon in its bulk form is the prevailing system that produces the electronic devices; it possesses poor optical features for optoelectronic devices, which can be significantly changed at nanoscale dimensions. For instance, the band gap in silicon would be blue-shifted from the infrared to the optical region versus the size. One of the first revealing effects detected in porous silicon illustrated notable visible (red) photoluminescence (PL) at room temperature [5–10]. This striking behavior assisted intensive research attempts into the examination of the optical properties of Si nanocrystals that have been investigated extensively at frequencies approaching the fundamental energy gap [11–14]. Then, both shape and dot size are mandatory for the modulation of the optical gap of these QDs. In this perspective, silicon quantum dots have recognized a great curiosity owing to the intriguing underlying physical aspects and the propitious applications in advanced nano-electronic and optoelectronic devices, such as solar cells [6, 15, 16]. The search of optical active devices based on silicon, namely, lasers has received a stimulating mission. The pure bulk silicon owns an indirect optical gap which is inappropriate material for active optical applications, whereas enormous low-dimensional silicon nanostructures are seemingly favorable for diverse technological applications. A substantial change is anticipated in the electronic and optical aspects of size-reduced Si nanostructures [6, 17, 18] because their bulk form is a semiconductor with an indirect band gap of 1.1 eV. It is inevitably recognized that the emission and absorption in silicon occurs in the visible light with the reduction of the size to nanometer scale beneath its exciton Bohr radius (5 nm). Then, the direct-gap semiconductor nanostructures pledge a luminous future. The usage of silicon in photonics has been also significantly triggered by the pragmatic aim such as the inexpensive price and the undemanding manufacturability of the material [19–21].

Since the mechanism of visible PL in Si was first detected by L. T. Canham [2] earlier in the 1990s, many researchers have brought experimental and theoretical works toward Si luminescence [22–25]. Importantly, the substantial enhancement of photoluminescence (PL) in porous silicon could be elucidated by the simple quantum confinement effects [26–29]. Eventually, this may drive to high-performance photoluminescent Si nanostructures. A substantial prospective for next-generation lighting has been manifested in Si quantum dots by engineering their band gaps, quantum efficiency (QE), high photoluminescence (PL), and intense color clarity [27]. The alleviation of the devices' size and boosting the performance would prevail the device mechanism, such as the nano-miniaturization by enhancing the optical characteristics of Si QDs. Over the last few decades, PL features of silicon nanomaterials, like emission and absorption levels, carrier dynamics, and spectral characteristics have been extensively scrutinized on ensemble of silicon nanomaterials [28, 29]. According to the previous research, PL internal quantum efficiency of single Si QDs was acquired around 10% at room temperature, for oxide-passivated nanocrystals [30]. However, the tunable size of the nanocrystal leads essentially to the viable shift of the band gap to the optimal energy gap (i.e., 1.5 eV) for supreme efficiency of single-junction solar cells, or for stacking cells with dissimilar band gap and then, joining various spectral response can establish tandem cells. In this respect, Si QDs have exhibited substantial PL in the visible red span of the solar spectrum, and external quantum efficiencies about 60% have been delineated in literature [15, 16]. This substantial PL can be employed in the photoluminescence downshifting to enhance the efficiency. Irrespective with the bulk Si counterpart, the relaxation of excited carriers is declined in quantum dots because of the enlarging of the distance between the energy states. It would be feasible to enhance the open-circuit voltage of the solar cells, and this could practically happen by the extraction of the carriers through the dot before the relaxation to the extremity of the band gap. The successful extraction of the multiple excitons through the dot may conduct to a boosted current of the solar cell. On the other hand, the recombination radiative of the multiple excitons in the QD conducts to the enhancement of the PL quantum efficiency. Other features, such as hot-carrier cells, are greatly favorable to boost the high thermodynamic efficiency limit (~85%), and however the hot-carrier solar cell has not been produced thus far. The characteristics of Si QDs concentrated exclusively on the properties pertinent for the usage in solar cells, like the band gap increase, PL characteristics, electrical transport, etc. In this respect, an extensive amount of scrutiny was performed in this arena since the detection of substantial PL from Si porous [24–30].

On the other hand, based on both underlying understanding and realistic application of Si QDs, a keen scrutiny in their optical properties is compulsory. Hence, it has been found via experimental measurements that the optical properties of Si quantum dots can drastically influenced by quantum confinement effects. Using theoretical techniques, researchers can predict also precise optical gaps in Si quantum dots [24, 25]. In this respect, the energy gap (Eg) of Si dots was determined to drastically rely upon the diameter and surface passivation. Substantial experimental and theoretical works were evidenced to display the quantum confinement effects on PL via quantum-confined excitons [30]. Several traditional experimental tools employed to investigate the bulk materials and bulk surfaces cannot be directly applicable to Si nanocrystals. Furthermore, theoretical and computational modeling represents a challenging task because many nanostructures are not stable due to their size, exhibiting an inherently quantum mechanical phenomenon. Providing a sizeable surface area of Si nanoparticles, in order to tune their nanostructures for particular applications, it is imperative to comprehend and rule the effect of surface properties on their optical gap and in overall their physical properties. The comprehension of the impact of surface reconstruction and passivation on the ground-state properties of Si nanodots represents a key prerequisite, not exclusively to design the technological applications of single dots but to manage the deposition of nanoparticles on aggregation and surfaces of multi-dots into new structures. Subsequently, the examination of visible photoluminescence in porous silicon and in silicon nanostructures [26, 27] demands a realistic control of the optical response of Si nanosized structures versus the size. This has conducted to one of the most exciting features of the contemporary silicon-based nanotechnologies [28].

According to the previous experimental and theoretical determinations, it has been found that hydrogen-terminated surface can mainly affect the electronic and optical aspects of silicon nanocrystals [26–28]. In these systems, nearly most of silicon atoms are lying on or in the vicinity of the surface and are presumably influenced by the surface or passivation, which can lead to a drastic change in their electronic structures. Although several models have been proposed to reveal the PL phenomenon in porous silicon (exposed to air), till now its emergence is still under debate. The experimental results have not clarified yet the observation of surface effects from the bulk state. Thereby, it was not possible to register the size relationship between the PL energies and radiative lifetimes in passivated silicon nanocrystals, and the luminescence aspects would be connected to the surface terminations with H atoms in Si QDs. The previous experimental works have found a difficulty with Si QDs possessing ultrasmall 1–2 nm-diameters [27, 28], albeit density functional theory (DFT) simulations are promising for supporting or predicting the electronic structure of these nanomaterials. Numerous problems, like hydrogen-covered Si QDs and size correlation on the absorption spectra are still under debates. Therefore, to evidently distinguish the effect of surface terminated by hydrogen atoms on the electronic and optical properties of host silicon QDs, our recent research based on theoretical examination toward these nanostructures with diameter up to 1.1 nm will be presented. The effect of morphological properties will be presented, by involving size and shape of hydrogenated and unsaturated silicon QDs on the electronic and optical properties versus diameter of dot. Accordingly, the previous simulations were employed by means of first-principles methodology within the framework of pseudopotential scheme [31]. The effect of hydrogen-passivated silicon surface will be essentially investigated by the elimination of dangling bonds. To clarify theoretically the size impact on the electronic structure of Si QDs terminated by H atoms and the mechanism of the observed red shifts in PL, the modeling of Si nanostructures topologies with small and moderate size of medium quantum dots will be addressed [32]. The dot size dependence and hydrogen-passivated surface impacts on the energy gap and absorption spectra of Si dots will be handled.

In this chapter, the recent work concerning the effect of quantum confinement on the energy gap will be reviewed in both terminated and non-terminated silicon quantum dots versus the change of size [32]. A short review of Si nanostructures will be reported in the next section. Additionally, the photoluminescence measurements of silicon nanostructures will be illustrated briefly. The important quantum confinement and its effects on the energy gap of nanostructures will be discussed after that. To control the electronic, bonding character, and optical properties that are relying on the dot size and surface termination by hydrogen on Si QDs, the recent results, such as the density of states, optical absorption spectra, and static dielectric constant versus the dot size will be presented [32].This is based on the estimation of the structure and stability for various hydrogenated and unsaturated Si quantum dots, which are compared with the data of the relevant literatures. To understand what is happening at microscopic level of the bonding of unsaturated and H-saturated Si QDs, the distribution of the highest occupied molecular orbitals (HOMO) and the lowest unoccupied molecular orbitals (LUMO) will be desalinated after that. Thus, to elucidate this mechanism, it is pertinent to examine how the correlation between hydrogen and size dot influences the quantum confinement on energy gap and optical absorption spectra in these nanomaterials. The computed optical absorption spectra of hydrogenated and unsaturated Si quantum dots are based on the dielectric function jointly together with dipole matrix elements of

HOMO–LUMO states. For hydrogenated and unsaturated Si QDs, the static dielectric constant was also calculated with different size of dots. At the end, this chapter will be finalized by a summary and conclusion. This will highlight the significance of the previous results [32] respecting the optical parameters of silicon quantum dots to comprehend the optical properties in the luminescence process.

## 2 Previous Review on Si Quantum Dots

Over the last years, the fabrication of silicon nanocrystals have had an enormous advance for the light-emitting devices, while sharper nanocrystal size distributions have been acquired [1–4]. A prominent purpose was about the revelation of a significant photoluminescence through the porous silicon, comprising a very simple and productive strategic for maintaining high-performance photoluminescent silicon structures [5]. After the realization of visible PL in Si, enormous investigations were conducted both theoretically and experimentally [8–12]. Employing the linear combination of atomic orbital scheme, the computed electronic structures of spherical Si crystallites having diameter up to 4.3 nm have been pointed out earlier in 1992 by Proot et al. [33]. Accordingly, the change in the optical band gap versus the cluster size was found to verify the experimental determinations [34]. An augmentation in the radiative efficiency of Si–PL was evaluated previously [33] because of the confinement yielding a spread of the wave functions in the reciprocal space. Lately, based on first-principles DFT, Delley et al. [35] examined the visible luminescence in Si nanoclusters varying up to 3 nm. Their results implied that the scaling of the band gap has a linear behavior with the cluster diameter. They also realized that the dipole transitions through the gap in Si nanoclusters are symmetry permitted, conversely to the dipole-forbidden transitions in a bulk Si lattice. Furthermore, the near-gap electronic structures of Si nanoclusters with the size up to 3.7 nm were analyzed hitherto [36, 37] by employing plane-wave pseudopotential framework. Hence, this theoretical finding was in consensus with both experiments and all electron calculations [38].

Albeit the aforementioned theoretical works [39–41] consistently illustrated that the gap in a Si nanocluster can be manipulated with the change of size; semiempirical modifications have to be implemented to the gap in comparison with experimental optical gaps. For instance, the value of 0.6 eV has been involved to theoretical gaps in the reference [40]. This is due to the electron self-energy and the screened electron-hole Coulomb attraction which are not exactly considered in the calculations, involving those determined with DFT scheme. With the above methods, researchers have computed exact optical gaps in Si clusters, and the gaps illustrate significant size dependence. Based on quantum confinement effects, the optical gap in Si cluster of 2.0 nm was approximately 2.6 eV, and the gap of the 1.5 nm cluster was nearby 3.5 eV. It has been stated that together with the quantum confinement, the optical gaps of Si clusters are exhibited to drastically alter across surface passivation. For instance, in a spherical Si cluster comprising 35 Si atoms, the DFT predicted an energy gap around 3.4 eV for $Si_{35}H_{36}$ (36 hydrogen atoms saturated surface dangling bonds). Their work indicated that hydrogen and other single-bonded atoms exhibit same gaps. Eventually, except the quantum confinement and surface passivation, researchers examined other factors that may in addition alter the energy gap [42]. Those key factors incorporate doping or defects. Since then, the optical properties of silicon nanocrystals have been explored substantially both experimentally and theoretically, but yet the correct atomic structures and their correspondence to the optical properties are unforeseeable [40–45]. Particularly, the cause of the photoluminescence in nano-silicon is still not clear and is under controversy.

In the other side, various phenomena have been suggested; the most feasible illustrates the quantum confinement effect, various defects in the surface like hydrogen or oxygen defects or dangling bonds, as well as suboxide formation. Then, the energy band gap could be opened up because of quantum confinement effect in the nanocrystallites and relaxation of the selection rules for radiative transmissions. This could lead to the emergence of photoluminescence in the visible regime for crystallites size beneath ~5 nm. A crucial characteristic for porous silicon is how the excitations are located in nanoscale dimensions. If the technological request can be clarified, silicon nanomaterial would be a good optical material. A keen comprehension about the optical aspects of matter is demanded to capitalize satisfactorily such mechanism. In this attitude, a sound comprehension of the electronic and optical properties of silicon nanocrystals is necessary. The quantum size effects show novel mechanisms in silicon, like optical absorption. All the key ingredients that turn Si quantum dots into prosperous for optoelectronic applications will be discussed in the next coming sections.

## 3 Photoluminescence Measurements

Luminescence from a semiconductor excited with photons (photoluminescence) or by electric current (electroluminescence) is a robust tool in the investigation of its properties. This technique can supply information on the optical band gap, carrier relaxation, viable defect configurations, and numerous of other properties. Enormous previous studies have been performed on light emission through Si QDs since they are likely to be the luminescence centers in porous Si (Psi) that had been detected in the visible PL and electroluminescence at room temperature. As it was stated in Ref. [46], it is so hard to exactly control the size and purity of the sample where the results of various experiments could differ from each other to some extent. The size dependence on the energy gap of Si nanostructures has been examined substantially [47–50] since the observation of room temperature visible photoluminescence in silicon nanocrystals [49] and porous silicon [47]. The quantum size impact, inducing in a blue shift of the energy gap versus the reduction of size, was extensively surmised to be prominent for the optical features of porous silicon (PSi) [47, 48]. The energy gap was observed to rise significantly with the

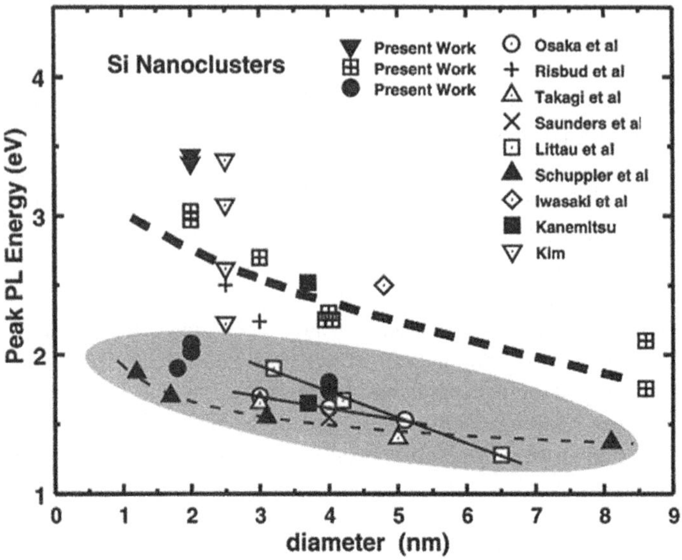


**Fig. 1** Summary of experimental data on peak PL energy versus Si QD size [46]

reduction of size. Very recently, many efforts have been directed to generate quasi-direct-gap semiconductor nanostructures and a valuable research attempts are concentrated on nanometer size crystallites or quantum dots fabricated from silicon or germanium [47, 48]. Significant PL from silicon nanostructures created by electrochemical anodization, frequently named porous silicon, has aroused much interest from basic physics perspective, specifically of the potential application in optical devices.

It has been outlined that the most of the observed experimental results for PL peak energies vary versus the diameter of the Si QDs, as depicted in Fig. 1. Note that the differences occur for the dependency on size, as reported by various authors [47, 48]. The entire PL peaks of Si QDs embedded in glass matrices or SiO2-capped Si QDs drop into the shaded zone of the diagram, although the experimental data for the oxygen-free samples (for similar value of the diameter) drop up this shaded zone. For this reason, it is evidently noticed that the interface case has a major contribution in the luminescence across Si QDs. Furthermore, an analogous conclusion has been acquired in Ref. [51] in comparison with the PL of PSi samples at various atmospheres, as displayed in Fig. 2. Firstly, the freshly etched samples (oxygen-free PSi) were examined at various porosities (equal to various sizes of Si QDs) and emitting across the visible spectrum. After that, the exposure of oxygen-free PSi samples to air was attained, and their PL energies were measured additionally. It is obviously detected that after the exposure of samples to air, a red shift of the PL is assigned, which is as big as 1 eV for blue luminescent samples comprising QDs lesser than 2 nm (see Fig. 2). Also, it has been reported that a PL red shift occurred as soon as the transfer of oxygen-free PSi samples has been

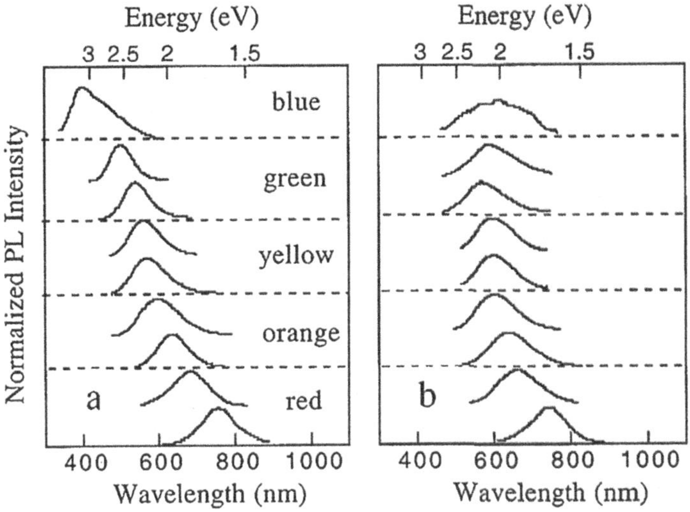


**Fig. 2** Room temperature PL spectra from PSi samples with different porosities kept under Ar atmosphere (**a**) and after exposure to air (**b**) [47]

obtained from Ar to a pure oxygen atmosphere, although no red shift was observed at all if the samples were remained in pure hydrogen atmosphere or in vacuum. Furthermore, an upper limit of the emission energy (2.1 eV) was acquired independently of size. In conclusion, the surface passivation has a crucial role, particularly the chemistry of oxygen at the surface can create an obvious alteration for the PL of Si QDs and an upper limit of energy gap. On the whole, in spite of some inevitable dissimilarities, the experimental results exhibit that for ultrapure Si QDs as their sizes reduce, a significant blue shift in the peak PL energy occurs which is owed to the augmentation of band gap. Therefore, if the Si QD is not pure enough, much more complex results will be acquired because of the effect of impurity contamination and complicated interface case.

## 4 Important Quantum Features for Nanoscale Si Dots

One strategy to ameliorate the optical properties of indirect semiconductors could be based on the decrease of their dimensionality, for instance, by the formation of nanocrystals (NCs). The quantum confinement effects are developed and might change the electronic structure if the size of a structure is approximated to the size of Bohr radius of an exciton in a material [7]. The most significant modifications correspond to the opening of the band gap. Then, the production of wider exciton energies happened in smaller NCs which possess wider band gap [8]. From experimental point of view, this can be noticed like a switch of luminescence to shorter wavelengths [9, 10]. Overall, three main recombination pavements represented by direct band-to-band transition, a transition including a phonon, and, ultimately, from a defect state (which is mostly a non-radiative procedure) of excited carriers to recombine in bulk Si. The first procedure relies upon the excess energy dissipation: a radiative recombination, in which the transfer of energy happened with the propagating wave and Auger recombination. Therefore, the absorption of energy is developed through another preexcited free carrier. The bulk silicon owns an intrinsic indirect band gap property, as referred in Ref. [13]. The direct band-to-band transition is of exceedingly low probability and considered as "forbidden" because of indirect band structure of Si [13]. The location of the valence-band maximum and conduction-band minimum is not positioned at similar momentum wave vector k. The assistance of carrier band gap transitions is based on phonon absorption or emission [14]. For nanoscale silicon dots, firstly, the electron and hole wave functions have an augmentation in the spatial localization and an augmentation of their spread in momentum space with the dot. Thus, an enhancement in their overlap probability will occur in the Brillouin zone. Secondly, nanodots possess a very insignificant number of atoms. However, imperfections, like point defects and dislocations which produce non-radiative recombination are implausible to occur. The carriers' confined nanocrystals induce the efficiency of luminescent recombination to substantially ameliorate effectively since the suppression of non-radiative recombination pathways is present, and the radiative recombination rate is increased. At room temperature, the excitonic luminescence in silicon nanocrystals can simply be detected in the near-infrared Raman (nearby 700–800 nm) which is hundreds of times brighter than the roughly 1100-nm bulk intrinsic luminescence at cryogenic temperatures.

The visible photoluminescence which has been detected in silicon nanostructures at room temperature has significantly aroused extensive attention in these specific nanoclusters and in small semiconductor particles [52, 53]. The most stimulating prospect of this discovery is connected to the light-emitting devices based on this effect, and it could emerge likely within the well-attainable silicon technology for optoelectronic applications [54, 55]. The well-recognized band structures of these nano-silicon structures are compulsory for the comprehension of this effect. In the last decade, valuable attempts have been devoted to analyses whether silicon can change inherently a direct semiconductor in the case of nanocrystalline clusters (or porous). It has been attained that quantum confinement can alter the energy gap in which the visible luminescence is generated as experimentally detected [54, 55]. Furthermore, the band gap in silicon is significantly small to get interacted efficiently with the visible spectrum. With the adjustment of the gap in silicon, then it could be employed for either electronic or optical application. Exploration of mechanism, as the Stokes shift (difference between absorption and emission energies), the radiative lifetimes, the nonlinear optical properties, the PL emission energy as a function of nanocrystals size, the quantum-confined Stark effect (QCSE), and so on can provide basic contribution to the comprehension of tunable optical response of such systems.

In the case of free-standing Si-NCs, significant luminescence is detected, the color relying upon the size of the Si-NCs, and the gap energy is grown with the decrease of their size. A variety of procedures have been attempted during the last few years for light generation and amplification in silicon. The most valuable one is founded on silicon nanocrystals (Si NCs) with the purpose to take advantage of the diminished dimensionality of the nanocrystalline structure (1–5 nm in size), while the band folding, quantum confinement, and surface effects have a significant contribution [55]. In fact, it has been procured that the Si NC band gap enhances versus the reduction of size and a photoluminescence (PL) external efficiency in excess of 23% is attained [56]. Si NCs-based LED with high efficiency have been acquired in Si NC active layers [52] and establishing separate injection of electrons and holes [56]. Furthermore, optical gain under optical pumping has been previously manifested in a wide diversity of experimental conditions [53–57].

## 5 Structure of Silicon Quantum Dots

The typical size of Si QDs lesser than ten nanometers is approximately near the exciton Bohr radius of bulk silicon. In the other side, silicon quantum dots show strong quantum confinement which induces a broaden band gap, and the electronic states may develop discrete levels because of the extreme small dimensions. Overall, at ideal conditions, it was accounted that the interior of the dot possesses the structure of crystalline silicon, whereas the surface of the dot is saturated with particular atoms relying upon the surrounding environment of the dot, like hydrogen, oxygen, and so on.

### *5.1 Ideal Structure*

Enormous experimental investigations have been performed on the electronic and optical properties of Si QDs. Therefore, various elements have major role in rendering the explanation of measurements an arduous task. For example, samples illustrate a considerable dispersion in the QD size that is hard to be established. Furthermore, Si-NCs have been synthesized using various techniques that usually illustrate distinctive properties in shape, size, and the interface structure [58]. For the justifications indicated above, most of the experimental investigations provide diverse results. Hence, theoretical model calculations for some ideal structures have been accounted very imperative to explore the properties of Si QDs. Overall, saturated-surface silicon nanoclusters are the ideal theoretical structure for us to study Si QDs [56–58].

# 6 Physical Properties of Unpassivated and Passivated Si Quantum Dots

The previous calculations were carried out by means of pseudopotential ab initio simulations [31] to explore the effect of hydrogen atoms terminating the surface of silicon QDs (the size change ~0.2–1.1 nm) and the unsaturated Si QDs. DFT methodology has a crucial beginning point in our recent study and can be useful for more complex materials by extracting pertinent microscopic information. In the previous study, ab initio calculations were performed by employing pwscf package [31], in which the basis set is a plane-wave pseudopotential framework. The structural calculations are carried out by applying DFT [59] with the conventional generalized gradient approximation (GGA). Specifically, PBE exchange and correlation functional [60] was used and a plane-wave approach with the super cell technique. For the exchange-correlation energy, the ultrasoft Vanderbilt pseudopotentials with semicore corrections [61] were employed for both Si and H atoms. The plane-wave kinetic energy cut-off is chosen to be 45 Ry for guarantying the convergence for the total energy to be less than $10^{-5}$ Ry/cell. All the atoms are thoroughly relaxed until the Hellmann–Feynman force acting on each atom is smaller than $10^{-4}$ Ry/a.u. The dots are then relaxed to minimize the total energy employing the conjugate gradient algorithm [31].

## 6.1 *Size Effect*

Beginning from a central Si atom, the construction of Si nanocrystals is done by a consecutive shells of Si atoms. Si QDs with bigger diameters were not taken into account in the previous work [32] because of the limitation of computational resources. For unpassivated Si QDs and Si QDs terminated with H atoms, the nanoclusters with spherical shapes are described by a wide periodic repeated supercell (20~a.u.). By considering these configurations, the interactions between the dot and its replica are not taken into account. The modeling of spherical dots is done by the generation of atomic positions at the bulk interatomic distances up to the limited diameter. The H-terminated Si QDs own tetrahedral clusters that can be relaxed with an average of Si-Si and Si-H bond lengths of 2.36 Å and 1.45 Å, respectively. Then, by minimizing the total energy of each configuration, the stable configurations of all dots with given size can be determined. A ball and stick models for silicon quantum dots at different diameters are displayed in Fig. 3. The structures with hydrogen termination are also depicted in Fig. 4 [32].

For example, the modeling of Si QDs is established by commencing from tetrahedral, Si-atom centered, hydrogen-saturated clusters, such as $Si_{28}H_{32}$, as illustrated in Fig. 4. Since DFT was fundamentally employed at systematic ground-state calculations, the aspects of excited states (relative to the optical gap) are not exactly predicted. However, DFT theory underestimates the energy gap

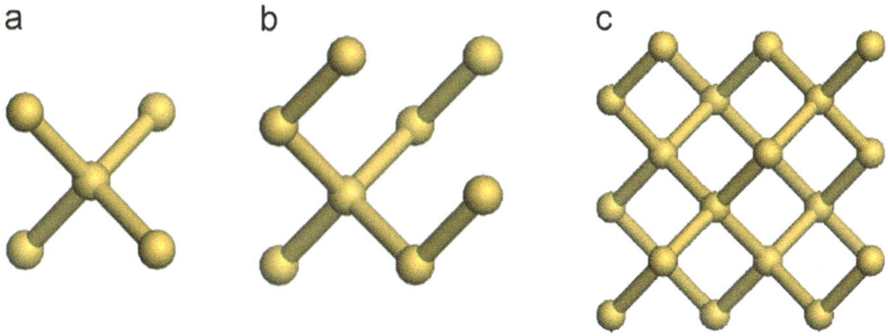


**Fig. 3** Snapshots of various unsaturated Si quantum dots. Yellow dots are Si atoms. The size of these quantum dots is also given. (**a**) $Si_5$, (**b**) $Si_8$, and (**c**) $Si_{28}$ [32]

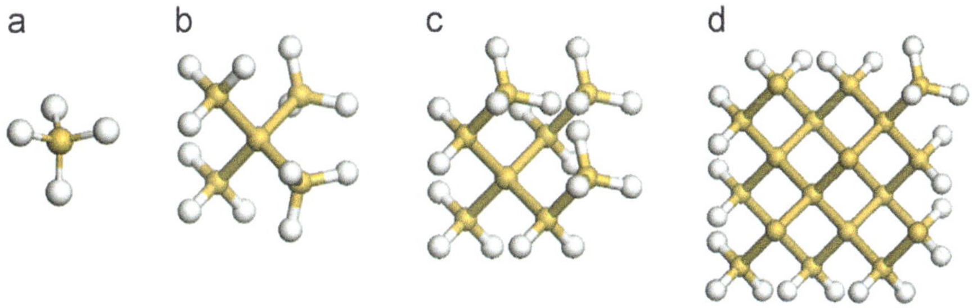


**Fig. 4** Snapshots of various passivated Si quantum dots by hydrogen atoms. Yellow dots are Si atoms and white H atoms. In all cases, the surface dangling bonds are saturated with hydrogen atoms. The size of these QDs is also given. (**a**) $SiH_4$, (**b**) $Si_5H_{12}$, (**c**) $Si_8H_{18}$, and (**d**) $Si_{28}H_{32}$ [32]

irrespective with the experimental results. Fortunately, the underestimation of the energy gap via DFT might be ameliorated using GGA-1/2 approximation, and the detailed calculations are provided elsewhere in Ref. [61]. In fact, all GGA-1/2 calculations are performed with GGA-relaxed structural parameters. Here, the GGA-1/2 method is founded on the half occupation (transition state) by involving hole self-energy in the Schrodinger equation. The GGA-1/2 technique can treat the elimination of the spurious electrostatic electron self-energy in the electronic structure calculations of material. Its technique can proceed a variational principle without using any fitting parameter. For this purpose, after carrying out DFT geometry optimizations and calculating the gap between the HOMO and LUMO, the electronic structures of the considered systems can be simulated with the energy gap corrections based on GGA-1/2 [61]. The HOMO–LUMO gaps are intimately connected to the size of the QDs and the localization of electronic states close to the gaps. By this way, adequate trends for the HOMO–LUMO gap can be procured versus the dot size besides to the hydrogen-terminated and unsaturated Si QDs.

To examine the electronic and optical properties of Si QDs in terms of surface termination, it is important to clarify their underlying physics by the elimination of the dangling bonds from the surface. Thereafter, to stabilize the system at different dots, the contribution of hydrogenated Si QDs is important. The saturated and unsaturated Si quantum dots are taken into account at different sizes and are compared to their corresponding band gap results. The states near the band gap will be presented next, since the termination of all dangling bonds at the surface is pertinent, or else, the band gap could be modified by the dangling bond states. Then, the localization of surface states will be removed from the energy region close to Fermi level when passivating the dangling bonds.

## 6.2 *Electronic Properties of Unpassivated and Passivated Si Quantum Dots*

The electronic structure results of different atomic configurations of Si QDs will be reviewed with the size variation between 4 and 60 atoms, as depicted in Figs. 3 and 4 [32]. To examine the quantum size effects and the effect of hydrogen-terminated surface on the energy gap, the recent results of the computed density of states (DOS) for the optimized Si QDs at various dot diameters will be presented. The DOS of unpassivated and passivated Si QDs are exhibited in Figs. 5 and 6 [32], where the diameters are altered between 0.2 and 1.1 nm. Apparently, the hydrogenated Si QDs possess larger energy gap than the unsaturated Si dots. These nanocrystals illustrate molecular characteristics with manifold peaks, and their components are increased more due to the surface termination from the Si-H bonding. The fluctuations are more considerable for smaller dots. The terminated $Si_{28}H_{32}$ quantum dot illustrates a band gap around 4.2 eV, in which the dangling bonds are saturated by 32 hydrogen atoms at surface. This is a good anticipation for the dots with this size span. Clearly, the energy gap of pristine dot emerges through the surface states, and the results are in consensus with the previous available theoretical reports [62, 63].

For unpassivated $Si_8$ dot, two main peaks are observed at two boundaries of gap. The first peak is only centered downward the highest valence states, and the second one is located upward the band gap. Because of the surface termination with hydrogen in $Si_8H_{18}$ QD, the two major peaks of pure dot disappeared. In this situation, the energy gap is enlarged due to the termination of the dangling bonds by hydrogen atoms. Hence, some changes in the inner states arise through Si-H bonding. Note that the occurrence of hydrogen pulls up principally the conduction states with the addition of new peaks, and its influence also changes the character of the valence states. It is notable that the optical response spectra will be modified in silicon terminated with hydrogen irrespective with the unsaturated dot. In the case of the dot having larger size, the system favors fourfold $sp^3$-like coordination surrounding the inert region, although the surface of the dot is over-coordinated through the core atoms. The DOS of Si QDs having small diameters are sparser than those with wider diameters, as displayed in Fig. 6. With the augmentation of dot

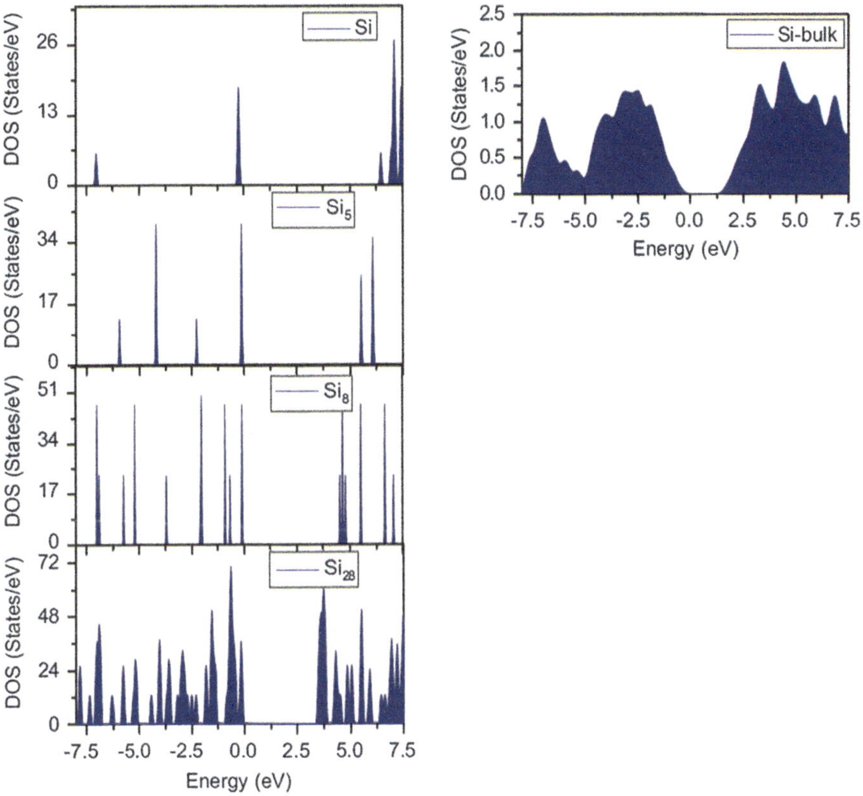


**Fig. 5** Density of states for various pure Si quantum dots as a function of dot size compared to Si bulk [32]

diameter, this behavior is the characteristic of the highest symmetry of Si QDs associated to the high degeneracy of the electronic states. The resulting DOS of Si QDs having larger diameters illustrate larger peaks and tend to be nearly like bulk system, whereas for smaller dots, the DOS have sharp peaks resembling molecular structures. Hence, it is anticipated that the optical spectra possess peaks (features) in smaller dots which are related to the separated electronic transitions than the narrower spectra with larger dots.

The termination of surface dangling bonds with hydrogen atoms besides to the impact of dot diameter on the energy gaps is also examined (see Fig. 7) [32]. The alteration of energy gap of different dot sizes for unsaturated and hydrogen-terminated surface is displayed in Fig. 7. The energy gap is found to increase with the reduction of the number of atoms in the dot. This behavior corroborates the stronger confinement for the smaller sizes [61–63]. Overall, the energy gaps of Si nanostructures saturated by H atoms and unsaturated QDs are reduced with the variation of number of atoms in the dot. For unsaturated dots, the energy gap was found to diminish (from 6.2 to 3.3 eV) as the dot size is augmented which is an

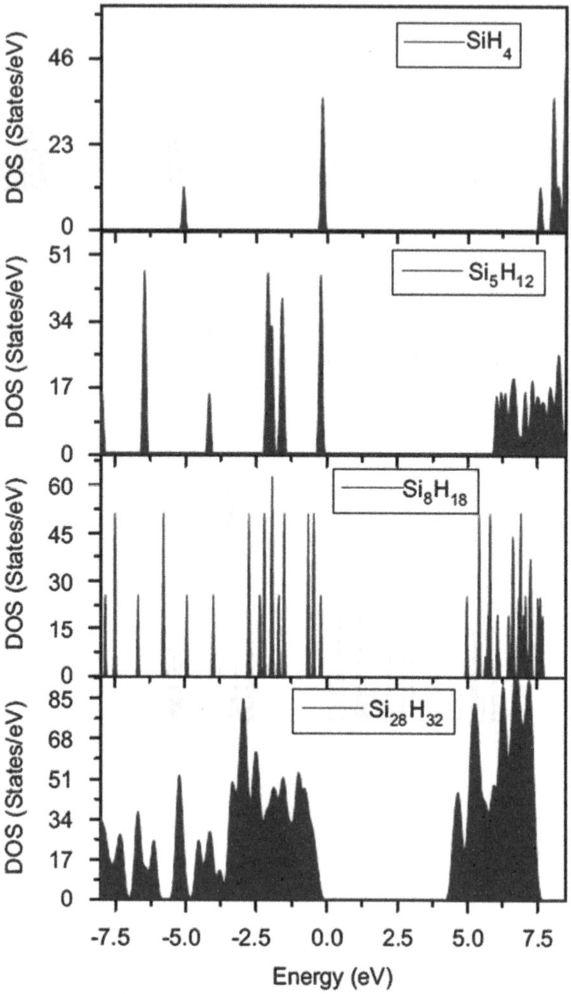


**Fig. 6** Density of states for various hydrogenated Si quantum dots as a function of dot size [32]

obvious evidence of the drastic confinement for the smaller diameters. The impact of surface termination by hydrogen atoms might conduct to an augmentation in the energy gap at the interval 0.4–1.2 eV for the concerned nanostructures. Thus, the bigger energy gap with the hydrogenation of surface corresponds to the localized electron cloud induced from the $sp^3$ hybridization of Si-dangling bond with H atom. These recent results corroborate the previous theoretical reports [64, 65]. In the hydrogenation situation, the larger value of energy gap is attributed to the strong Si-H bonding which arises from the confinement impact approximately at the surface region. For saturated surface of all dots with hydrogen atoms, the filled and empty electronic states are varied, and the resulting value of the energy gap is changed. As apparent, for smaller dots, the degeneracy of energy states is also considerable. Therefore, the termination of the surface dangling bonds with

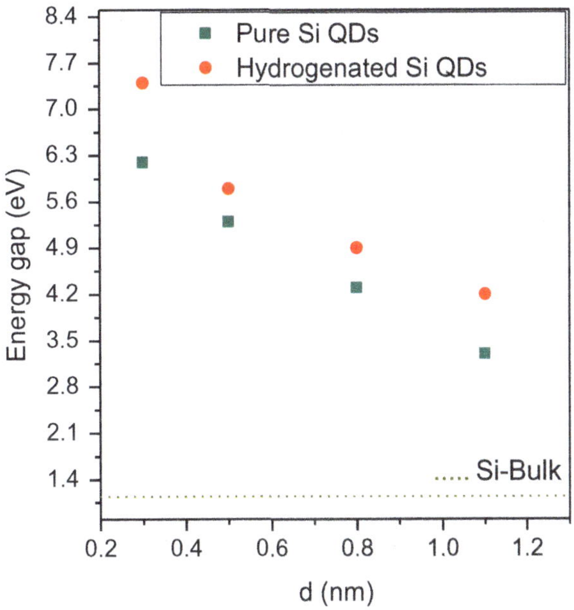


**Fig. 7** Variation of HOMO–LUMO gap as a function of QD diameter. Squares are pure quantum dots, circles for a hydrogen-passivated surface. The horizontal dotted line is the energy gap of silicon bulk [32]

hydrogen has an influence of enlarging the energy gap and lifting the degeneracy of energy states. For wider dots, the mixed $sp^2$- and $sp^3$-hybridized states are localized at the surface and core zone, respectively. Hence, the electrons are considerable, and the energy spectra are more feasibly non-degenerated. In contrast, the unsaturated Si QDs may get smaller energy gap due to the quantum size effects. Then, both the surface passivation and quantum confinement effects supply the electronic properties of Si quantum dots.

## *6.3 Bonding Character in Unpassivated and Passivated Si Quantum Dots*

To get more insight into the absorption spectra, we present now the recent work concerning the spatial distribution of both HOMO and LUMO states for unsaturated and hydrogenated Si dots at various sizes. The computed three-dimensional spatial distribution for the bonding–antibonding states versus the dot size are illustrated in Figs. 8 and 9 [32]. For bigger sizes, it is well noticed that the distribution of HOMO states is essentially located over the core part of the dot, although the LUMO states are concentrated on the surface. In saturated dots, it is apparent that the surface of LUMO states is contributed differently from the unsaturated Si dots (as noticed in Figs. 8 and 9). For the larger sizes, the behavior of the localized HOMO and LUMO states is dissimilar from the smaller sizes which clearly reflect the impact of the structural configuration. For unpassivated Si QDs, it is clearly seen in Fig. 8 that the

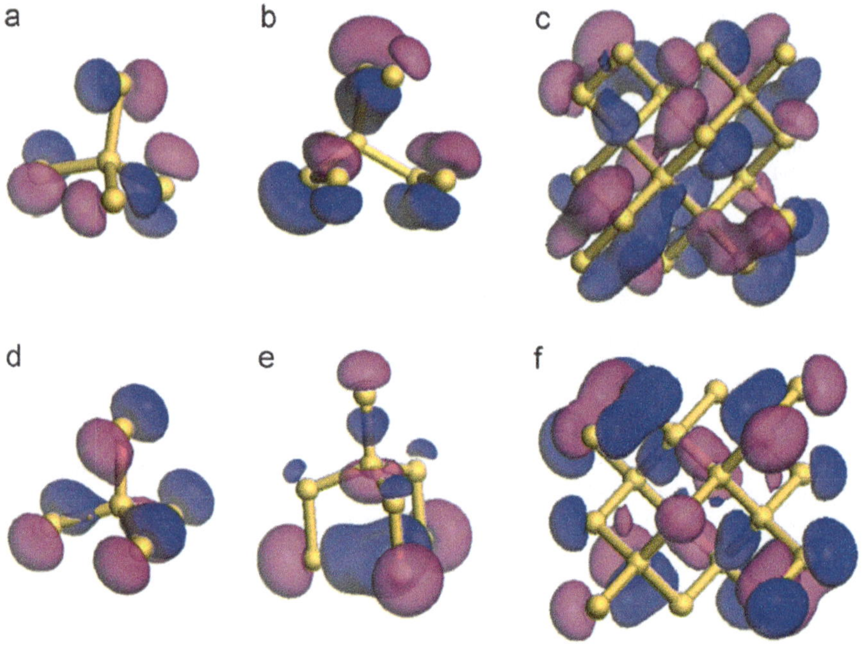


**Fig. 8** Isosurface plots of HOMO and LUMO for $Si_5$, $Si_8$, and $Si_{28}$ QDs. HOMO and LUMO spatial distribution plots for $Si_5$, $Si_8$, and $Si_{28}$ QDs at the same numerical value in the same species. The structures show strong bonding characters in HOMO. Antibonding feature of LUMO is dominated in $Si_5$, while the LUMO of $Si_{28}$ includes a certain mixture between the bonding and antibonding characters. (**a**) $Si_5$-HOMO, (**b**) $Si_8$-HOMO, (**c**) $Si_{28}$-HOMO, (**d**) $Si_5$-LUMO, (**e**) $Si_8$-LUMO, and (**f**) $Si_{28}$-LUMO [32]

localization of both HOMO and LUMO orbitals is surrounding silicon, and their overlap is augmented for smaller dots. In larger QDs, the HOMO density is more significant on the interior Si atoms, although the density is less considerable on the surface of the dot (as noticed in Fig. 8). In this situation, the density of LUMO has less support in the interior of the central tetrahedral cavity. This behavior is anticipated because the bonding and the antibonding states own dissimilar symmetry.

In saturated Si QDs, it is visible that the bonding of hydrogen is tight with silicon atoms even though its valance is one. In the case of Si5H12, the electron spatial distribution topology of HOMO and LUMO illustrates that the HOMO possesses a bonding character (as displayed in Fig. 8), because the electron cloud is mainly located in the intermediate parts that are shared with Si atoms. In contrast, the LUMO owns an antibonding behavior, i.e., the charge density is essentially distributed nearby the atoms. For large size QDs, LUMO acquires a level of bonding behavior additionally to its antibonding character (as referred in Figs. 8 and 9). It is remarkable from the DOS that the HOMO and LUMO states illustrate the occurrence of the bonding interaction of 1 s orbital of hydrogen atoms in conjunction

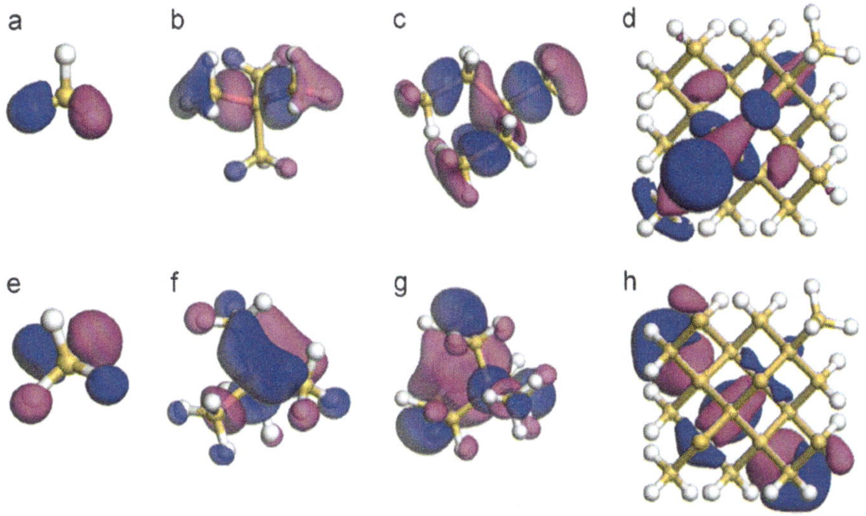


**Fig. 9** Isosurface plots of HOMO and LUMO for $SiH_4$, $Si_5H_{12}$, $Si_8H_{18}$, and $Si_{28}H_{32}$ QDs. HOMO and LUMO spatial distribution plots for $SiH_4$, $Si_5H_{12}$, $Si_8H_{18}$, and $Si_{28}H_{32}$ QDs at the same numerical value in the same species. The structures show strong bonding characters in HOMO. Antibonding feature of LUMO is dominated in $SiH_4$ and $Si_5H_{12}$, while the LUMO of $Si_8H_{18}$ and $Si_{28}H_{32}$ includes a certain mixture between the bonding and antibonding characters. (**a**) HOMO-$SiH_4$, (**b**) HOMO-$Si_5H_{12}$, (**c**) HOMO-$Si_8H_{18}$, (**d**) HOMO-$Si_{28}H_{32}$, (**e**) LUMO-$SiH_4$, (**f**) LUMO-$Si_5H_{12}$, (**g**) LUMO-$Si_8H_{18}$, and (**h**) LUMO-$Si_{28}H_{32}$ [32]

with the LUMO of Si-H (as seen in Fig. 7). The HOMO of hydrogen-saturated atom illustrates a section of high density in the interior of the QD as expected for 1 s kind quantum size wave functions. It was surmised from the isosurface plots of HOMO and LUMO hydrogenated Si QDs, the upper gap is attributed to the localized electron cloud at surface which originates through the $sp^3$ hybridization of Si-dangling bond with hydrogen. The localization of dangling bond states occurs between the HOMO state and the lower edge of the energy gap. In the case of the LUMO of the unsaturated Si QDs having smaller diameter, the participation of the density is less remarkable inside Si atoms, although the density is more significant in the hydrogen terminated with silicon. Also, the density is more pronounced inside with the increment of the size. Additionally, it is inferred that the inner region of these dots owns a bulk-like tetrahedral structure for larger size. However, the majority remnant dangling bonds are removed, and the hydrogen atoms are terminated at the surface. This can lead to a widening of energy gap. The localization of both HOMO and LUMO orbitals on hydrogen atoms displays a remarkable energy which can be discerned by hydrogen bonds, at least for smaller sizes. Overall, the occurrence of the strong electronegativity difference between Si and the terminated atoms in QDs conducts to the tendency of the concentration of the HOMO (by preserving its delocalization) on the weakened surface Si–Si bonds. Thus, the surface states possess different participation for both HOMO and LUMO states.

Also, the LUMO states illustrate a more localization for the electrons around the hydrogen atom. This behavior is expected due to the occurrence of a small charge transfer from silicon atoms to hydrogen [62, 63].This could be reflected by the slightly excessive electronegativity of H atom irrespective with Si atom. In fact, H atom favors to bond with silicon atom having more number of electrons. In the case of silicon QDs, it can be at the price of giving up the $sp^3$ hybridization. So, the $sp^3$ nature of diamond-like bulk might be absent in these small dots. The hydrogen terminated with Si QDs conduct to the diminishing of surface dangling bonds, and the bonding in these structures is tightened with the surrounding atoms of the host and by the termination of the dangling bonds. Indeed, the interaction between Si and H has polar covalent bonding [64]. Hydrogen-terminated QDs can roughly proceed the simple quantum size model in the spatial distribution of HOMO and LUMO and the size relationship with the band gap. This is a notable elucidation for the interaction between hydrogen and silicon not only to terminate the dangling bonds but also to modify electronic structure of QDs.

## 6.4 *Optical Properties of Unpassivated and Passivated Si Quantum Dots*

Now, the recent results about the optical absorption spectra of both unsaturated and saturated Si QDs as a function of the variation of dot diameter will be reviewed [32]. Note that the impact of both quantum confinement and hydrogen termination on the optical spectra profiles is also elucidated. In the beginning, the imaginary dielectric function is obtained. The dielectric function depends on the electronic band structure of a crystal, and its investigation by optical spectroscopy is a powerful tool for the determination of the overall band behavior of a solid. It has two parts, real and imaginary:

$$\varepsilon(\omega) = \varepsilon_1(\omega) + i\varepsilon_2(\omega). \tag{1}$$

It describes the material response to the photon spectrum. Other optical properties can be derived through the dielectric function. The imaginary part of the dielectric function $\varepsilon_2(\omega)$ denotes the optical absorption in the crystal, which can be obtained from the momentum matrix elements between the filled and the vacant wave functions. It designates the change of inter-band transitions in a semiconductor.

$\varepsilon_2(\omega)$ is the imaginary part of the dielectric function which is calculated by the following equation:

$$\mathrm{Im}\left[\varepsilon^{ij}(\omega)\right] = \frac{e^2\hbar^2}{\pi m^2\omega^2}\sum_{\mathrm{v,c}} \left|\psi_{\mathrm{c}}|\widehat{e}_j.P|\psi_{\mathrm{v}}\right|^2 \delta(E_{\mathrm{c}} - E_{\mathrm{v}} - \hbar\omega), \tag{2}$$

where $\hat{e_j}$ represents the unitary vector onward the direction of the external electromagnetic field of energy $\hbar\omega$. $\psi_v$ and $\psi_c$ designate the vacant and occupied level eigenfunctions of the system, respectively, and $E_v$ and $E_c$ represent their accompanied energies. $e$ and $m$ are the charge and mass of the bare electron, and p denotes the momentum operator. Utilizing the spectra of real and imaginary parts of dielectric functions, it is feasible to compute the spectra of remaining optical components like the refractive index, reflectivity, absorption coefficient, and so forth.

The real part $\varepsilon_1(\omega)$ can be calculated from the imaginary part $\varepsilon_2(\omega)$ by using Kramers–Kronig relation:

$$\varepsilon_1(\omega) = 1 + \frac{2}{\pi} P \int_0^{\infty} \frac{\varepsilon_2(\omega')\omega'}{\omega'^2 - \omega^2} d\omega' . \tag{3}$$

The electron transitions up to 15 eV are involved, to ensure that the real part of dielectric function is reasonable up to 12 eV, and the static dielectric constant is computed as zero-photon energy.

The absorption coefficient, $\alpha$, is a property to characterize a material which designates the amount of light absorbed through it. Accordingly, note that the inverse of the absorption coefficient $\alpha^{-1}$ represents the average path to be traveling over a photon before its absorption. For photon energy less than the energy gap, the creation of electron-hole pairs will not appear. Then, the material is transparent and $\alpha$ is insignificant. For photon energy more than the energy gap, the absorption is substantial. Afterward the optical absorption spectra are computed, although the summation is over the filled and empty states. From the dielectric function, the absorption coefficient is given by:

$$\alpha(\omega) = \sqrt{2}\omega\left[\left\{\varepsilon_1^2(\omega) + \varepsilon_2^2(\omega)\right\}^{1/2} - \varepsilon_1(\omega)\right]^{1/2} . \tag{4}$$

The optical absorption of bulk Si is also portrayed in order to compare with both unpassivated and passivated Si QDs. The search of optical excitations in hydrogen-saturated Si QDs is essential to clarify the absorption and emission of visible light in porous silicon [11–13]. The impact of size increment of silicon QDs on the optical spectra is discussed. Hence, the size effect on passivated and unpassivated Si QDs is substantially affected in the optical absorption spectra.

The optical absorption spectra of unpassivated Si QDs and Si QDs passivated by H atoms, respectively, are plotted in Figs. 10 and 11 [32]. The absorption energy is related to the peak location of the optical absorption as a function of photon energy curve with the change of dot sizes in Si QDs (refer to Figs. 10 and 11). The structures in absorption spectra are discerned by the transitions from filled to empty levels. Apparently, the optical absorption gaps for smaller dots can be noticed directly via the energy of the first dipole-permitted transition based on their absorption spectra. Therefore, the peaks are merged for larger dots, and the absorption spectra turn undoubtedly into quasi-continuous and have a tendency to

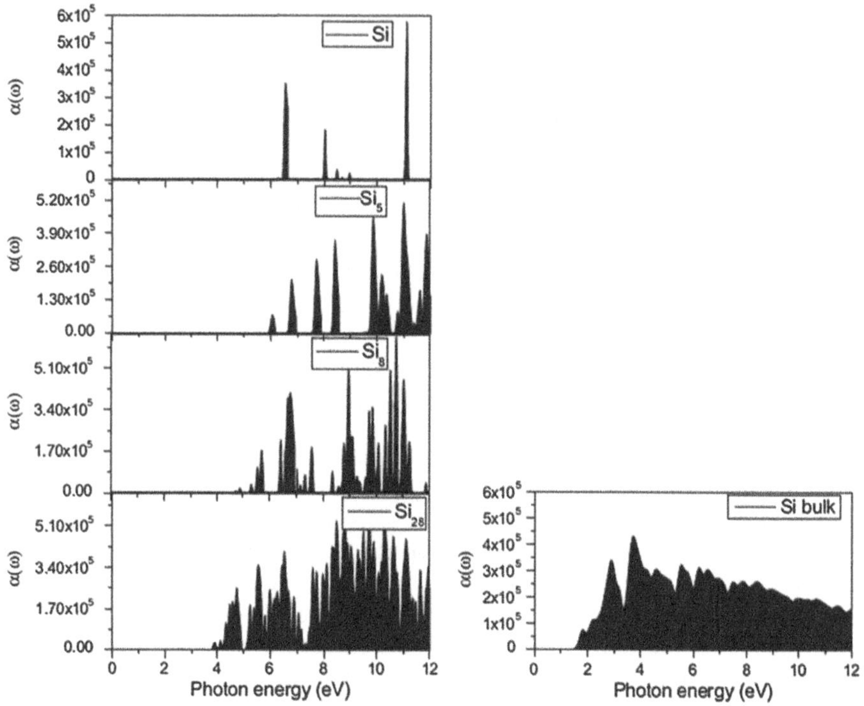


**Fig. 10** Plots of absorption spectra for various pure Si quantum dots as a function of dot size in comparison with the bulk spectrum [32]

be almost like bulk profile (refer to Figs. 10 and 11). Additional shoulders emerged into peaks below 6 eV are obtained for larger sizes, and they are mostly affected by quantum confinement effects. So, the number of these states is increased as a function of the dot diameter. The difference between smaller and larger dots is mostly associated to the intensities that are visible in the regime of the first main absorption peak of Si together with the corresponding shoulder at the underneath-energy part (as shown in Fig. 10). As apparent, the quantum size effects illustrate a significant effect on the results of absorption spectra. In the unpassivated QDs, the shoulder at downward-energy region close to the HOMO–LUMO gap shifts mostly in the direction of smaller energies versus the increase of QD size, and the LUMO is nearby the onset of absorption. Notably, only insignificant difference occurs between the spectra of two dots with five and eight atoms. The behavior of absorption energies is monotonous between 6.4 and 3.8 eV, with the increase of the size of the QD from 0.2 to 1.1 nm due to the reduction of quantum confinement effects. In the case of smaller dots, the peak locations are evidently blue-shifted against to their bulk positions. The downward-energy edges of the absorption spectra are red-shifted with the increase of QD size. A large number of low-intensity transitions occur in the proximity of the absorption edge. The absorption gap is diminished slowly as the dot diameter grows, and the discrete spectra for

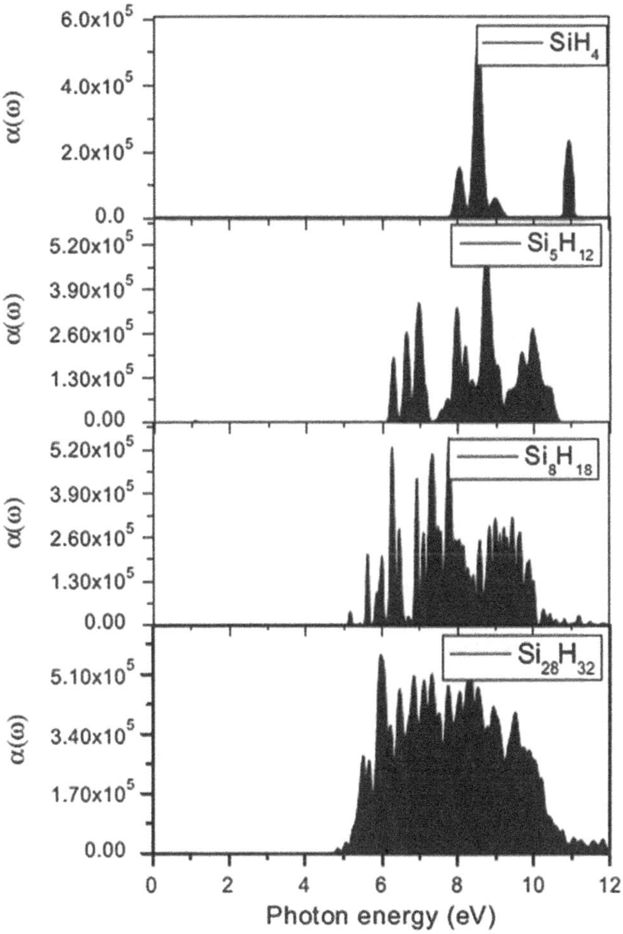


**Fig. 11** Plots of absorption spectra for various hydrogenated Si quantum dots as a function of dot size [32]

small QDs are emerged into quasi-continuous spectra. The optical absorption of Si bulk illustrates electronic transitions beginning around 1.6 eV, with two major peaks at 3.3 and 4.2 eV, respectively. These transitions verify well those detected experimentally at 3.4 and 4.5 eV, respectively [13]. For larger diameter, almost same peak is remarkable at 4.5 eV in the optical absorption, inferring that the peak for the QDs with bigger diameter is bulk like.

In saturated Si QDs, the higher occupied states emerge from the hybridized Si and H states, whereas the lower vacant states are essentially governed by Si states, as shown in Fig. 10. For the dot with larger diameter, a sharp peak is detected nearby 5.6 eV (refer to Fig. 10). An edge peak is more visible on the left side of the major peak for the smaller dots, and this behavior does not occur in the bulk Si (in Figs. 10 and 11), which is an important feature for Si nanostructures. This peak is located around 7.75 and 6.2 eV for $SiH_4$ and $Si_5H_{12}$ nanostructures, respectively. The optical absorption curves as a function of photon energy indicate an existence of significant absorption nearby the downward energy, showing a marked peak. This important characteristic surmises that a drastic absorption of photons at the underneath-energy regime for Si nanostructures, and this situation is in sharp contrast with bulk Si. Aside from the occurrence of this peak at downward energy, also other interesting features related to its shift toward downward energy for the surface saturated QDs and moderate decrease in the intensity with the increment of QD size. This is in consensus with the experimental realization for the absence of any absorption of photons in bulk Si with energy lower than 3.8 eV, and this is in gratifying agreement with the previous theoretical reports [65–67]. The computed optical gaps for $Si_5H_{12}$, $Si_8H_{18}$, and $Si_{28}H_{32}$ nanostructures occur around 6.2 eV, 5.4 eV, and 5.0 eV, respectively. The recent results illustrated that the surface effects induced by the single-bonded passivants contribute to the optical transitions. It is evident that with the increase of dot size, both the energy gap and optical gap are diminished, and the energy gap would be nearby the value of bulk Si with bigger dot size. This particular tendency of the energy variation validates the quantum size impact. Moreover, for all concerned quantum dots, the absorption energy is more significant than the energy gap and the intensity of photon absorption is essentially connected to the DOS close to the HOMO–LUMO energy. Notably, the silicon terminated with hydrogen is associated to the change of electron degeneracy which is significantly reflected in the optical response of the dots. Therefore, the rest of spectra for larger dots will preserve a similar shape but not the magnitude of intensity. For smaller dots, the absorption spectra exhibit a mixture of many peaks and mimic isolated atoms. In bigger Si QDs, the overlapped electronic wave functions lift the degeneracy of the energetic states emerging in the associating of energy states in a narrow energy regime. This will produce an enlarged absorption spectrum for bigger nanostructures. A blue shift of the optical absorption spectrum occurs with the decrease in the dot diameter of Si QDs. For larger dots, the main feature in the absorption spectra is positioned nearby 6.7 eV and an inconsiderable characteristic starts to emerge around 9 eV. For smaller QDs, the absorption spectrum varies substantially with supplementary of hydrogen atoms, although for bigger QDs, this modification is almost negligible. The occurrence of these important features evidently discriminates the nanostructures. Then, additional hydrogen atoms saturated in Si dots will induce new transitions which are absent in the unsaturated nanostructures. This evidence is attributed to the major change produced by the atomic configuration and the electronic structure. Remarkably, the surface termination hydrogen atoms illustrated that the intensity of optical absorption is more marked than the one of the unsaturated Si dots at underneath energies. This would develop an increase in the optical absorption at lesser excitations. Indeed, hydrogen presents a principal contribution for obtaining the optical properties of Si QDs. As the size of the dot is extended, the band gap is diminished, and the hydrogen peaks slowly overlap with the valence and conduction states.

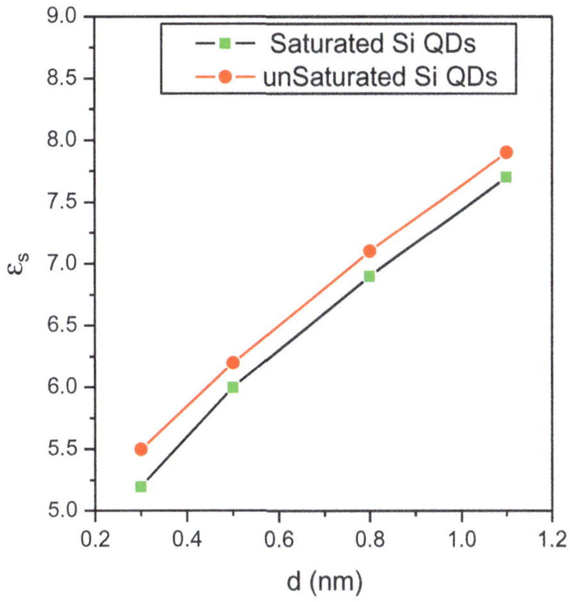


**Fig. 12** Variation of static dielectric constant as a function of the dot diameter. Circles are for QDs without surface passivation and squares for a hydrogen-passivated surface [32]

In nanostructures with bigger radius, hydrogen which is employed as a simple passivant has an inconsiderable role in their optical properties.

Recently, enormous reports have been focused on the exploration of Si nanostructures screening properties [68–70] that represent prominent information for the calculation of electron-hole interaction energy. The computed real part of the dielectric function enables us to find the electronic dielectric constant of Si nanostructures. Then, we computed the static dielectric constants for unsaturated and saturated Si QDs. The recent new results concerning the static dielectric constant $\varepsilon_S$ as a function of dots diameter are illustrated in Fig. 12, and a clear increase appears with the growing of dot size [32]. This behavior is more or less considerable for passivated surface than the unpassivated one and is close to the bulk value with the increase of the dot diameter. The computed value of the static dielectric constant for larger dots has a tendency to be in the vicinity of the bulk value ($\varepsilon(0) = 11.16$). This value is slightly higher than the experimental one [69]. The recent results concerning the dielectric constant infer that the smaller dots majorly support the luminescence than the bigger dots, which validated the previous theoretical studies [69, 70]. Our recent theoretical report is anticipated to support the elucidation of the discrepancies between the absorption and luminescence energies as well as the remarkable contribution of quantum confinement and the surface states concerning a better microscopic discerning of the visible luminescence in Si nanostructures.

# 7 Conclusion

In order to examine the quantum size and surface impacts on the energy gap and optical absorption, our recent works concerning the electronic structure and optical properties of Si quantum dots versus the variation of sizes were reviewed. A fruitful comparison between all properties has illustrated a different behavior between small and large quantum dots. In Si QDs, the energy gap is diminished as the number of atoms in the dot is incremented due to the drastic quantum confinement. Interesting results about the HOMO–LUMO gap, density of states, absorption spectra, and static dielectric constants versus the dot size and surface passivation were presented. The variation in the density of states and absorption spectra are more pronounced for hydrogenated Si QDs with bigger dots. The energy gap grows in the dot terminated by H atoms. The computed DOS for bigger dots emerge majorly as its bulk counterpart, although the smaller dots have analogous behavior as the atomic-like structures. With more hydrogen, the degeneracies are lifted in the DOS spectra, and this could lead to the increment in the energy gaps. It is expected that the HOMO and LUMO states are shifted versus the diminution in the size of Si QDs because of the quantum confinement. The intriguing characteristics for all quantum dots related to the wave functions show the location of HOMO states at the core of QD, albeit the LUMO states govern the surface. Our recent study illustrated that the addition of hydrogen in the surface eliminates the surface states by the augmentation of the energy gap and the confinement effects that are in gratifying agreement with the previous experimental determinations. Nevertheless, the phenomena at the fundamental of the photoluminescence have not been utterly understood. In fact, it is inferred that the surface saturation and the quantum confinement effect may have a crucial role on the optical properties and electronic structure of silicon QDs. Our recent results showed that the hydrogenated and unpassivated Si QDs own different optical response with salient dependence on the dot size. The absorption spectra results illustrated that the absorption properties are intimately related to the surface passivation and dot size. The saturated surface states have spectacular consequences as far as the photoluminescence of finite-size dots are incorporated. The previous theoretical determinations are mostly in consensus with the available experimental and theoretical reports of the optical spectra of Si QDs. However, based on the computed energy gap, absorption spectra, and static dielectric constant, we surmise that Si QDs are convenient for illustrating a high luminescence in the visible regime as detected experimentally.

# References

1. M. Forcales, N.J. Smith, R.G. Elliman, J. Appl. Phys. B **100**, 014902 (2006)
2. L.T. Canham, Appl. Phys. Lett. **57**, 1046 (1990)
3. Z. Deng, X.D. Pi, J.J. Zhao, D. Yang, J. Mater. Sci. Technol. **29**, 221 (2013)
4. L.W. Wang, A. Zunger, Phys. Chem. **98**, 2158 (1994)

5. L. Koponen, L. Tunturivuori, et al., Phys.Rev. B **79**, 2353321 (2009)
6. Z.Y. Ni et al., J. Phys. D: Appl. Phys. **48**, 314006 (2015)
7. M. Luppi, S. Ossicini, Phys. Rev. B **71**, 035340 (2005)
8. D. Li et al., Appl. Phys Lett. **104**, 204101 (2014)
9. L.E. Ramos, H.-C. Weissker, J. Furthm¨uller, F. Bechstedt, Phys. Status Solidi B **242**, 3053 (2005)
10. D. Melnikov, J.R. Chelikowsky, Phys. Rev. B **69**, 113305 (2004)
11. A. Tsolakidis, R.M. Martin, Phys. Rev. B **71**, 125319 (2005)
12. F. Trani, G. Cantele, D. Ninno, G. Iadonisi, Phys. Rev. B **72**, 075423 (2005)
13. L. Yao, T. Yu, L.X. Ba, H. Meng, X. Fang, Y.L. Wang, L. Li, X. Rong, S. Wang, X.Q. Wang, G.Z. Ran, X.D. Pi, G.G. Qin, J. Mater. Chem. C **4**, 673 (2016)
14. B.B. Sahu, Y. Yin, J.G. Han, M. Shiratanib, Phys. Chem. Chem. Phys. **18**, 15697 (2016)
15. S. Askari, M. Macias-Montero, T. Velusamy, P. Maguire, V. Svrcek, D. Mariotti, J. Phys. D: Appl. Phys. **314002**, 48 (2015)
16. L. Eleonora, I. Federico, M. Rita, P. Olivia, O. Stefano, D. Elena, O. Valerio, Phys. Rev. B **75**, 033303 (2007)
17. F. Sangghaleh, I. Sychugov, Z. Yang, J.G.C. Veinot, J. Linnros, J. ACS Nano **9**, 7097 (2015)
18. B. Pejova, Semicond. Sci. Technol. **29**, 045007 (2014)
19. S. Ossicini, L. Pavesi, F. Priolo, *Light Emitting Silicon for Microphotonics* (Springer, Berlin, 2004)
20. M.B. Gongalsky, L.A. Osminkina, A. Pereira, A.A. Manankov, A.A. Fedorenko, A.N. Vasiliev, V.V. Solovyev, A.A. Kudryavtsev, M. Sentis, A.V. Kabashin, V.Y. Timoshenko, Sci. Rep. **6**, 24732 (2016). https://doi.org/10.1038/srep24732
21. M. Miyano, S. Endo, H. Takenouchi, S. Nakamura, Y. Iwabuti, O. Shiino, T. Nakanishi, Y. Hasegawa, J. Phys. Chem. C **118**, 19778 (2014)
22. G. Shen, D. Chen, K. Tang, Y. Qian, S. Zhang, Chem. Phys. Lett. **375**, 177 (2003)
23. D. Zhang, A. Alkhateeb, H. Han, H. Mahmood, D.N. Mcllroy, M. Grant Norton, Nano Lett. **3**, 983 (2003)
24. G.W. Ho, A.S.W. Wong, D.J. Kang, M.E. Welland, Nanotechnology **15**, 996 (2004)
25. R. Rurali, Phys. Rev. **B71**, 205405 (2005)
26. B. Tian et al., Nature **449**, 885 (2007)
27. D.V. Melnikov, J.R. Chelikowsky, Phys. Rev. Lett. **92**, 046802 (2004)
28. X. Liu, Y. Zhang, Y. Ting, X. Qiao, R. Gresback, X. Pi, D. Yang, Part. Part. Syst. Charact. **33**, 44 (2016)
29. S. Ossicini, M. Amato, R. Guerra, M. Palummo, O. Pulci, Nanoscale. Res. Lett **5**, 1637 (2010)
30. S. Askari et al., D. Appl. Phys. Lett. **104**, 163103 (2014)
31. P. Giannozzi et al., J. Phys.: Condens. Matter **21**, 395502 (2009)
32. A. Laref, N. Alshammari, S. Laref, S.J. Luo, Sol. Energy Mater. Sol. Cells **120**, 622 (2014)
33. J.P. Proot, C. Delerue, G. Allan, Appl. Phys. Lett. **61**(16), 1948 (1992)
34. J.S. Biteen, D.P. Pacifici, N.S. Lewis, H.A. Atwater, Nano Lett. **5**, 1768 (2005)
35. B. Delley, E. F. Steigmeier, Phys. Rev. B **47**, 1397 (1993); Appl. Phys. Lett. **67**, 2370 (1995)
36. T. Trupke, J. Zhao, A. Wang, R. Corkish, M. Green, Appl. Phys. Lett. **82**, 2996 (2003)
37. L.-W. Wang, A. Zunger, J. Chem. Phys. **100**, 2394 (1994)
38. A. Liu, R. Jones, L. Liao, D. Samara-Rubio, D. Rubin, O. Cohen, R. Nicolaescu, M. Paniccia, Nature **427**, 615 (2004)
39. F. Bruneval, F. Sottile, V. Olevano, R.D. Sole, L. Reining, Phys. Rev. Lett. **94**, 186402 (2005)
40. I. Vasiliev, S. Ogut, J.R. Chelikowsky, Phys. Rev. Lett. **86**, 1813 (2001); A.J. Williamson, J.C. Grossman, R.Q. Hood, A. Puzder, G. Galli, Phys. Rev. Lett. **89**, 196803 (2002)
41. C. Delerue, M. Lannoo, G. Allan, Phys. Rev. Lett. **84**, 2457 (2000)
42. C.S. Garoufalis, A.D. Zdetsis, S. Grimme, Phys. Rev. Lett. **87**, 276402 (2001)
43. A. Zunger, Phys. Status Solidi A **190**, 467 (2002)
44. R.M. Martin, *Electronic Structure* (Cambridge University Press, Cambridge, 2004)
45. K.W. Kolasinski, Curr. Opin. Solid State Mater. Sci. **9**, 73 (2005)

46. J. Wilcoxon, G. Samara, et al. Phys. Rev. B **60**, 2704 (1999)
47. M. Wolkin, J. Jorne, et al., Phys. Rev. Lett. **82**, 197 (1999)
48. A.D. Zdetsis, C.S. Garoufalis, S. Grimme, in *NATO Advanced Research Workshop on "Quantum Dots: Fundamentals, Applications, and Frontiers" (Crete 2003)*, ed. by B.A. Joyce et al. (Springer, Heidelberg, 2005), pp. 317–332
49. S.Z. Bisri et al., Adv. Mater. **26**, 5639–5645 (2014)
50. Y. Liu, Z.Y. Zhang, Y.F. Hu, C.H. Jin, L.-M. Peng, J. Nanosci. Nanotechnol. **8**, 252 (2008)
51. Y.-Y. Noh, X. Cheng, H. Sirringhaus, J.I. Sohn, M.E. Welland, D.J. Kang, Appl. Phys. Lett. **91**, 043109 (2007)
52. B. Ghosh et al., Adv. Funct. Mater. **24**, 7151 (2014)
53. Y. Du Y et al., Acs Nano **8**, 10019–10025 (2014)
54. N.J. Thompson et al., Nat. Mat. **13**, 1039 (2014)
55. G. Conibeer, M. Green, et al. Thin Solid Films **511-512**, 654 (2006)
56. G. Conibeer, M. Green, M. Cho, et al., Thin Solid Films. 516, 6748 (2008); C.S. Garoufalis, A.D. Zdetsis, J. Math. Chem. 46, 952 (2009)
57. S.Z. Bisri, et al., Adv. Mater. **26**, 5639–5645 (2014); R. Guerra, E. Degoli, et al., Phys. Rev. B. **80**, 155332-1—155332-5 (2009)
58. D. König, J. Rudd, et al., Sol. Energy Mater. Sol. Cells **93**, 753 (2009)
59. W. Kohn, L.J. Sham, Phys. Rev. **140**, 1133 (1965)
60. J.P. Perdew, K. Burke, M. Ernzerhof, Phys. Rev. Lett. **77**, 3865 (1996)
61. M. Ribeiro Jr., L.R.C. Fonseca, L.G. Ferreira, Phys. Rev. B **241312**(R), 79 (2009)
62. J. von Behren et al., Solid State Commun. **105**, 17 (1998)
63. M. Stupca, M. Alsalhi, T. Al-Saud, A. Almuhanna, M. Nayfeh, Appl. Phys. Lett. **91**, 063107 (2007)
64. F.A. Reboredo, A. Franceschetti, A. Zunger, Phys. Rev. B **61**, 13073 (2000)
65. G. te Velde et al., J. Comput. Chem. **22**, 931 (2001)
66. L.C. Lew-Yan-Voon, L.R. Ram-Mohan, Phys. Rev. B **47**, 15500 (1993)
67. M. Virgilio, G. Grosso, Nanotechnology **18**, 075402 (2007)
68. G. Pizzi, M. Virgilio, G. Grosso, Nanotechnology **21**, 055202 (2010)
69. O. Lehtonen, D. Sundholm, Phys. Rev. B **72**, 085424 (2005)
70. C. Tserbak, H.M. Polatoglou, G. Theodorou, Phys. Rev. B **47**, 7104 (1993)

# Absorption by Particulate Silicon Layer: Theoretical Treatment to Enhance Efficiency of Solar Cells

**Alexander A. Miskevich and Valery A. Loiko**

**Abstract** Absorption of light by single by crystalline silicon spherical particle and 2D and 3D layers from such particles is theoretically investigated in the wavelength range from 0.28 to 1.12 μm. The range of particle diameters from 0.05 to 1000 μm is covered. Absorption coefficient of monolayer of small- and wavelength-sized particles is calculated in the quasicrystalline approximation of the theory of multiple scattering of waves. For monolayer of large particles, the single scattering approximation is used. Absorption by multilayer is examined under the transfer matrix method. The spectral and integral over the sun spectrum absorption coefficients are studied. The results are compared with the data for homogeneous plane-parallel silicon plates of the equivalent volume of material (equivalent plates). The monolayer and multilayer consisting of silicon particles with sizes significantly smaller than the wavelength absorb lesser than the equivalent silicon plates. The absorption coefficient of the monolayer of large particles is smaller than the one of equivalent plate. Absorption by three- and more monolayer systems of such particles is larger than the one of the equivalent plates. Absorption by monolayer of wavelength-sized particles can be significantly larger than the one of the equivalent plate. It is caused by strong resonance scattering by individual silicon particles and strong multiple scattering in particle arrays. The narrow wavelength intervals (up to 10 nm) of the resonance peak spectral absorption coefficient of monolayer can be more than 100 times larger than the one of the equivalent plate. In the wavelength range from 0.8 μm to 1.12 μm, integral absorption coefficient of monolayer can be more than 20 times higher than the one of the plate. Enhancement of light absorption due to tuning of the multilayer parameters is considered. The sketch of the solar cell based on gradient particulate structure of active layer is presented.

A.A. Miskevich (✉) · V.A. Loiko
Institute of Physics of National Academy of Sciences of Belarus, Minsk, Belarus
e-mail: miskevic@ifanbel.bas-net.by; loiko@ifanbel.bas-net.by; http://loiko.org



S. Ikhmayies (ed.), *Advances in Silicon Solar Cells*,
https://doi.org/10.1007/978-3-319-69703-1_3

# 1 Introduction

In the last decades, photovoltaic (solar) cells are the object of intensive theoretical and experimental investigations. That is caused by increasing importance of "green," renewable, and nonfossil energy sources. However, relatively low efficiency of solar cells is one of the restrictive factors for wide usage of such sources [1].

The significant problem in the enhancement of the solar cell (SC) performance is optimization of its structure to trap more amount of the incident light. To attain this end, the plasmonic and diffractive nanostructures, down-converting particles, surface texturing, nanohole patterning, etc. are widely studied [1–18]. In this investigation, we examine the possibility to enhance light absorption by a particulate active layer.

Currently the most used material for solar cell production is silicon. Because it is an indirect gap semiconductor, the probability of an electron transition from valence to conduction band due to photon absorption is small [1, 2]. If we deal with a homogeneous active layer of SC, we need to increase the layer thickness to increase light absorption. That can reduce the efficiency of solar cell due to the diffusion length of the minority carrier [16–18] limitation. Only carriers produced in the space-charge (depletion) region and adjacent areas, which are determined by the diffusion length of the minority carriers, are separated by the electric field of the *p–n* junction and, thus, contribute in photoelectromotive-force (photo-emf) generation [1, 11–17].

The amount and efficiency of light absorption can be enhanced using the particulate structure of an active layer. Such a layer consists of plane-parallel monolayers of silicon particles with diameter of the diffusion length order. That enables one to decrease reflection and simultaneously increase light absorption in comparison with the homogeneous plane-parallel silicon plate [19, 20]. Silicon particles of such sizes and the layers of these particles can be fabricated by chemical [21, 22], lithographic [23] methods, plasma synthesis [24], laser ablation in air or liquids [25, 26], laser-induced pyrolysis [27], and the recently proposed method based on the laser-induced transfer of molten droplets [28, 29].

In crystalline silicon (c-Si), the light absorption in the long-wavelength part of visible and near-infrared spectral regions is small, while in the ultraviolet region and at the short-wavelength part of visible spectrum it is large (Fig. 1). An important problem for c-Si solar cells is to enhance light absorption in the wavelength range of small imaginary part $\kappa$ (absorption index) of the complex refractive index $m = n + i\kappa$ of semiconductor. Absorption by the particulate structure can be increased in conditions of the pronounced multiple scattering of waves. Such conditions occur when the particle sizes and distances between them are comparable with wavelengths of the incident light. Therefore, as one can expect, the systems of submicron- and micrometer-sized particles should satisfy the mentioned conditions. Moreover, individual submicron spherical silicon particles exhibit strong peaks of scattering efficiency in the spectral range of interest due to optical

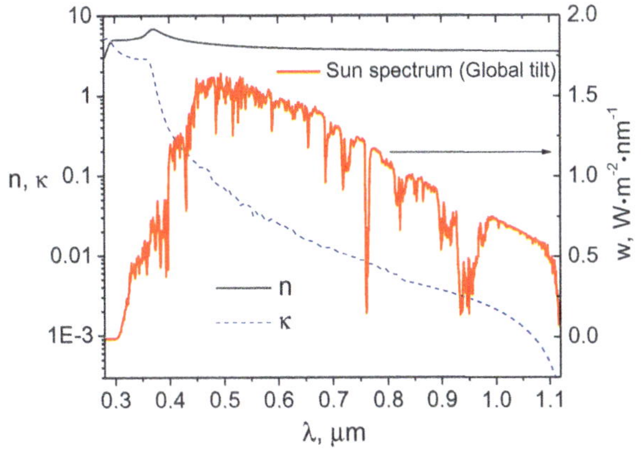


**Fig. 1** Spectra of real $n$ and imaginary $\kappa$ parts of the complex refractive index of crystalline silicon (c-Si) by the data of [48] and the terrestrial solar spectral irradiance "Global tilt" ASTM G173–03 by the data of [47]

resonances (also known as Mie resonances) [30–37]. This can be an important factor of overall absorption enhancement.

In the present chapter, we theoretically examine absorption by the single spherical crystalline silicon particle and by the layers consisting of large, wavelength-sized and small silicon particles. We use the single scattering approximation (SSA) [38–40] for monolayers of large particles [19] and the quasicrystalline approximation (QCA) [41–43] of the theory of multiple scattering of waves for monolayers of wavelength-sized and small ones [20]. For multilayers consisting of particulate monolayers, we utilize the transfer matrix method (TMM) [44–46]. The wide range of particle diameters which correspond to the diffusion lengths of the minority carriers is studied. The wavelength range from 0.28 to 1.12 μm is covered. The lower and upper bounds of the range are determined by the power of solar spectral irradiance [47] which we have chosen and by the c-Si bandgap, respectively. Figure 1 shows spectra of real $n$ and imaginary $\kappa$ parts of the complex refractive index $m = n + i\kappa$ of crystalline silicon [48] and the terrestrial solar spectral irradiance "Global tilt" ASTM G173–03 [47].

# 2 Characteristics of Single Scattering

## *2.1 Absorption by a Single Silicon Particle*

To describe scattering and absorption by a single particle, the dimensionless parameters are used: scattering ($Q_{sc}$) and absorption ($Q_{abs}$) efficiency factors [31, 39, 49–63]. They characterize removal of radiation from the parallel beam of radiation illuminating a particle due to scattering and absorption, respectively. Their sum yields the extinction efficiency factor $Q_{ext}$:

$$Q_{\mathrm{ext}} = Q_{\mathrm{sc}} + Q_{\mathrm{abs}} \tag{1}$$

It characterizes total extinction (attenuation) of the parallel beam resulting from the effects of scattering and absorption. The factors depend on the size, shape, internal structure of particle, roughness of their surface, particle orientation, and the polarization state of the incident radiation. Generally the efficiency factors are determined via elements of the amplitude scattering matrix [31, 49, 50]. The value of $Q_{\mathrm{abs}}$ according to (1) is the difference between the extinction and scattering efficiency factors:

$$Q_{\mathrm{abs}} = Q_{\mathrm{ext}} - Q_{\mathrm{sc}} \tag{2}$$

We consider only systems with homogeneous spherical particles for which these factors are determined only by the first element of the scattering matrix called scattering function and

$$Q_{\mathrm{ext}} = \frac{2}{x^2}\mathrm{Re}\sum_j (2j+1)(a_j + b_j) \tag{3}$$

$$Q_{\mathrm{sc}} = \frac{2}{x^2}\sum_j (2j+1)\left(|a_j|^2 + |b_j|^2\right) \tag{4}$$

Here, $x = \pi D/\lambda$ is the size parameter of particle with diameter $D$, $\lambda$ is the wavelength of the incident light, and $a_j$ and $b_j$ are the Mie coefficients. They depend on diameter and complex refractive index of particle [39, 40].

The spectra of efficiency factors of spherical c-Si particle calculated by Eqs. (3), (4), and (2) are shown in Fig. 2. Data for complex refractive index are taken from [48] (Fig. 1). Linear interpolation is used for the wavelength intervals, where the values of $n$ or $\kappa$ are not presented in [48]. As can be seen, small particle ($D = 0.05$ μm) possesses noticeable extinction and absorption only in the wavelength range of large absorption index ($\lambda \leq \sim 0.4$ μm) of c-Si. When particle diameters are comparable with the wavelength ($D = 0.1$, 0.2, 0.5 μm), the sharp peaks on the absorption efficiency factor spectra occur in the range of middle and small absorption indices of material ($\lambda > \sim 0.4$ μm). They are caused by strong scattering efficiency of particle with high refractive index ($n_{\text{c-Si}} > 3.5$ (Fig. 1)) due to optical resonances [30–37, 39, 49, 58]. Note that refractive indices of dielectrics are typically within one to two [48, 64]. For large particle diameters ($D = 2$, 50, 200 μm), the spectra are more uniform. As a whole, the absorption efficiency factor increases for small and decreases for large absorption indices with particle size increasing.

Spectral dependences of the $Q_{\mathrm{abs}}/Q_{\mathrm{ext}}$ ratio for spherical c-Si particle are shown in Fig. 3. This ratio determines the probability of photon absorption and characterizes contribution of absorption to the extinction. One can see that, as a whole, the dependences have character similar to the absorption efficiency factor $Q_{\mathrm{abs}}$ (see Fig. 2).

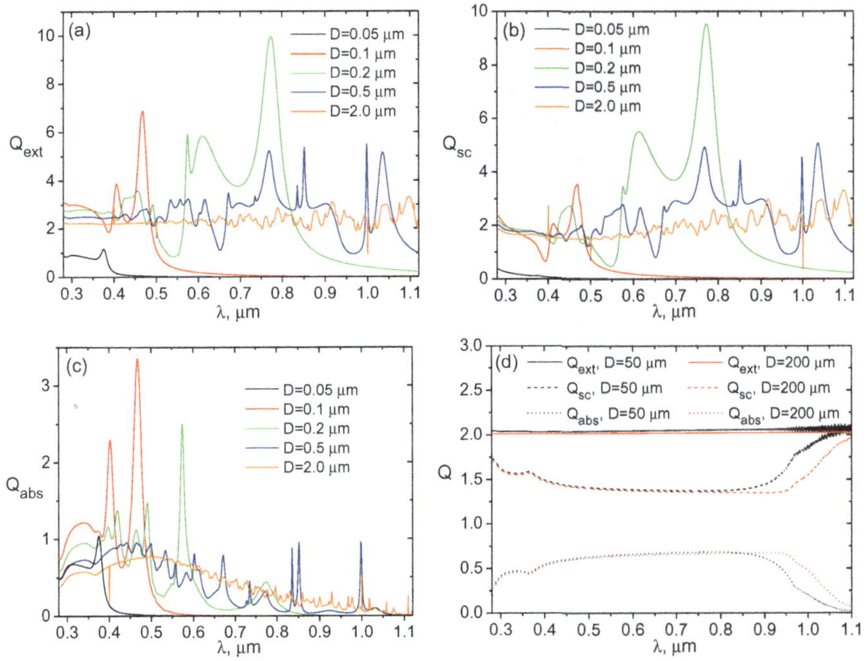


**Fig. 2** Spectra of extinction $Q_{ext}$, scattering $Q_{sc}$, and absorption $Q_{abs}$ efficiency factors of spherical c-Si particle at different diameters $D$

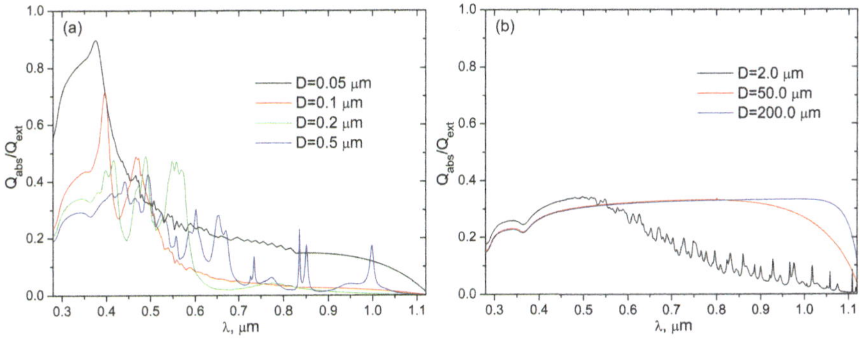


**Fig. 3** Dependences of $Q_{abs}/Q_{ext}$ of spherical c-Si particle on the wavelength at different diameters $D$

The map and surface plot of $Q_{abs}(\lambda, D)$ are shown in Fig. 4. The data allow one to find particle diameters with most efficient absorption in the considered range of wavelengths.

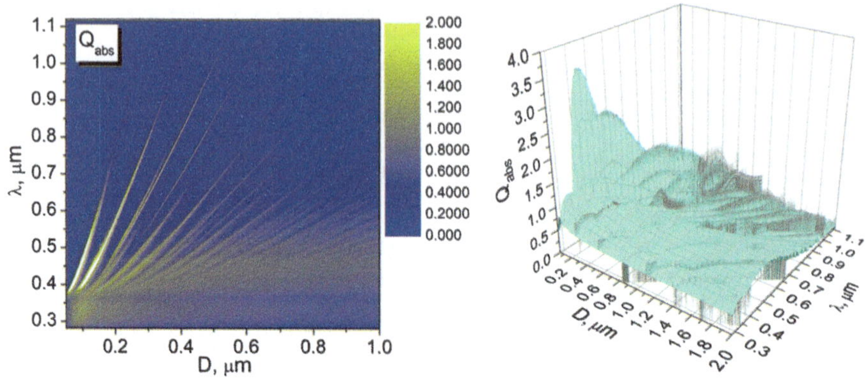


**Fig. 4** The map (left) and surface plot (right) of $Q_{abs}(\lambda, D)$ for spherical c-Si particles

## 2.2 Absorption by Ensemble of Silicon Particles

Let $N$ monodisperse spherical particles with diameter $D_1$ are distributed in a volume $V$ and the single scattering of light is implemented, i.e. reillumination by particles does not occur. It means that the amount of scattered and absorbed radiation is proportional to the volume size. In this case, absorption index of the particulate medium, $k_{abs}$, is determined as follows:

$$k_{abs} = N\frac{\pi D_1^2}{4}Q_{abs} \tag{5}$$

If we hold the volume $V_p$ of the material concluded in all particles as a constant ($V_p = \text{const}$) and change the size of particles, we find that absorption index of the medium consisting of particles with diameter $D_2$, which differs from the initial one ($D_1$), is

$$k_{abs}(x_2) = Nx_1^3\frac{\lambda^2}{4\pi}\frac{Q_{abs}(x_2)}{x_2} \tag{6}$$

Here $x_1 = \pi D_1/\lambda$ and $x_2 = \pi D_2/\lambda$ are the size parameters of particles with diameter $D_1$ and $D_2$, respectively, and $\lambda$ is the wavelength of the incident wave. Equation (6) shows that at $V_p = \text{const}$ absorption coefficient of medium depends on the ratio $Q_{abs}/x$.

The map and surface plot of the function $Q_{abs}(\lambda, D)/x$ are shown in Fig. 5.

Using these data, one finds conditions to provide maximum absorption by an ensemble at constant volume of the material. In appearance, they are similar to the ones for $Q_{abs}(\lambda, D)$ (see Fig. 4), but maxima of the dependences occur at different particle sizes.

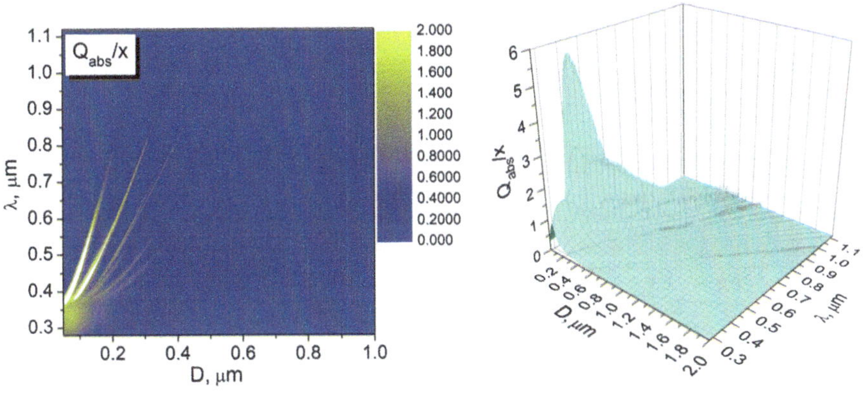


**Fig. 5** The map (left) and surface plot (right) of the $Q_{abs}(\lambda, D)/x$ ratio for spherical c-Si particles

## 2.3 *Integral Efficiency Factors*

Let $f(\lambda)$ be a function which describes a spectral characteristic of an object. Then the value of $\langle f \rangle$ which is integrated over the illumination spectrum of the source is

$$\langle f \rangle = \frac{\int_{\lambda_1}^{\lambda_2} f(\lambda) w(\lambda) d\lambda}{\int_{\lambda_1}^{\lambda_2} w(\lambda) d\lambda} \tag{7}$$

where $\lambda$ is the wavelength of incident light, $w(\lambda)$ is the power of the illumination spectrum of the source, and wavelengths $\lambda_1$ and $\lambda_2$ specify the considered spectral range.

Consider the integrated (over the solar spectral illumination) absorption $\langle Q_{abs} \rangle$, scattering $\langle Q_{sc} \rangle$, and extinction $\langle Q_{ext} \rangle$ efficiency factors (we will refer these quantities as the "integral efficiency factors") of a particle. The dependences of these quantities on the particle diameter are shown in Fig. 6. They are calculated by Eq. (7), where $f$ stands for $Q_{abs}$, $Q_{sc}$, and $Q_{ext}$, respectively, and $w(\lambda)$ is the terrestrial solar spectral irradiance "Global tilt" ASTM G173–03 [47] (see Fig. 1). Two spectral ranges are studied. The first one (0.28 μm $\leq \lambda \leq$ 1.12 μm) encloses all spectral range indicated in Fig. 1. The second one (0.8 μm $\leq \lambda \leq$ 1.12 μm) is narrower and encloses wavelengths where absorption index of c-Si is small. For the range of 0.28 μm $\leq \lambda \leq$ 1.12 μm, the value of $\langle Q_{abs} \rangle$ initially monotonically grows with particle size and then saturates at diameter close to 0.1 μm. More complex oscillating dependence is observed for the spectral range of 0.8 μm $\leq \lambda \leq$ 1.12 μm.

In Fig. 6, the dependences of the $\langle Q_{abs} \rangle / \langle Q_{ext} \rangle$ ratio, which determines the relative contribution of absorption to the extinction, are depicted as well. One can see that the ratio approaches unity for small particles, i.e., the extinction efficiency

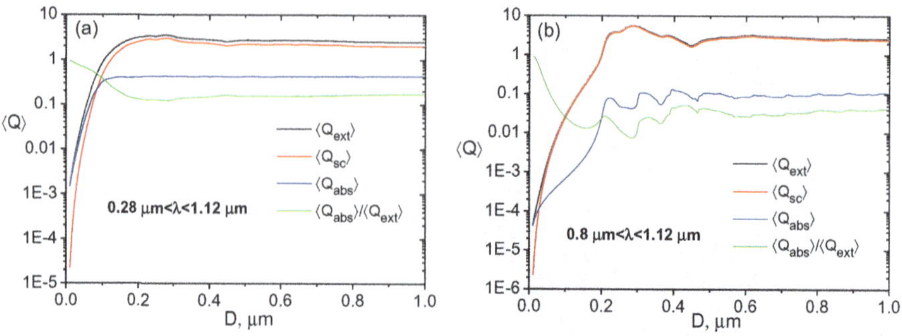


**Fig. 6** Dependences of integral absorption $\langle Q_{abs}\rangle$, scattering $\langle Q_{sc}\rangle$, and extinction $\langle Q_{ext}\rangle$ efficiency factors of spherical c-Si particle and the $\langle Q_{abs}\rangle/\langle Q_{ext}\rangle$ ratio on particle diameter for 0.28 μm $\leq \lambda \leq$ 1.12 μm (**a**) and 0.8 μm $\leq \lambda \leq$ 1.12 μm (**b**) wavelength ranges

factor of these particles is determined mainly by absorption. With particle size increase, the value of $\langle Q_{abs}\rangle/\langle Q_{ext}\rangle$ at first decreases monotonically and then oscillates. The more pronounced oscillations occur when the wavelength range of small absorption index of c-Si is considered (Fig. 6b). The oscillations attenuate with diameter increasing.

The results obtained in this section are used in the following sections at consideration of absorption by the particulate structures of c-Si.

# 3 Layers Composed of Large Particles: Single Scattering Approximation

## *3.1 Spectral Absorption, Transmission, and Reflection Coefficients of Particulate Monolayer: Basic Equations*

Consider a monolayer of large monodisperse spherical particles of crystalline silicon. Let us write equations to calculate absorption, coherent (directional) and incoherent (diffuse) transmission, and reflection coefficients of such a layer in the single scattering approximation (SSA) [19, 38–40, 65].

Absorption coefficient $A_{ml}$ of a monolayer of particles in this approximation is

$$A_{ml} = \eta Q_{abs} \tag{8}$$

Here $\eta$ is the monolayer surface filling factor (the ratio of the area of particle projections onto the monolayer plane to the layer area where they are distributed), and absorption efficiency factor $Q_{abs}$ of a single particle is calculated by Eq. (2).

Equations for the coherent transmission $T_c$ and coherent reflection $R_c$ coefficients of monolayer of spherical particles in the SSA can be written as follows [19, 40]:

$$T_c = |t_c|^2 = \left|1 - \frac{\eta}{x^2}\sum_j (2j+1)(a_j + b_j)\right|^2 \tag{9}$$

$$R_c = |r_c|^2 = \left|-\frac{\eta}{x^2}\sum_j (-1)^j(2j+1)(a_j - b_j)\right|^2 \tag{10}$$

where $t_c$ and $r_c$ are the amplitude coherent transmission and reflection coefficients and $a_j$ and $b_j$ are the Mie coefficients.

To calculate incoherent transmission $T_{\text{inc}}$ and reflection $R_{\text{inc}}$ coefficients (they characterize the parts of light diffusely scattered in forward and backward hemispheres, respectively) of a monolayer, we write the following equations:

$$T_{\text{inc}} = (1 - T_c - R_c - A_{\text{ml}})F_{\text{fs}} \tag{11}$$

$$R_{\text{inc}} = (1 - T_c - R_c - A_{\text{ml}})F_{\text{bs}} \tag{12}$$

Here functions $F_{\text{fs}}$ and $F_{\text{bs}}$ determine parts of light scattered in the forward and backward hemispheres, respectively:

$$F_{\text{fs}} = \frac{\int_0^{\pi/2} i(\theta)\sin\theta d\theta}{\int_0^{\pi} i(\theta)\sin\theta d\theta} \tag{13}$$

$$F_{\text{bs}} = \frac{\int_{\pi/2}^{\pi} i(\theta)\sin\theta d\theta}{\int_0^{\pi} i(\theta)\sin\theta d\theta} \tag{14}$$

where $i(\theta) = [i_1(\theta) + i_2(\theta)]/2$, $\theta$ is the polar scattering angle (angle between wave vectors of incident and scattered waves), and $i_1(\theta)$ and $i_2(\theta)$ are the dimensionless Mie intensities [39]:

$$i_1(\theta) = \left|\sum_j \frac{(2j+1)}{j(j+1)}\left(a_j\tau_j(\cos\theta) + b_j\pi_j(\cos\theta)\right)\right|^2 \tag{15}$$

$$i_2(\theta) = \left|\sum_j \frac{(2j+1)}{j(j+1)}\left(a_j\pi_j(\cos\theta) + b_j\tau_j(\cos\theta)\right)\right|^2 \tag{16}$$

Angular functions $\pi_j(\cos\theta)$ and $\tau_j(\cos\theta)$ are

$$\pi_j(\cos\theta) = \frac{P_j^{(1)}(\cos\theta)}{\sin\theta} \tag{17}$$

$$\tau_j(\cos\theta) = \frac{dP_j^{(1)}(\cos\theta)}{d\theta} \tag{18}$$

where $P_j^{(1)}(\cos\theta)$ are the associated Legendre functions.

## 3.2 *Absorption by Homogeneous Plane-Parallel Plate: Basic Equations*

Absorption coefficient $A_{\text{pl}}$ of the homogeneous plane-parallel plate is calculated by the equation:

$$A_{\text{pl}} = 1 - T_{\text{pl}} - R_{\text{pl}} \tag{19}$$

where $T_{\text{pl}}$ and $R_{\text{pl}}$ are transmission and reflection coefficients of the plate.

In the case of thin plate and normal illumination, taking into account the multiple-beam interference, the transmission $T_{\text{pl}}$ and reflection $R_{\text{pl}}$ coefficients are determined by the relations:

$$T_{\text{pl}} = \left|t_{\text{pl}}\right|^2 = \left|\frac{t_{01}t_{12}e^{ikh}}{1 - r_{12}r_{10}e^{2ikh}}\right|^2 \tag{20}$$

$$R_{\text{pl}} = \left|r_{\text{pl}}\right|^2 = \left|r_{01} + \frac{t_{01}t_{10}r_{12}e^{2ikh}}{1 - r_{12}r_{10}e^{2ikh}}\right|^2 \tag{21}$$

Here $t_{\text{pl}}$ and $r_{\text{pl}}$ are the amplitude transmission and reflection coefficients of the plate; $t_{01}$, $t_{10}$ and $r_{01}$, $r_{10}$ are the amplitude transmission and reflection coefficients for a wave incoming (subscript "01") and outgoing (subscript "10") the plate through the top interface; $t_{12}$ and $r_{12}$ are the amplitude transmission and reflection coefficients for a wave outgoing the plate (subscript "12") through the bottom interface; $h$ is the plate thickness; $k = 2\pi m/\lambda$, $m = n + i\kappa$ is the complex refractive index of the plate material; and $\lambda$ is the wavelength of the incident light.

In the case of thick plate, taking into account multiple reflections between the interfaces, we write equations for the $T_{\text{pl}}$ and $R_{\text{pl}}$ as follows:

$$T_{\text{pl}} = \frac{(1 - R_1)(1 - R_2)e^{-\alpha h}}{1 - R_1R_2e^{-2\alpha h}} \tag{22}$$

$$R_{\text{pl}} = R_1 + \frac{R_2(1 - R_1)^2e^{-2\alpha h}}{1 - R_1R_2e^{-2\alpha h}} \tag{23}$$

Here $R_1$ and $R_2$ are the energy reflection coefficients of the top and bottom interfaces of the plate with thickness $h$, $\alpha = 4\pi\kappa/\lambda$, $\kappa$ is the absorption index (imaginary part of the complex refractive index) of the plate, and $\lambda$ is the wavelength of the incident light.

The $t_{\text{ij}}$, $r_{\text{ij}}$, $R_1$, and $R_2$ coefficients are calculated by the Fresnel formulae [66].

### 3.3 Spectral Absorption by a Monolayer of Silicon Particles: Comparison with the Equivalent Plane-Parallel Plate

We compare the absorption by the particulate monolayer and plane-parallel homogeneous layer under condition of the equality of the silicon volume per unit surface area in both systems. Such homogeneous layer is called in our consideration as an "equivalent plate." Its thickness $h$ is associated with the particle diameter $D$ and the monolayer filling factor $\eta$ by the relationship

$$h = \frac{2}{3}\eta D \tag{24}$$

In Fig. 7, spectra of calculated absorption coefficient $A_{\mathrm{ml}}$ of monolayer of c-Si particles and absorption coefficient $A_{\mathrm{pl}}$ of the homogeneous plane-parallel equivalent plate are presented. Surface filling factor of monolayers $\eta = 0.6$. Absorption coefficient of the thin plate (see Fig. 7a) is calculated by Eq. (19) taking into account the multiple-beam interference (Eqs. (20) and (21)). The coefficient of the thick plate (see Fig. 7b) is calculated taking into account the multiple reflections (Eqs. (22) and (23)). In Fig. 7a, the Mie resonances (sharp peaks) are observed in the spectra of monolayers of "small" particles, and the interference structure (oscillations) is observed for the thin plates at small absorption index of material. For "large" particles and the thick homogeneous plane-parallel plates, these features disappear (Fig. 7b).

As follows from the results displayed in Fig. 7, in the spectral range of large absorption index of silicon ($\lambda \leq 0.4$ μm (see Fig. 1)), the plane-parallel homogeneous plate absorbs more light than a monolayer of particles. This is caused by absorption

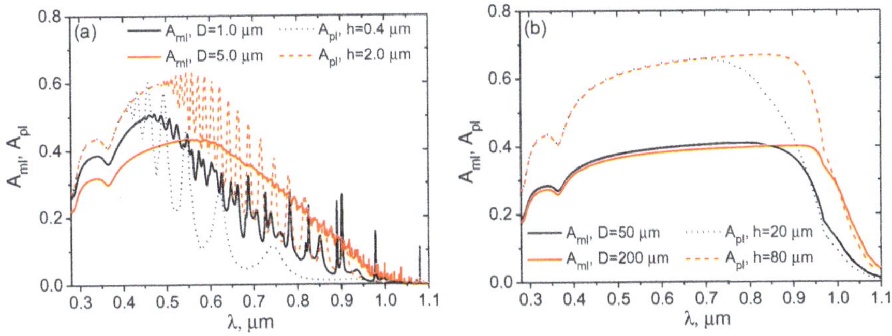


**Fig. 7** The spectra of absorption coefficient $A_{\mathrm{ml}}$ of the monolayer of monodisperse c-Si particles and absorption coefficient $A_{\mathrm{pl}}$ of a homogeneous plane-parallel plate of the same material. (**a**) Black solid line, $A_{\mathrm{ml}}$, $D = 1$ μm; black dotted line, $A_{\mathrm{pl}}$ of the equivalent plate, $h = 0.4$ μm; red solid line, $A_{\mathrm{ml}}$, $D = 5$ μm; red dashed line, $A_{\mathrm{pl}}$ of the equivalent plate, $h = 2$ μm. (**b**) Black solid line, $A_{\mathrm{ml}}$, $D = 50$ μm; black dotted line, $A_{\mathrm{pl}}$ of the equivalent plate, $h = 20$ μm; red solid line, $A_{\mathrm{ml}}$, $D = 200$ μm; red dashed line, $A_{\mathrm{pl}}$ of the equivalent plate, $h = 80$ μm. Filling factor $\eta = 0.6$ for all monolayers

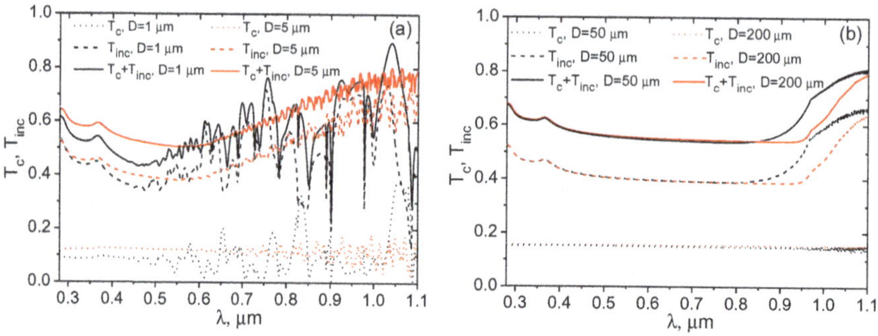


**Fig. 8** Spectral dependences of the coherent $T_c$ and incoherent $T_{\mathrm{inc}}$ transmission coefficients and their sum $T_c + T_{\mathrm{inc}}$ of monolayer. Filling factor $\eta = 0.6$. Diameter of particles $D = 1$ μm, $D = 5$ μm (**a**), $D = 50$ μm, $D = 200$ μm (**b**)

of the incoming into the plate radiation in a very thin surface layer. A similar situation occurs for a monolayer of particles. In the region of strong absorption of the material, the light is absorbed by a thin surface layer of the particles. However, the filling factor of the monolayer is much less than the "filling factor of the homogeneous plate," which is equal to 1. Therefore, in this region the absorption coefficient of a homogeneous plate exceeds the one of the monolayer for all considered particle sizes. For the moderate and small values of the absorption index of silicon (wavelength range $\lambda > 0.4$ μm), the absorption coefficient of a monolayer can exceed the one of the homogeneous plate. This is due to the fact that radiation propagates through a homogeneous silicon layer without significant absorption. Absorption by a monolayer of particles is determined by scattering and absorption of individual particles and their concentration. As follows from the results presented in Fig. 7, the $A_{\mathrm{ml}} > A_{\mathrm{pl}}$ in a wider spectral range for monolayers of smaller considered particles, i.e., the absorption efficiency of such monolayers, is higher.

Let us consider some results for coherent (directly transmitted and specularly reflected) and incoherent (diffusely scattered in the forward and backward hemispheres) components of radiation. Figure 8 shows spectral dependences of the coherent $T_c$ and incoherent $T_{\mathrm{inc}}$ transmission coefficients of monolayers and $T_c + T_{\mathrm{inc}}$ sums. Figure 9 shows spectral dependences of the coherent $R_c$ and incoherent $R_{\mathrm{inc}}$ reflection coefficients of monolayers and $R_c + R_{\mathrm{inc}}$ sums. The transmission and reflection spectra of monolayer of "small" particles have large oscillations [see the spectra for $D = 1$ μm in Figs. 8a and 9a]. They decrease with particle sizes increasing. Growing the particle sizes leads to increasing the $T_c$ and $T_{\mathrm{inc}}$ in the spectral range of large absorption index and decreasing the $R_c$ and $R_{\mathrm{inc}}$ in the spectral range of small absorption index of silicon.

As follows from the calculation results, the major part of radiation incident on a monolayer of particles is scattered in the forward hemisphere. Consequently, it does not contribute to generation of the photo-emf. Therefore, to increase absorption per unit surface area of the particulate layer, it is reasonable to use a multilayered system.

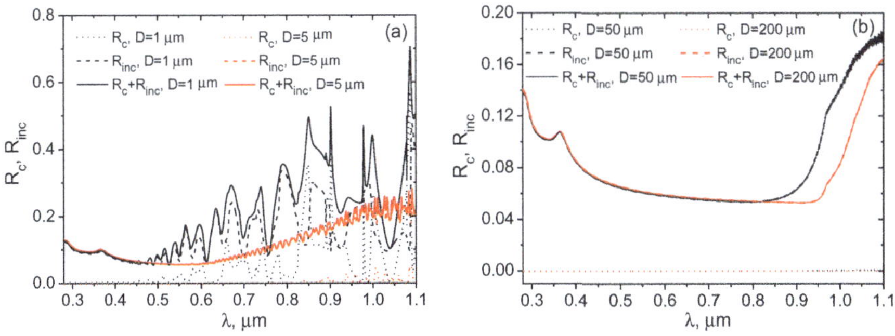


**Fig. 9** Spectral dependences of the coherent $R_c$ and incoherent $R_{\mathrm{inc}}$ reflection coefficients and their sum $R_c + R_{\mathrm{inc}}$ of monolayer. Filling factor $\eta = 0.6$. Diameter of particles $D = 1$ μm, $D = 5$ μm (**a**), $D = 50$ μm, $D = 200$ μm (**b**)

## 3.4 *Spectral Absorption by a Particulate Multilayer and an Equivalent Plane-Parallel Plate*

Consider a multilayer as a stack of identical plane-parallel monolayers of silicon particles. Such a system is schematically shown in Fig. 10.

Let us simulate absorption by the structure and compare it with the results for a homogeneous plane-parallel plate with the equivalent material volume per unit area. Absorption coefficient of multilayer consisting of $N_{\mathrm{ml}}$ plane-parallel monolayers is calculated by the equation

$$A_{N_{\mathrm{ml}}} = 1 - T_{N_{\mathrm{ml}}} - R_{N_{\mathrm{ml}}} \tag{25}$$

where $T_{N_{\mathrm{ml}}} = (T_c + T_{\mathrm{inc}})_{N_{\mathrm{ml}}}$ and $R_{N_{\mathrm{ml}}} = (R_c + R_{\mathrm{inc}})_{N_{\mathrm{ml}}}$ are the energy coefficients of the total transmission and reflection of the system consisting of $N_{\mathrm{ml}}$ monolayers.

We determine coefficients $T_{N_{\mathrm{ml}}}$ and $R_{N_{\mathrm{ml}}}$ (at $N_{\mathrm{ml}} > 1$) in the framework of the transfer matrix method (TMM) [44–46], which takes into account multiple reflection between the interfaces. In this method, the monolayers are considered as "interfaces," which are characterized by their transmission and reflection coefficients, and the spaces between the monolayer planes (planes where the particle centers are located) as "layers" (see Fig. 10) [19, 20, 67–69].

Suppose that spacings between monolayers (i.e., thicknesses of "layers") significantly exceed the coherence length of the radiation. In this case, phase relationships between the waves scattered by monolayers can be neglected, and the energy coefficients $T_{N_{\mathrm{ml}}}$ and $R_{N_{\mathrm{ml}}}$ are calculated as follows [44, 46]:

$$T_{N_{\mathrm{ml}}} = \frac{1}{T_{11}^{\mathrm{int}}} \tag{26}$$

$$R_{N_{\mathrm{ml}}} = \frac{T_{21}^{\mathrm{int}}}{T_{11}^{\mathrm{int}}} \tag{27}$$

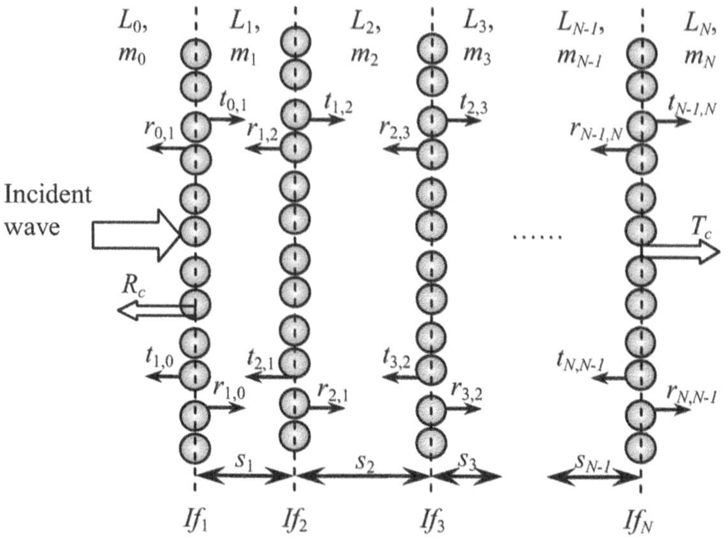


**Fig. 10** Schematic representation of multilayer consisting of plane-parallel particulate monolayers (side view, monolayer planes are indicated by dashed lines). $m_i$ and $s_i$ are complex refractive index and thickness of the $i$-th layer $L_i$, respectively; $If_i$ are numbers of interfaces (monolayers); $t_{i,j}$ and $r_{i,j}$ ($t_{j,i}$ and $r_{j,i}$) are amplitude coherent transmission and reflection coefficients of monolayers for the waves propagated forward (backward); $T_c$ and $R_c$ are coherent transmission and reflection coefficients of multilayer

where $T_{ij}^{\text{int}}$ are elements of the intensity transfer matrix $\mathbf{T}^{\text{int}}$ of the multilayer. For a system of $N$ interfaces (monolayers), the intensity transfer matrix $\mathbf{T}_{0,N}^{\text{int}}$ can be written as [44]

$$\mathbf{T}_{0,N}^{\text{int}} = \begin{bmatrix} T_{11}^{\text{int}} & T_{12}^{\text{int}} \\ T_{21}^{\text{int}} & T_{22}^{\text{int}} \end{bmatrix} = \begin{bmatrix} \dfrac{1}{T_{0,N}} & \dfrac{-R_{N,0}}{T_{0,N}} \\ \dfrac{R_{0,N}}{T_{0,N}} & \dfrac{T_{0,N}T_{N,0} - R_{0,N}R_{N,0}}{T_{0,N}} \end{bmatrix} \tag{28}$$

Here $T_{0,N}$ and $R_{0,N}$ ($T_{N,0}$ and $R_{N,0}$) are the energy total transmission and reflection coefficients of the multilayer for the forward (backward) propagating light.

The transfer matrix of a multilayer is calculated by the sequential multiplying of the interface transfer matrices $\mathbf{T}_j^{\text{int}}$ with the layer propagation matrices $\mathbf{P}_j^{\text{int}}$:

$$\mathbf{T}_{0,N}^{\text{int}} = \left( \prod_{j=1}^{N-1} \mathbf{T}_j^{\text{int}} \mathbf{P}_j^{\text{int}} \right) \mathbf{T}_N^{\text{int}} \tag{29}$$

where

$$\mathbf{T}_j^{\text{int}} = \begin{bmatrix} \dfrac{1}{T_{j-1,j}} & \dfrac{-R_{j-1,j}}{T_{j-1,j}} \\ \dfrac{R_{j-1,j}}{T_{j-1,j}} & \dfrac{T_{j-1,j}T_{j,j-1} - R_{j-1,j}R_{j,j-1}}{T_{j-1,j}} \end{bmatrix} \tag{30}$$

is the transfer matrix of $j$-th interface,

$$\mathbf{P}_j^{\text{int}} = \begin{bmatrix} \left|\exp(-ik_j s_j)\right|^2 & 0 \\ 0 & \left|\exp(ik_j s_j)\right|^2 \end{bmatrix} \tag{31}$$

is the propagation matrix of $j$-th layer, $T_{j-1,j}$ and $R_{j-1,j}$ ($T_{j,j-1}$ and $R_{j,j-1}$) are the energy total transmission and reflection coefficients of $j$-th interface for the forward (backward) propagating wave, wave number $k_j = 2\pi m_j/\lambda$, $m_j$ and $s_j$ are the complex refractive index and thickness of $j$-th layer, and $\lambda$ is the wavelength of incident light.

For the system in homogeneous nonabsorbing medium (air), the $\mathbf{P}_j^{\text{int}}$ is unity matrix and $T_{j-1,j} = T_{j,j-1} \equiv T_j$, $R_{j-1,j} = R_{j,j-1} \equiv R_j$. In this case, Eqs. (28) and (30) can be written as

$$\mathbf{T}_{0,N}^{\text{int}} = \begin{bmatrix} \dfrac{1}{T_{0,N}} & \dfrac{-R_{N,0}}{T_{0,N}} \\ \dfrac{R_{0,N}}{T_{0,N}} & \dfrac{T_{0,N}^2 - R_{0,N}^2}{T_{0,N}} \end{bmatrix} \tag{32}$$

$$\mathbf{T}_j^{\text{int}} = \begin{bmatrix} \dfrac{1}{T_j} & \dfrac{-R_j}{T_j} \\ \dfrac{R_j}{T_j} & \dfrac{T_j^2 - R_j^2}{T_j} \end{bmatrix} \tag{33}$$

Here $T_j$ and $R_j$ are the energy total transmission and reflection coefficients of $j$-th monolayer. Remind that they are calculated in this section in the single scattering approximation (SSA).

Data for spectral absorption coefficients of one monolayer ($A_{N_{\text{ml}} = 1}$) and multilayers consisting of two ($A_{N_{\text{ml}} = 2}$) and three ($A_{N_{\text{ml}} = 3}$) identical monolayers of silicon particles are shown in Fig. 11. Absorption coefficients $A_{\text{pl}}$ of the equivalent plane-parallel plates are presented here as well. Note that thicknesses $h$ of the plane-parallel plate indicated in the figure legends correspond to volume of material per unit area for one, two, and three monolayers of particles, respectively. The thickness of a plane-parallel plate equivalent to a single monolayer and to multilayer composed of $N$ identical monolayers is determined, respectively, by Eq. (24) and

$$h = \frac{2}{3}\eta DN \tag{34}$$

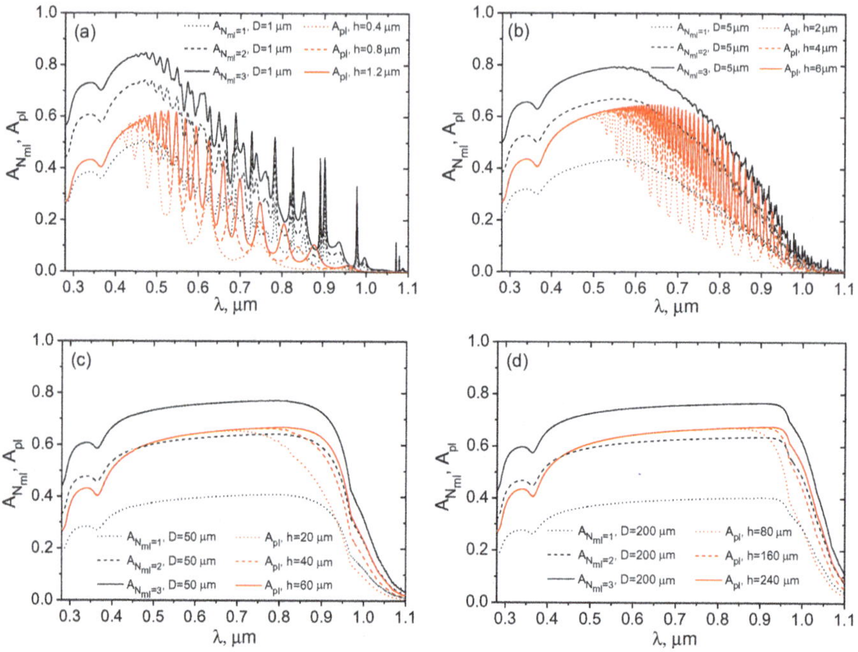


**Fig. 11** Spectral absorption coefficient of one monolayer of silicon particles, the system consisting of two and three identical monolayers ($A_{N_{\mathrm{ml}}=1}$, $A_{N_{\mathrm{ml}}=2}$, and $A_{N_{\mathrm{ml}}=3}$, respectively) and the plane-parallel plate of the equivalent amount of material ($A_{\mathrm{pl}}$). The filling factor of monolayers $\eta = 0.6$, particle diameter $D = 1$ μm (**a**), $D = 5$ μm (**b**), $D = 50$ μm (**c**), $D = 200$ μm (**d**)

The absorption coefficients are calculated using Eqs. (19), (20), and (21) for a thin plate (Fig. 11a, b); Eqs. (19), (22), and (23) for a thick plate (Fig. 11c, d); and Eqs. (25), (26), (27), (28), (29), (30), (31), (32), and (33) for the system of monolayers.

As follows from the results presented in Fig. 11, absorption coefficient of the three-monolayer system as a whole exceeds the one of the equivalent plane-parallel plate. In the range of large absorption index of silicon, the two- and more monolayer system absorbs more light than the equivalent plate. As in the case of single monolayer, increasing of particle size results in significant increasing of the absorption coefficient of the layered system at moderate and small values of the absorption index of material.

Figure 12 shows spectral absorption coefficient of the system consisting of various numbers of identical monolayers of silicon particles and the plane-parallel silicon plate with a volume equal to the particle volume of ten monolayers. The absorption spectra are calculated using Eqs. (19), (22), and (23) for the plate and Eqs. (25), (26), (27), (28), (29), (30), (31), (32), and (33) for the system of monolayers. The absorption coefficient of a system increases with the number of

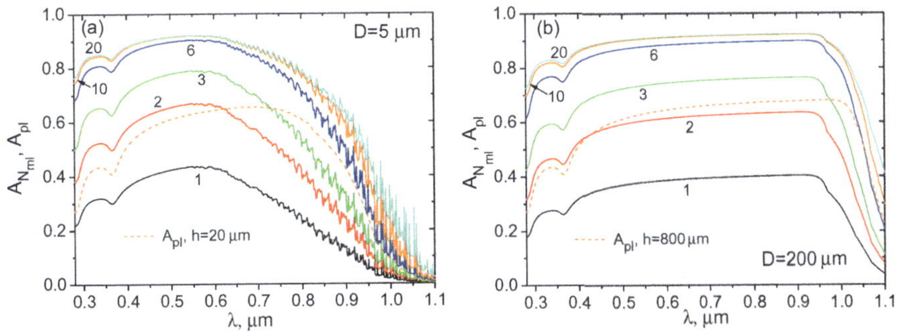


**Fig. 12** Spectral absorption coefficient $A_{N_{ml}}$ of single monolayer of silicon particles ($N_{ml} = 1$), systems of $N_{ml} = 2$, 3, 6, 10, and 20 identical monolayers and $A_{pl}$ of silicon plane-parallel plate. The numbers near the curves indicate the number $N_{ml}$ of monolayers. The filling factor of monolayers $\eta = 0.6$. Particle diameter $D = 5$ μm (a), $D = 200$ μm (b). Dashed line: spectral absorption coefficient $A_{pl}$ of the homogeneous plane-parallel plate with a volume equal to the volume of particles in ten monolayers (plate thickness $h = 20$ μm (a), $h = 800$ μm (b))

monolayers. The coefficient can exceed 0.8 for large and 0.9 for moderate and small values of the absorption index of silicon (see Fig. 12). Absorption coefficient is larger for the systems of larger particles in the range of moderate and small values of the absorption index of c-Si.

Although the absorption increases with the number of monolayers, the fraction of energy absorbed by each subsequent monolayer is reduced relative to the incident energy flux due to absorption by the previous monolayers. As a result, the absorption coefficients of the systems consisting of ten and more monolayers of particles are differed insignificantly. Therefore, for practical use of the proposed approach, the balance between the number of monolayers and the value of the absorption coefficient increasing should be found.

Note that the results described in this section are obtained on the assumption that each monolayer is illuminated by a plane wave. In the considered model, it occurs only for the upper (first) monolayer of the system. The other monolayers are illuminated by the directly transmitted (coherent) and scattered (incoherent) light. But we consider them as illuminated by the plane waves. As a whole, it has to be resulted in the underestimation of the amount of light scattered and absorbed by the multilayer. The approach gives best accuracy for multilayer composed of particles with the highly peaked in the forward direction scattering functions (phase functions) [31].

Examples of the angular distributions of scattered light for individual c-Si particles with diameters $D = 5$ μm and $D = 50$ μm are shown in Fig. 13.

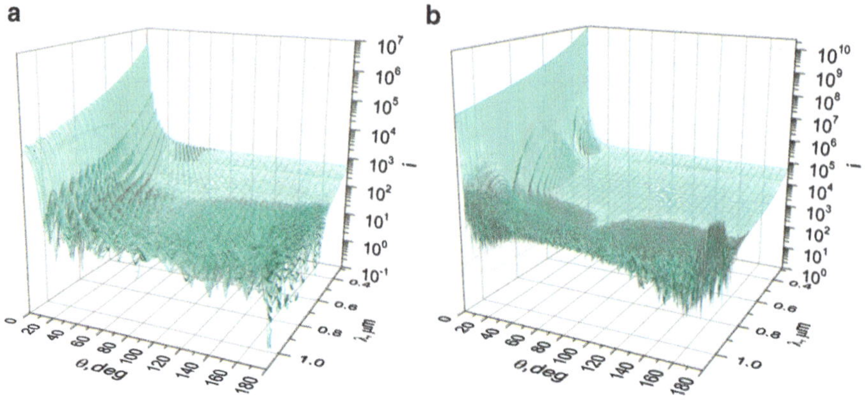


**Fig. 13** Angular distributions of scattered light $i(\theta,\lambda) = [i_1(\theta,\lambda) + i_2(\theta,\lambda)]/2$ versus polar scattering angle $\theta$ and wavelength $\lambda$ of the incident light for spherical c-Si particle with diameter $D = 5$ μm (**a**) and $D = 50$ μm (**b**)

## 3.5 *Integral Absorption by a Particulate Multilayer and an Equivalent Plane-Parallel Plate*

Let us consider the integral over the solar illumination spectrum (Fig. 1) absorption coefficient $\langle A_{N_{\mathrm{ml}}} \rangle$ of the layered system consisting of $N_{\mathrm{ml}}$ monolayers and compare it with the integral absorption coefficient $\langle A_{\mathrm{pl}} \rangle$ of the equivalent plane-parallel plate. We calculate them by Eq. (7), where $f(\lambda)$ is the spectral absorption coefficient of the corresponding system.

To compare absorption of different structures, we use the relative absorption coefficient

$$A_{\mathrm{rel}} = \left( \frac{\langle A_{L_1} \rangle}{\langle A_{L_2} \rangle} - 1 \right) \times 100\% \tag{35}$$

which characterizes difference in integral absorption coefficients of the layers under consideration. Here indices $L_1$ and $L_2$ designate the particulate system and the equivalent plane-parallel plate, respectively. The positive values of $A_{\mathrm{rel}}$ correspond to an increase, while the negative ones correspond to a decrease of light absorption by the particulate system in comparison with absorption by the plate.

Figure 14a shows dependences of the integral absorption coefficient $\langle A_{N_{\mathrm{ml}}} \rangle$ of the systems of c-Si particles with different diameters on the number $N_{\mathrm{ml}}$ of monolayers and the $\langle A_{\mathrm{pl}} \rangle$ of the equivalent plane-parallel plates. Dependence of $A_{\mathrm{rel}}(N_{\mathrm{ml}})$ is displayed in Fig. 14b. The integral absorption coefficient of single monolayer consisting of particles with diameter $D = 1$ μm is larger than the one of the equivalent plane-parallel plate. With the particle sizes increasing, the coefficient of a single monolayer becomes smaller than the one of the equivalent plate. For small- and medium-sized particles, integral absorption coefficient of the

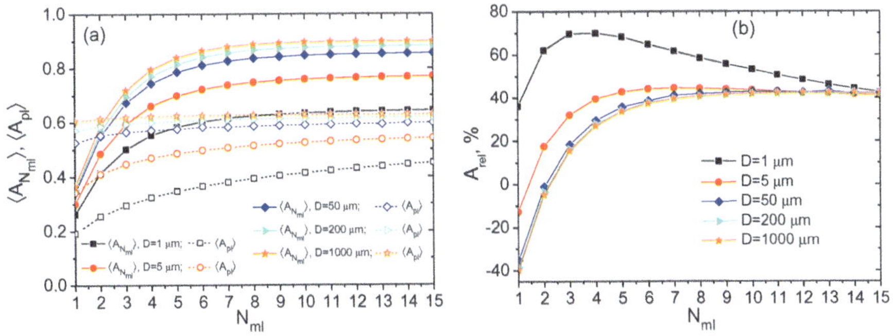


**Fig. 14** Dependences of integral absorption coefficient $\langle A_{N_{ml}} \rangle$ of multilayer on the number of monolayers $N_{ml}$, $\langle A_{pl} \rangle$ of the equivalent plane-parallel plates (**a**) and relative absorption coefficient $A_{rel}$ (**b**). Particle diameters are indicated in the legends. Values of $\langle A_{N_{ml}} \rangle$, $\langle A_{pl} \rangle$, and $A_{rel}$ are calculated for $N_{ml} = 1, 2, \ldots, 15$. The results are indicated by symbols. Filled and empty identical symbols for each $N_{ml}$ correspond to the same volume of material per unit surface area for the multilayer and plate, respectively

two-monolayer system is larger than the one for the equivalent plane-parallel plate. Integral absorption coefficient of three- and more-monolayer systems is larger than the one of the equivalent plane-parallel plates for all considered particle sizes. The values of $A_{rel}$ are larger for smaller particles. Maximum of the $A_{rel}(N_{ml})$ dependence shifts to larger $N_{ml}$ with the particle size increasing (Fig. 14b). At particle diameter $D = 1$ μm, absorption by the system consisting of three and four monolayers is about 1.7 times larger than the one for the equivalent plate. However, the values of $\langle A_{N_{ml}} \rangle$ are greater for systems of larger particles (Fig. 14a). Pay attention that light absorption by the system increases with the particle size increasing, but gradient of $\langle A_{N_{ml}}(D) \rangle$ function decreases. Therefore, for creating the solar cells based on the layered particulate structure, one should optimize the size of the particles. For silicon particles, this diameter is up to ~200 μm. Note that for a sufficiently large number of monolayers the integral absorption coefficient of the particulate structure can exceed the one of the plane-parallel homogeneous plate with volume much greater than the volume of particles (see Fig. 14a).

From the obtained results, it follows that a layered particulate system consisting of three or more monolayers of spherical silicon particles absorbs more light than the equivalent homogeneous plane-parallel silicon plate.

It is worth noting that in the considered range of particle diameters in this section, the monolayer with larger particles reflects smaller part of incident radiation than the monolayer of the smaller particles (see Fig. 9). Therefore, it is advisable to create the front monolayer in the stack from larger particles. The sketch of such a cell is given in [19, 70, 71].

Recall that the results for particulate system presented in this section are obtained in the framework of the TMM with characteristics of individual monolayers simulated in the SSA. More thorough investigation requires taking into account multiple scattering of waves. That is made in the next section.

# 4 Multiple Scattering and Absorption of Light by Layers of Spherical Submicron Particles

## 4.1 Monolayers

### 4.1.1 Basic Equations for Direct Transmission, Specular Reflection, and Absorption Coefficients

The thorough analysis of light absorption by particulate structure requires taking into account multiple scattering of waves especially at high particle concentration and particle sizes comparable with the wavelengths. Such a case is considered in this section.

An important problem for solar cells is to enhance light absorption in the wavelength range of small absorption index of semiconductor. Absorption by the particulate structure can be increased under conditions of the pronounced multiple scattering of waves. Such conditions occur when the particle sizes and distances between them are comparable with wavelengths of the incident light. The small values of absorption index of silicon occur at the longer-wave part of the visible spectrum and in the near-infrared region (Fig. 1). Therefore, as one can expect, the systems of submicron- and micrometer-sized particles should satisfy the abovementioned conditions. Moreover, individual submicron spherical silicon particles exhibit strong peaks of scattering efficiency (see, for example, Fig. 2b) in the spectral range of interest due to optical resonances [30–37]. This is an important factor of overall absorption enhancement. In the present section, we examine absorption by the monolayer and multilayer of submicron crystalline silicon particles in air at normal illumination. The multiple scattering of waves in individual monolayers and multiple reflections between them are taken into account in the quasicrystalline approximation (QCA) [41–43] and in the framework of the TMM [44–46], respectively. The main difference of this approach from considered in previous subsections is that here we take into consideration multiple scattering of waves in individual monolayers and phase relations between the waves scattered by them in multilayer.

Absorption coefficient $A$ of an arbitrary particulate structure can be determined as

$$A = 1 - T_t - R_t \tag{36}$$

where $T_t$ and $R_t$ are the total transmission and reflection coefficients, respectively. Generally, it is a complex problem to simulate these coefficients for the systems under consideration. The complexity essentially increases in the case of concentrated ensembles of particles with sizes of the wavelength order when interference effects, and multiple scattering of waves must be taken into account.

The solution can be simplified when there is only coherently scattered light [41, 42]. In this case, the absorption coefficient can be found as follows:

$$A = 1 - T_c - R_c \tag{37}$$

where $T_c$ and $R_c$ are the coherent transmission and coherent reflection coefficients including all diffraction orders. When only zero diffraction order is implemented, these coefficients characterize only directly transmitted and specularly reflected light, respectively.

The inherent property of the ordered systems to scatter light coherently allows one to obtain relatively simple solution of the scattering problem and use the well-developed methods to simulate coherent transmittance and reflectance. In this section we investigate light absorption by ordered monolayers of spherical particles and layered systems (multilayers) of such monolayers. The consideration is limited by the conditions when only zero diffraction order is implemented.

Here we use the recently developed by us approach to model spatial distribution of particles forming the ordered monolayers (we designate such monolayers as planar crystals (PC)) [72, 73] with the imperfect two-dimensional lattices. The majority of the known methods to model such systems is related to the PCs with perfect lattices [35, 74–82]. The approach inherently takes into account deviations of particle positions from the perfect lattice sites in the monolayer plane and enables one to simulate such type of imperfection (positional imperfection) of real planar crystals simply and naturally. The sketch showing deviations of particle centers from the sites of the perfect two-dimensional (2D) lattices is given in Fig. 15.

To calculate coherent transmission $T_c$ and reflection $R_c$ coefficients of monolayers, we use formalism of the quasicrystalline approximation [42] of the theory of multiple scattering of waves developed by Hong [43] for monolayers with the short-range order (partially ordered monolayers).

Let the infinite statistically uniform monolayer of monodisperse homogeneous isotropic spherical particles is normally illuminated by a plane wave. The particle centers are located in the same plane (monolayer plane). The coherent transmission $T_c$ and reflection $R_c$ coefficients (describing the directly transmitted and specularly reflected components of light in the far-field zone) can be written as

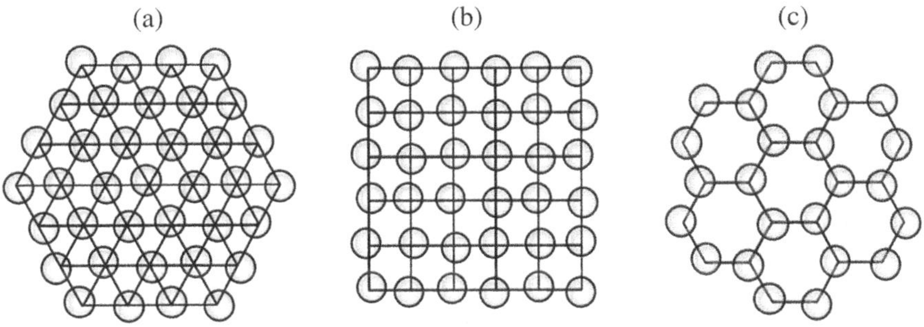


**Fig. 15** The sketch showing deviations of particle centers from the sites of the perfect triangular (**a**), square (**b**), and hexagonal (**c**) lattices

$$T_c = |t_c|^2 = \left| 1 - \frac{\eta}{x^2} \sum_{j=1}^{N} (2j+1)(z_j + y_j) \right|^2 \tag{38}$$

$$R_c = |r_c|^2 = \left| -\frac{\eta}{x^2} \sum_{j=1}^{N} (-1)^j (2j+1)(z_j - y_j) \right|^2 \tag{39}$$

Here $t_c$ and $r_c$ are amplitude coherent transmission and reflection coefficients, $\eta$ is the filling factor, $x = \pi D/\lambda$ is the size parameter of particle with diameter $D$, $\lambda$ is the wavelength of incident light, and $N = x + 4.05x^{1/3} + 2$ is the number of coefficients which is taken into account in the expansions of waves by vector spherical wave functions (harmonics) for given $x$ [39]. The $z_j$ and $y_j$ coefficients are found from the solution of the system of equations:

$$\begin{cases} z_l = b_l + \rho_0 b_l \sum_{j=1}^{N} \left( A_{lj} z_j + B_{lj} y_j \right) \\ y_l = a_l + \rho_0 a_l \sum_{j=1}^{N} \left( B_{lj} z_j + A_{lj} y_j \right) \end{cases} \tag{40}$$

where $\rho_0$ is the average numerical concentration of particles in the monolayer, $a_l$ and $b_l$ are the Mie coefficients (see, for example, [39]) for scattered light

$$a_l = \frac{m\psi_l(mx)\psi_l'(x) - \psi_l(x)\psi_l'(mx)}{m\psi_l(mx)\chi_l'(x) - \chi_l(x)\psi_l'(mx)} \tag{41}$$

$$b_l = \frac{\psi_l(mx)\psi_l'(x) - m\psi_l(x)\psi_l'(mx)}{\psi_l(mx)\chi_l'(x) - m\chi_l(x)\psi_l'(mx)} \tag{42}$$

$m = n + i\kappa$ is the complex refractive index of particle, $\psi_l(x) = x\mathrm{j}_l(x)$ and $\chi_l(x) = x\mathrm{h}_l^{(1)}(x)$ are the Riccati–Bessel functions, $\mathrm{j}_l(x)$ is spherical Bessel function, and $\mathrm{h}_l^{(1)}(x)$ is spherical Hankel function of the first kind. The subscript indicates the order of the functions.

The $A_{lj}$ and $B_{lj}$ coefficients in (40) are found from the solution of equations:

$$A_{lj} = \frac{2j+1}{2} [l(l+1)j(j+1)]^{-\frac{1}{2}} \sum_{p=0,2,\ldots}^{N} i^{-p}(2p+1)[l(l+1) + j(j+1) - p(p+1)] \mathrm{P}_p(0) \times \begin{pmatrix} l & j & p \\ 0 & 0 & 0 \end{pmatrix} \begin{pmatrix} l & j & p \\ 1 & -1 & 0 \end{pmatrix} H_p$$

$$B_{lj} = \frac{2j+1}{2} [l(l+1)j(j+1)]^{-\frac{1}{2}} \sum_{p=0,2,\ldots}^{N} i^{-p}(2p+1)$$

$$[(p+l-j)(p-l+j)(l+j+1+p)(l+j+1-p)]^{\frac{1}{2}} \mathrm{P}_p(0) \times \tag{43}$$

$$\times\begin{pmatrix} l & j & p-1 \\ 0 & 0 & 0 \end{pmatrix}\begin{pmatrix} l & j & p \\ 1 & -1 & 0 \end{pmatrix}H_p \tag{44}$$

where

$$H_p = 2\pi\int_D^\infty g(R)h_p^{(1)}(kR)R\mathrm{d}R \tag{45}$$

g($R$) is the radial distribution function (RDF) [83, 84] of particles in the layer, $R$ is distance in the monolayer plane relatively to the coordinate origin located in the center of an arbitrary particle, $k = 2\pi/\lambda$ is the wavenumber, and $\begin{pmatrix} l & j & p \\ 0 & 0 & 0 \end{pmatrix}$, $\begin{pmatrix} l & j & p \\ 1 & -1 & 0 \end{pmatrix}$, and $\begin{pmatrix} l & j & p-1 \\ 0 & 0 & 0 \end{pmatrix}$ are the Wigner 3j symbols [85], which satisfy the relations:

$$\begin{pmatrix} l & j & p \\ 1 & -1 & 0 \end{pmatrix} = -\frac{1}{2}\left[\frac{(p+l-j)(p-l+j)(l+j+1+p)(l+j+1-p)}{l(l+1)j(j+1)}\right]^{\frac{1}{2}}\begin{pmatrix} l & j & p-1 \\ 0 & 0 & 0 \end{pmatrix} \tag{46}$$

if $l+j+p$ is odd and

$$\begin{pmatrix} l & j & p \\ 1 & -1 & 0 \end{pmatrix} = -\frac{1}{2}\frac{l(l+1)+j(j+1)-p(p+1)}{[l(l+1)j(j+1)]^{\frac{1}{2}}}\begin{pmatrix} l & j & p \\ 0 & 0 & 0 \end{pmatrix} \tag{47}$$

if $l+j+p$ is even.

To calculate 3j symbols, we use the expression [85]

$$\begin{pmatrix} j_1 & j_2 & j_3 \\ m_1 & m_2 & m_3 \end{pmatrix} = (-1)^{j_3+m_3+2j_1}\frac{1}{\sqrt{2j_3+1}}C_{j_1-m_1 j_2-m_2}^{j_3 m_3} \tag{48}$$

where $C_{j_1-m_1\ j_2-m_2}^{j_3 m_3}$ are the Wigner coefficients. For $m_1 = m_2 = m_3 = 0$,

$$C_{a0b0}^{c0} = \begin{cases} 0, a+b+c=2q+1 \\ \dfrac{(-1)^{q-c}\sqrt{2c+1}q!}{(q-a)!(q-b)!(q-c)!}\left[\dfrac{(2q-2a)!(2q-2b)!(2q-2c)!}{(2q+1)!}\right]^{\frac{1}{2}}, a+b+c=2q \end{cases} \tag{49}$$

where $q$ is an integer.

Summations in Eqs. (43) and (44) are carried out over even $p$, because $P_p(0)$ polynomials are zeros at odd $p$, and $P_p(0) = i^p(p-1)!!/p!!$ at even $p$ [86], where $p!! = p(p-2)(p-4)\cdots$.

Let us write Eq. (45) as $H_p = 2\pi(H1_p + H2_p)$, where

$$H1_p = \int_D^\infty \mathrm{h}_p^{(1)}(kR)R\mathrm{d}R \tag{50}$$

$$H2_p = \int_D^\infty (\mathrm{g}(R) - 1)\mathrm{h}_p^{(1)}(kR)R\mathrm{d}R \tag{51}$$

Using the recurrence relations for the spherical Bessel functions [86], we can write integral (50) in the form

$$H1_p = k^{-2}\left\{-kD\mathrm{h}_{p+1}^{(1)}(kD) + \sum_{q=0,2,\ldots}^{p} [2(p-q)+1]\frac{p!!(p-q-1)!!}{(p-1)!!(p-q)!!}\mathrm{h}_{p-q}^{(1)}(kD)\right\} \tag{52}$$

Let us transform the integral (51) into the more convenient for the calculation form. Introduce the dimensionless integration variable $u = R/D$, which is the distance in the monolayer plane expressed in the particle diameters $D$. Then

$$H2_p = D^2\int_1^\infty (\mathrm{g}(u) - 1)\mathrm{h}_p^{(1)}(2xu)u\mathrm{d}u \tag{53}$$

Integral (53) can be calculated numerically when integrand tends to zero at finite $u$. This is provided by convergence of $\mathrm{g}(u)$ to unity. The distance where the RDF attains unity is designated as the correlation length $l_c$, i.e., $\mathrm{g}(u\rightarrow l_c)\rightarrow 1$. Then in (53) upper limit of integration is replaced by $l_c$:

$$H2_p = D^2\int_1^{l_c} (\mathrm{g}(u) - 1)\mathrm{h}_p^{(1)}(2xu)u\mathrm{d}u \tag{54}$$

Correlation length characterizes the spatial scale of ordering of the system. When $u > l_c$ the function $\mathrm{g}(u) = 1$. In this chapter, we consider the monodisperse particles. Note that the above relations are applicable not only for monodisperse but for polydisperse systems as well. In this case, the partial radial distribution functions [83, 87–90] are used.

### 4.1.2 Radial Distribution Functions of Imperfect Planar Crystal

The radial distribution function $\mathrm{g}(u)$ characterizes the probability of finding any particle at distance $u$ relative to another one. There are a number of methods to calculate the RDFs of the partially ordered monolayers [91–95]. In this chapter, we use the iteration method [96] of the numerical solution of the Ornstein–Zernike integral equation [95] with the Percus–Yevick approximation [95].

It is a challenge to find the RDFs for the monolayer with the imperfect long-range order (planar crystal with imperfect lattice) as applied to the QCA. To solve this problem, we use the method based on blurring the coordination circles of perfect lattice [72, 73]. In Fig. 16 are shown the PCs with a perfect triangular, square, and hexagonal lattices and the numbers of particle centers on the coordination circle as functions N($u$) of distance $u$ from the center of a selected particle.

The equation for the radial distribution function of the planar crystal with an imperfect lattice can be written as follows [72, 73]:

$$g(u) = \rho_0^{-1} \sum_i \frac{N_i}{2\pi u_i} \frac{1}{\sqrt{2\pi}\sigma(u)} \exp\left( -\frac{(u - u_i)^2}{2\sigma^2(u)} \right) \tag{55}$$

Here $\sigma(u)$ is a continuous function of distance $u$ characterizing the blurring of coordination circles. We name the $\sigma(u)$ as a *blurring function.*

The simulation has shown that it is reasonable to use the linear blurring function:

$$\sigma(u) = \sigma_0(au + b) \tag{56}$$

In Eqs. (55) and (56), $\rho_0$ is the average numerical particle concentration in the monolayer; $N_i$ is the number of particle centers on the coordination circle with radius $u_i$ in an ideal lattice; and $\sigma_0$, $a$, and $b$ are the coefficients characterizing the ordering degree of the modeled crystal and the distance where the RDF converges to unity.

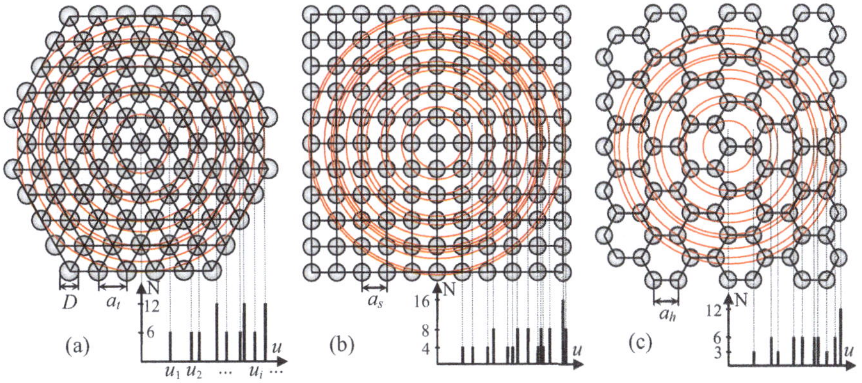


**Fig. 16** Schematic representation of PC with triangular (**a**), square (**b**), and hexagonal (**c**) lattice (top) and function of the number of particle centers N on distance $u$ from coordinate origin (bottom). $u_i$ are the radii of coordination circles (red circles), and $a_t$, $a_s$, and $a_h$, are the lattice constants

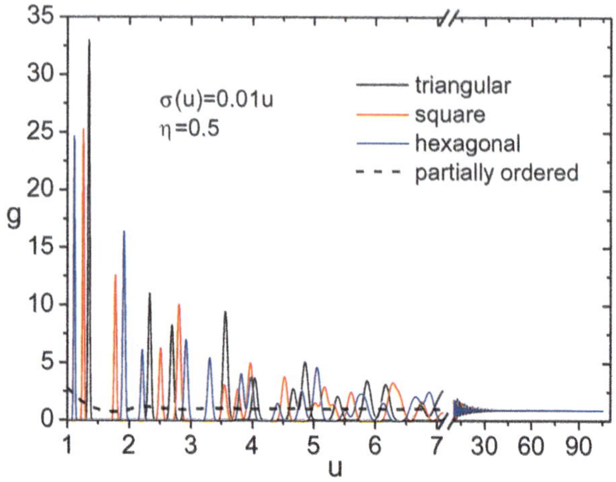


**Fig. 17** Radial distribution functions for monolayers with triangular, square, and hexagonal imperfect lattices and partially ordered monolayer. $\sigma(u) = 0.01u$, $\eta = 0.5$

Function (55) satisfies the normalization condition:

$$\frac{1}{\pi u_2^2} \int_{u_1=0}^{u_2\to\infty} g(u) 2\pi u du = 1 \tag{57}$$

The RDFs obtained using (55) and (56) have good agreement with experimental results [97, 98] and enable one to calculate the $T_c$ and $R_c$ coefficients of imperfect planar crystals with various degrees of ordering.

Shown in Fig. 17 are the RDFs g($u$) describing the monolayers with triangular, square, and hexagonal lattices. They are calculated by Eqs. (55) and (56). Function g($u$) for the partially ordered monolayer is shown as well. It is calculated by iteration method [96] of numerical solution of Ornstein–Zernike integral Eq. [94] in the Percus–Yevick approximation [95]. The data illustrate the difference in radial functions of monolayers with short- and imperfect long-range ordering.

One can see that near the coordinate origin, the RDF of the PC is a sequence of sharp peaks. It is typical for the long-range ordered systems (e.g., crystals). With $u$ increasing, the peaks are blurred and RDF becomes oscillating relative to unity and converging to unity function. The RDF of partially ordered monolayer is an oscillating function. It converges to unity at distances of several particle diameters. Such dependence is typical for the systems with a short-range order (e.g., liquids).

### 4.1.3 Transmission and Reflection Coefficients of Planar Crystal and Partially Ordered Monolayer

The dependences of coherent transmission $T_c$, reflection $R_c$ coefficients, and $T_c + R_c$ sum of planar crystals with triangular, square, and hexagonal lattices and partially ordered monolayer on the particle size parameter $x$ are shown in Fig. 18. They are calculated by Eqs. (38), (39), (40), (41), (42), (43), (44), (45), (46), (47), (48), (49),

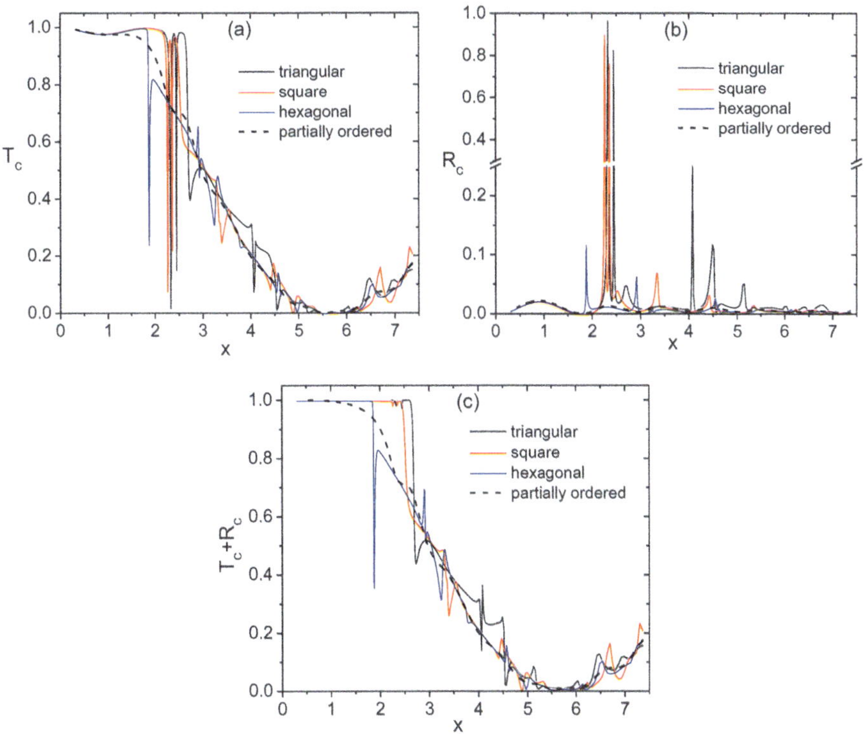


**Fig. 18** Dependences of coherent transmittance $T_c$ (**a**), reflectance $R_c$ (**b**), and the $T_c + R_c$ sum (**c**) of planar crystals with triangular, square, and hexagonal lattices ($\sigma(u) = 0.01u$) and partially ordered monolayer on the particle size parameter $x$. Filling factor $\eta = 0.5$ for all monolayers, particle refractive index $m = 1.4 + 5 \times 10^{-5}i$

(50), (51), (52), (53), and (54) with the RDFs shown in Fig. 17 and illustrate influence of the spatial organization of particles on the spectra of coherent components of light.

One can see that $T_c$ and $R_c$ dependences of the partially ordered monolayer are smooth. The functions $T_c(x)$ and $R_c(x)$ of PC are more complicated. The sharp resonance peaks occur when particle diameters are comparable with the wavelength (see in Fig. 18 the range of ~1.5 < $x$ < ~2.5). For the triangular and square lattices, the peaks are more pronounced than for the hexagonal one and the fine structure is observed.

The $T_c + R_c$ sum of partially ordered monolayer equals to unity only for small particles. With $x$ growing (up to the value of $x \approx 5.5$) the sum monotonically decreases. For monolayer with a long-range order, the sum is equal to unity in much wider region of size parameters than for the partially ordered monolayer. It increases with increasing the coordination number (the number of particle centers on the first coordination circle). Intervals of size parameters where $T_c + R_c = 1$ and $T_c + R_c < 1$ are separated by the resonances. If the sum $T_c + R_c = 1$ only zero diffraction order occurs. If $T_c + R_c < 1$, the higher diffraction orders are implemented along with the zero diffraction order. This causes sharp decrease of $T_c + R_c$ sum.

Therefore long-range order provides much more interval of size parameters where only directly transmitted and specularly reflected light exists than short-range order. In this interval the absorption coefficient of PC consisting of absorption particles can be calculated by Eq. (37).

The results described are used in the next subsections to calculate the absorption spectra of the layers of c-Si particles taking into account multiple scattering of waves.

### 4.1.4 Spectral Absorption Coefficient of Planar Crystal

To calculate the absorption spectra of the monolayer, at first we should find the conditions wherein only directly transmitted and specularly reflected light exist (i.e., $T_c + R_c$ is unity; see Fig. 18). To do that, we first consider monolayer of the model nonabsorbing particles with refractive index of crystalline silicon $n = n_{\text{c-Si}}$ and absorption index $\kappa = 0$. Spectra of coherent transmission $T_c$ and coherent reflection $R_c$ coefficients of the monolayer with imperfect triangular lattice of such particles at different diameters are shown in Fig. 19. The $T_c + R_c$ sum is shown also in the figure. The range of wavelengths where the value of $T_c + R_c$ is unity expands to the shorter wavelengths with the particle size decreasing and the filling factor increasing.

Effect of particle ordering on the directly transmitted and specularly reflected light is demonstrated in Fig. 19c. As one can see, the $T_c + R_c$ sum of the planar crystal consisting of 0.1-μm-diameter particles is practically unity over all considered spectrum (0.28 μm $\leq \lambda \leq$ 1.12 μm). The incoherent component [96, 99] is practically absent in this wavelength range (the incoherent component of scattered light intensity is determined as difference between intensities of total and coherent field). The $T_c + R_c$ sum of the partially ordered monolayer is less than unity in all considered spectral range. The range of 0.6 μm $\leq \lambda \leq$ 1.12 μm is presented in the inset to show the difference from unity more clear.

Above we considered the technique of finding the wavelength range where $T_c + R_c = 1$ for monolayers of nonabsorbing particles ($\kappa = 0$). If the particles with the same real part of the complex refractive index become absorbing ($\kappa > 0$), then in this range we can find the absorption coefficient $A$ of monolayer using Eq. (37).

Let us consider the highly ordered monolayer with triangular lattice from monodisperse spherical silicon particles. To describe the high degree of ordering, we should specify small value of the parameter $\sigma_0$. Let $\sigma_0 = 0.001$. At the chosen value of $\sigma_0$ and the blurring function parameters $a = 1$ and $b = 0$, the correlation length $l_c$ (i.e., the distance where RDF converges to unity) [72, 73, 83] for monolayer with filling factor $\eta = 0.9$ is 900. For monolayers with other $\eta$, we determined $a$ and $b$ values so that correlation length keeps the same value ($l_c = 900$).

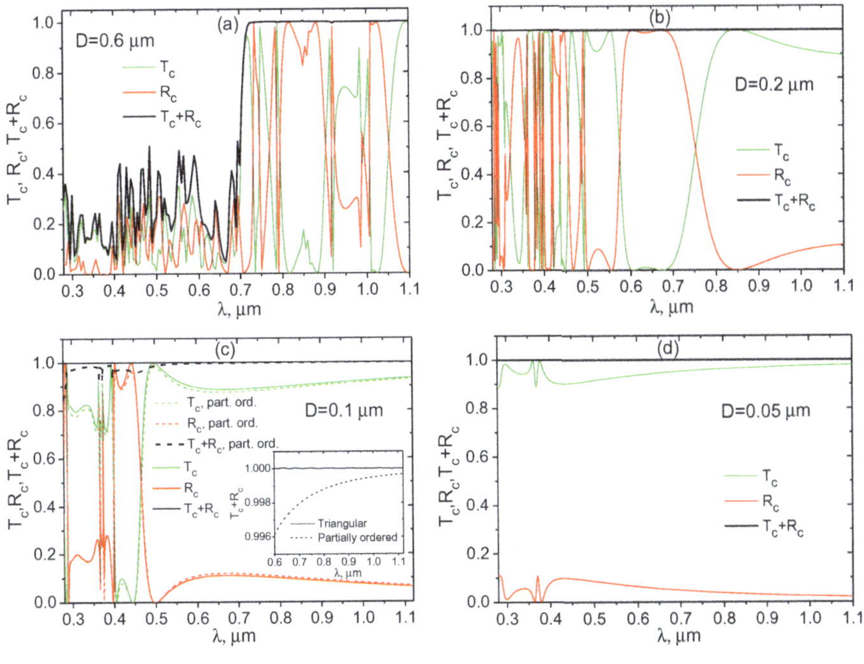


**Fig. 19** Calculated spectra of coherent transmission $T_c$, coherent reflection $R_c$ coefficients, and $T_c + R_c$ sum of monolayer with triangular lattice (solid lines) and partially ordered monolayer (dashed lines, (c)) of spherical particles at different particle diameters $D$. The inset in (c) shows the large scale of $T_c + R_c$ sums in the spectral range from 0.6 μm to 1.12 μm. Refractive index of particles is $n = n_{\text{c-Si}}$, $\kappa = 0$. Filling factor $\eta = 0.5$. Blurring function $\sigma(u) = 0.01u$, correlation length $l_c = 85$

Compare the results for absorption coefficient of monolayer obtained in the QCA (Eqs. (37), (38), (39), (40), (41), (42), (43), (44), (45), (46), (47), (48), (49), (50), (51), (52), (53), (54), (55), and (56)) with the ones obtained in the SSA (Eqs. (2), (3), (4), and (8)) and with the data for the equivalent homogeneous plane-parallel plate. The calculations for the plate are performed taking into account multiple-beam interference (Eqs. (19), (20), and (21)).

Figure 20 shows the spectral absorption coefficient $A_{\text{ml}}$ of monolayer with triangular lattice of c-Si particles with diameter $D = 0.05$ μm at different filling factors $\eta$ and spectral absorption coefficient $A_{\text{pl}}$ of the equivalent plane-parallel plate at different thicknesses $h$. One can see that, as a whole, the absorption coefficient of the equivalent plate is larger than the one of the monolayer of the 0.05-μm-diameter particles. As follows from the analysis of the results, the absorption coefficient of monolayer grows with particle concentration increasing. In the case of small filling factors ($\eta \leq 0.2$), the results obtained in the quasicrystalline and single scattering approximations practically coincide. The difference becomes more pronounced with concentration increasing.

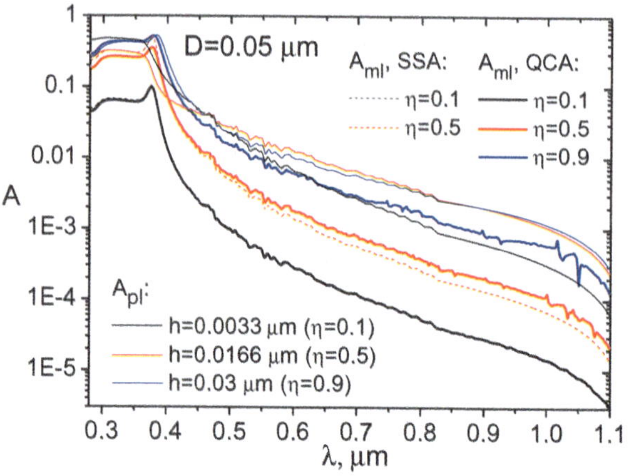


**Fig. 20** Spectral absorption coefficients $A_{ml}$ and $A_{pl}$ of monolayer with triangular lattice of c-Si particles at different filling factors $\eta$ and the equivalent plates with thicknesses $h$. The filling factors of monolayers to which the plates are equivalent are indicated in parentheses after the $h$-values. $D = 0.05$ μm, $\sigma_0 = 0.001$, $l_c = 900$

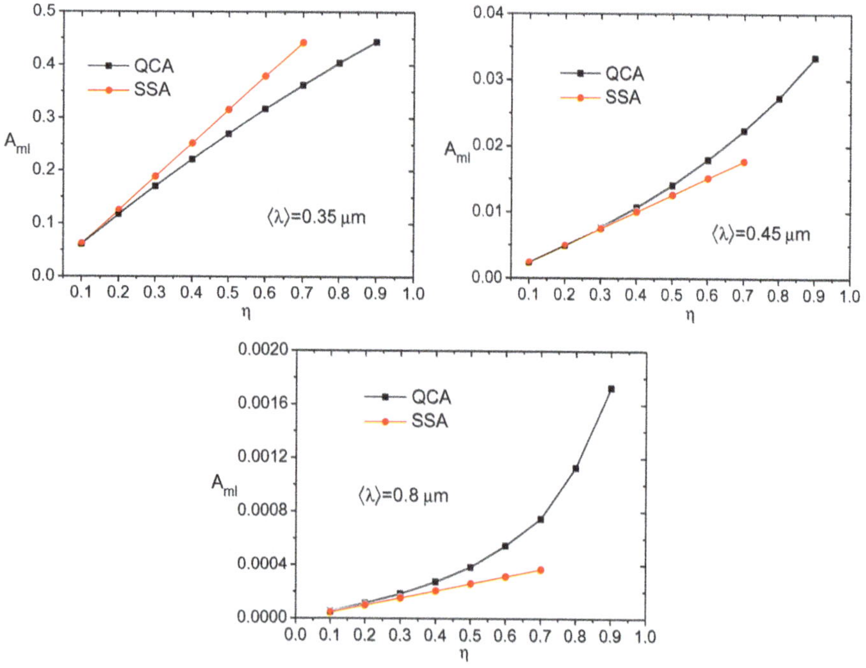


**Fig. 21** Dependences of absorption coefficient $A_{ml}$ of monolayer on filling factor $\eta$ at different $\langle\lambda\rangle$ obtained in the quasicrystalline approximation (QCA) and in the single scattering approximation (SSA). $D = 0.05$ μm, triangular lattice, $\sigma_0 = 0.001$, $l_c = 900$

The dependences of absorption coefficient $A_{ml}$ on the monolayer filling factor $\eta$ at different wavelengths $\langle\lambda\rangle$ are shown in Fig. 21. The $A_{ml}$ values are averaged in the wavelength range from $\langle\lambda\rangle - 10$ nm to $\langle\lambda\rangle + 10$ nm to smooth the absorption spectra oscillations. The data for three $\langle\lambda\rangle$ corresponding to large ($\langle\lambda\rangle = 0.35$ μm),

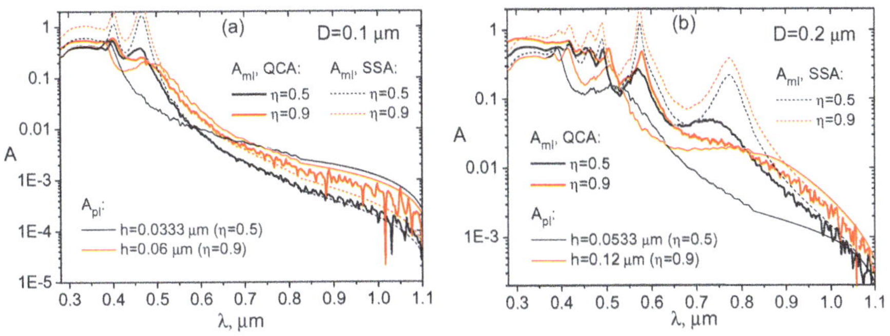


**Fig. 22** Spectral absorption coefficients $A_{ml}$ and $A_{pl}$ of the monolayers with triangular lattice of c-Si particles at different filling factors $\eta$ and equivalent plates with thicknesses $h$. The filling factors of monolayers to which the plates are equivalent are indicated in parentheses after the $h$-values. $\sigma_0 = 0.001$, $l_c = 900$. (**a**) $D = 0.1$ μm. (**b**) $D = 0.2$ μm

moderate ($\langle\lambda\rangle = 0.45$ μm), and small ($\langle\lambda\rangle = 0.8$ μm) absorption indices of c-Si are presented.

As follows from the results, the absorption coefficient increases with $\eta$. The $A_{ml}(\eta)$ dependence calculated in the SSA is linear. The $A_{ml}(\eta)$ dependence calculated in the QCA is close to linear for large and moderate absorption indices of material and nonlinear for the small ones. The difference between the results is more pronounced in the range of small absorption index. As one can see from Fig. 21, absorption coefficient calculated in the SSA can be larger or smaller than the one calculated in the QCA. This is due to multiple scattering of waves which is not taken into account in the SSA.

Consider the monolayers of larger particles. Figure 22 displays the calculated spectra of absorption coefficient of monolayers of 0.1-μm-diameter and 0.2-μm-diameter particles at different monolayer filling factors. The results for equivalent plates are shown as well.

As follows from the presented data, the SSA can give incorrect results ($A > 1$) for monolayer with high particle concentrations. Thus, this approximation cannot be used to obtain quantitative results for highly concentrated monolayers.

The absorption coefficient of monolayer of c-Si particles calculated in the QCA increases with concentration for the most part of the considered spectral range. However, as can be seen in Fig. 22, in some spectral ranges absorption decreases with $\eta$ growing (see the range 0.445 μm $< \lambda <$ 0.475 μm in Fig. 22a and 0.66 μm $< \lambda <$ 0.81 μm in Fig. 22b). To interpret the origin of such behavior, consider the spectra of extinction $Q_{ext}$, scattering $Q_{sc}$, and absorption $Q_{abs}$ efficiency factors of individual spherical particle and calculated in the QCA coherent transmission $T_c$ and reflection $R_c$coefficients of monolayers. Shown in Fig. 23a–c are spectra of these quantities for particle diameter $D = 0.1$ μm.

As can be seen from Fig. 23, in the ranges of 0.38 μm $< \lambda <$0.425 μm and 0.445 μm $< \lambda <$0.475 μm, individual c-Si particle exhibits strong peaks in

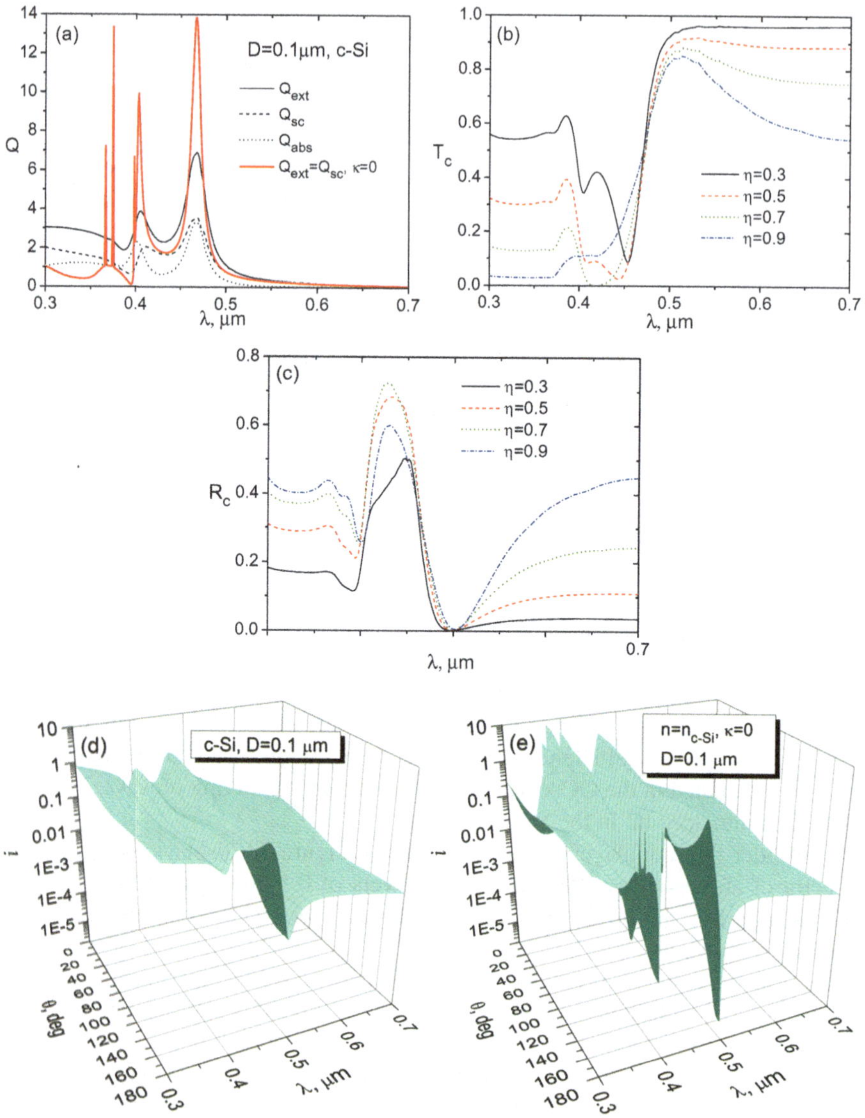


**Fig. 23** Spectra of extinction $Q_{ext}$, scattering $Q_{sc}$, absorption $Q_{abs}$ efficiency factors of spherical c-Si particle and $Q_{ext}$ of the model nonabsorbing particle with the real part of refractive index $n = n_{c\text{-}Si}$ and imaginary part $\kappa = 0$ (**a**). Spectra of coherent transmission $T_c$ (**b**) and reflection $R_c$ (**c**) coefficients of monolayer with triangular lattice, $\sigma_0 = 0.001$, $l_c = 900$. Angular distribution of intensity $i(\theta,\lambda) = [i_1(\theta,\lambda) + i_2(\theta,\lambda)]/2$ (see Eqs. (15) and (16)) of light scattered by c-Si (**d**) and model nonabsorbing ($n = n_{c\text{-}Si}$, $\kappa = 0$) (**e**) particle versus polar scattering angle $\theta$ and wavelength $\lambda$ of the incident light. $D = 0.1$ μm

extinction and scattering efficiencies due to Mie resonances [30–37]. They lead to peaks in absorptance and to sharp changing in coherent transmittance and reflectance of the monolayer. At that, the gradients of $T_c(\lambda)$ and $R_c(\lambda)$ are larger for monolayers with smaller filling factors. In the ranges of 0.455 μm < $\lambda$ < 0.465 μm and 0.47 μm < $\lambda$ < 0.475 μm, the decreasing monolayer absorption coefficient (with concentration growing, see Fig. 22) is accompanied by increasing the coherent transmittance and coherent reflectance, respectively. The similar results are observed for spectra of the 0.2-μm-diameter particle monolayer in the range of 0.66 μm < $\lambda$ < 0.81 μm.

In Fig. 23a, the spectrum of $Q_{ext}$ for model nonabsorbing particle ($n = n_{c\text{-Si}}$, $\kappa = 0$, $D = 0.1$ μm) is shown as well. In this case, $Q_{ext} = Q_{sc}$. Comparison of the spectrum with the one of the real c-Si particle shows that growing the absorption index can result in increasing as well as decreasing the amount of light scattered by individual particle. In considered case, the increasing occurs in the wavelength range of large absorption index of c-Si. The decreasing occurs in the wavelength range of moderate absorption index of c-Si where peaks are observed. One can see these features in the angular distribution of scattered light intensity (Fig. 23d, e) as well. We would like to pay attention that, although the single particle scatters light in all directions ($0 \le \theta \le \pi$, Fig. 23d, e), the ordered monolayer of these particles scatters only in strictly forward and strictly backward directions (see Fig. 19c).

Although the peaks on the absorption coefficient spectra of the monolayer caused by Mie resonances can be diminished with concentration increasing, these resonances as a rule lead to peaks in absorption by monolayer due to strong scattering efficiency, and therefore they are one of the factors of overall absorption enhancement.

As it was noted above, in the solar cells, it is important to provide high absorption of light, especially in the wavelength range of small absorption index of material. Figure 24 shows the calculated spectral absorption coefficient of monolayer of silicon particles in such a range (0.8 μm < $\lambda$ <1.1 μm).

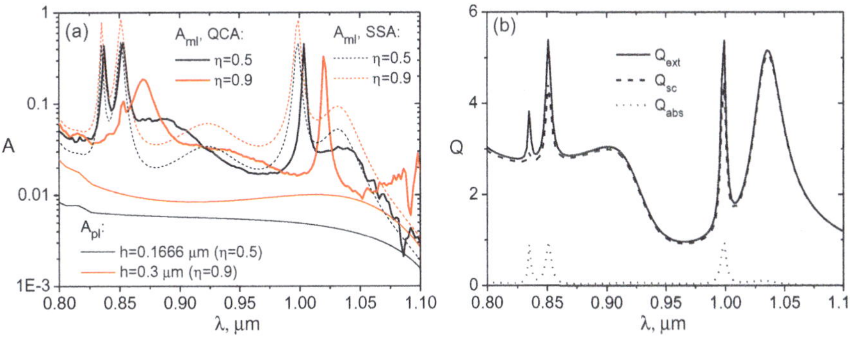


**Fig. 24** Spectral absorption coefficients $A_{ml}$ and $A_{pl}$ of monolayer with triangular lattice of c-Si particles and the equivalent plates. The filling factors $\eta$ of monolayers to which the plates are equivalent are indicated in parentheses after the values of plate thicknesses $h$. $\sigma_0 = 0.001$, $l_c = 900$ (**a**). Spectral dependence of extinction $Q_{ext}$, scattering $Q_{sc}$, and absorption $Q_{abs}$ efficiency factors of silicon spherical particle (**b**). $D = 0.5$ μm

As one can see, the peak positions of $A_{ml}$ obtained in the SSA coincide with the ones of $Q_{abs}$ of a single particle (Fig. 24b). The multiple scattering results in the long wavelength shift of the peaks and in formation of the additional ones. Thus, the absorption spectra of monolayers calculated with accounting the multiple scattering are more complicated than the ones obtained in the SSA. They are influenced by resonances caused by individual silicon particles as well as by their spatial arrangement.

Consider in more detail the monolayers with filling factors $\eta = 0.5$ and $\eta = 0.9$ at different particle sizes in the wavelength range of small absorption index of silicon. Calculated in the QCA, spectral absorption coefficients of such monolayers are presented in Fig. 25a, b. The results for equivalent plates are shown as well. As can be seen from the figure data, the number of peaks in the absorption spectra of monolayer increases with particle size. In spectral ranges of peaks the absorption efficiency significantly increases. Shown in Fig. 25c, d are spectral dependences of the ratio $A_{ml}/A_{pl}$, characterizing the difference of absorption coefficients of the monolayer and the equivalent plate. In the peaks, the $A_{ml}$ can be more than 100 times greater than the $A_{pl}$. Therefore, particulate monolayers can be used as effective light absorbers in the narrow wavelength intervals in the vicinities of the maxima of the peaks. It is worth noting that absorption gain due to the peaks can be

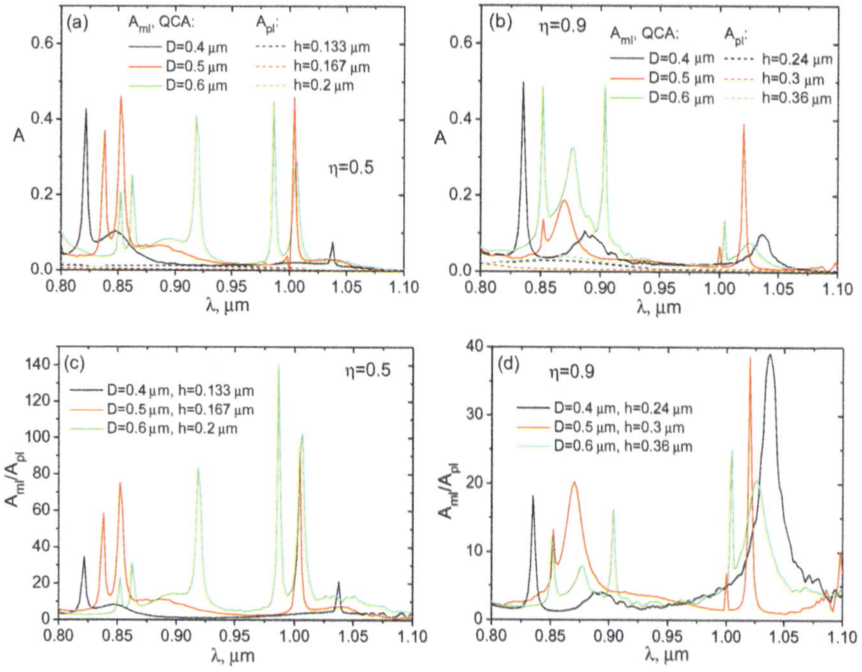


**Fig. 25** Spectral absorption coefficients $A_{ml}$ and $A_{pl}$ of monolayer with triangular lattice of c-Si particles with different diameters $D$ and the equivalent plates with thicknesses $h$ (**a**, **b**). Spectral dependences of the $A_{ml}/A_{pl}$ ratio (**c**, **d**). $\sigma_0 = 0.001$, $l_c = 900$. $\eta = 0.5$ (**a**, **c**), $\eta = 0.9$ (**b**, **d**)

reduced when their spectral positions coincide with the terrestrial solar spectral irradiance deeps caused by absorption bands of atmospheric water vapor, aerosols, etc. (see Fig. 1).

Analysis of the simulation results shows that absorption coefficient of monolayer can increase or decrease with concentration. As a whole, it increases with the particle size and can be larger than the one of the corresponding equivalent plate (see Figs. 20, 22, 24, and 25).

The monolayer absorption coefficient can be increased by optimization of particle diameters and filling factor. The maps and surface graphs of spectral absorption coefficient of monolayer of c-Si particles at different filling factors are shown in Fig. 26.

The outlined results allow one to choose parameters to optimize the monolayer spectral absorption coefficient and increase efficiency of the solar cell due to more efficient light harvesting.

### 4.1.5 Integral Absorption Coefficient

Compare the integral absorption coefficients of the monolayer, $\langle A_{ml} \rangle$, and the equivalent plane-parallel plate, $\langle A_{pl} \rangle$. They are calculated by the Eq. (7), where $f$ is the appropriate spectral absorption coefficient. The relative absorption coefficient is calculated by Eq. (35), where $L_1$ and $L_2$ indices indicate the monolayer and equivalent plane-parallel plate, respectively.

The dependences of $\langle A_{ml} \rangle$, $\langle A_{pl} \rangle$, and $A_{rel}$ on the monolayer filling factor are shown in Fig. 27. The calculations are made for the terrestrial solar spectral irradiance "Global tilt" ASTM G173–03 [47] at the wavelength range from $\lambda_1 = 0.28$ μm to $\lambda_2 = 1.12$ μm (see Fig. 1). The results are obtained using data presented in Figs. 20 and 22.

As follows from Fig. 27a, $\langle A_{ml} \rangle$ coefficient grows with the size of particles and the monolayer filling factor. For monolayer of the 0.05-μm-diameter particles, $\langle A_{ml} \rangle < \langle A_{pl} \rangle$. For other considered particle sizes, $\langle A_{ml} \rangle > \langle A_{pl} \rangle$. As can be seen from Fig. 27b, the relative absorption coefficient and, consequently, the efficiency of light absorption by the monolayer increase with the particle size in comparison with the equivalent plate. This result and the ones obtained in Sect. 3 show that maximum of $A_{rel}$ occurs, as one might expect, for monolayers of particles with sizes comparable with the wavelength.

Integral absorption coefficients of monolayer as functions of particle diameters at various filling factors for the wavelength range of $0.28\ \mu m \le \lambda \le 1.12\ \mu m$ are presented in Fig. 28. One can see that $\langle A_{ml} \rangle$ grows with $D$ and $\eta$ in this range. The dependences are close to the linear ones.

Consider the spectral range of small absorption index of silicon ($0.8\ \mu m \le \lambda \le 1.12\ \mu m$).

Shown in Fig. 29 are dependences of the integral absorption coefficient $\langle A_{ml} \rangle$ of monolayer of c-Si particles on their diameter at different filling factors. One can see that the dependences are more complicated in this range as compared to

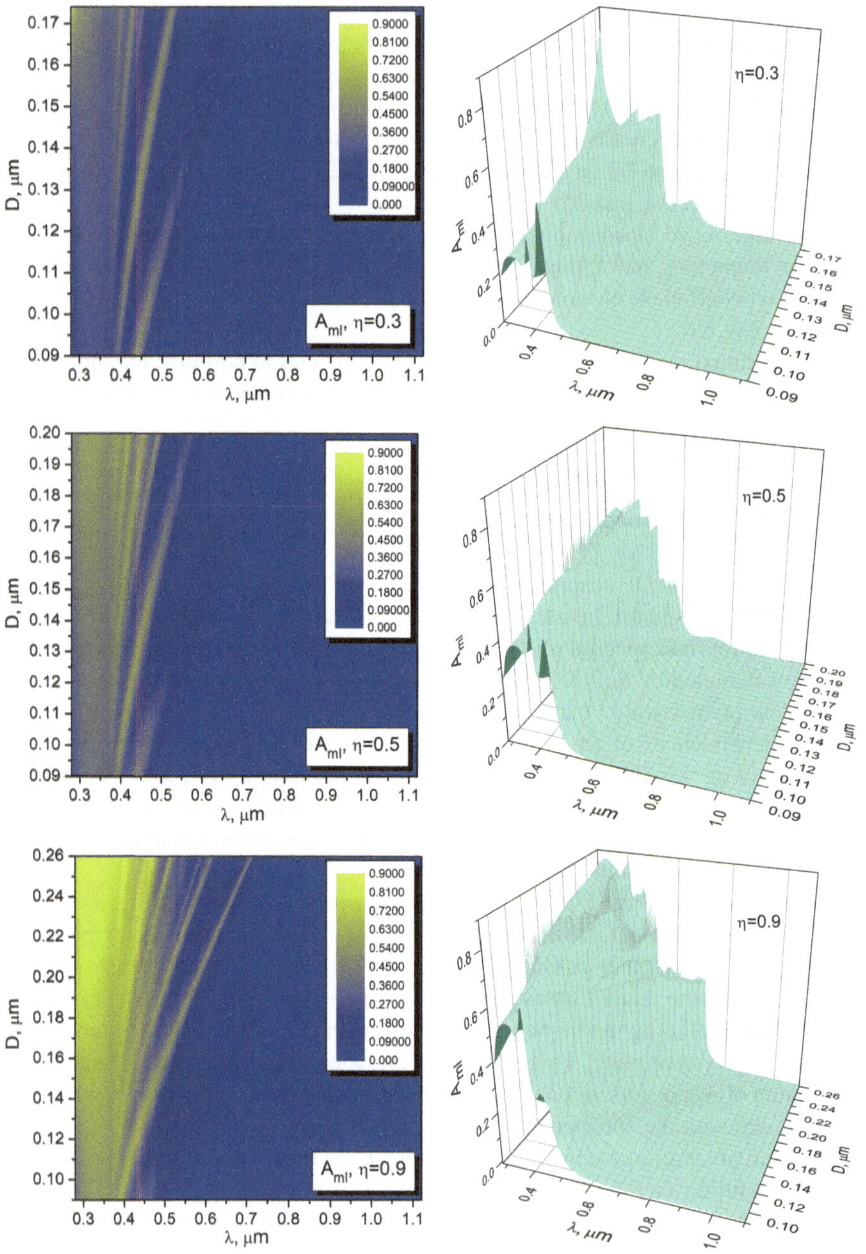


**Fig. 26** Maps (left) and surface graphs (right) of spectral absorption coefficient $A_{\mathrm{ml}}$ of monolayer of c-Si particles. Triangular lattice, $\eta = 0.3$, $\eta = 0.5$, $\eta = 0.9$

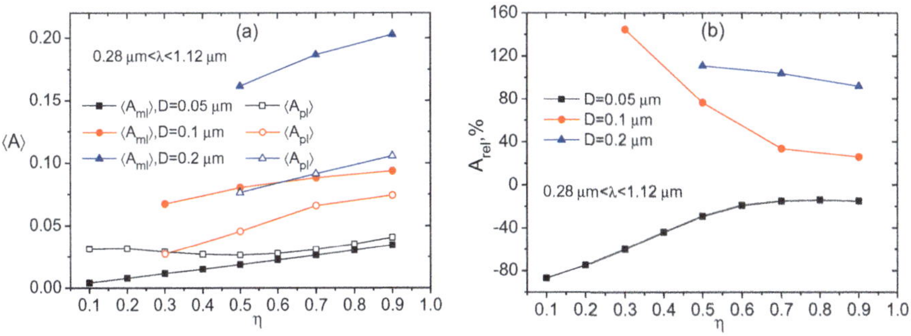


**Fig. 27** Integral absorption coefficient $\langle A_{\rm ml}\rangle$ of monolayer, integral absorption coefficient $\langle A_{\rm pl}\rangle$ of the equivalent plate (**a**), and relative absorption coefficient $A_{\rm rel}$ (**b**) versus monolayer filling factor $\eta$. Monolayers have triangular lattice, $\sigma_0 = 0.001$, $l_c = 900$, $0.28\ \mu\text{m} \leq \lambda \leq 1.12\ \mu\text{m}$

**Fig. 28** Dependences of integral absorption coefficient $\langle A_{\rm ml}\rangle$ of monolayer of c-Si particles on particle diameter at different monolayer filling factors $\eta$. Triangular lattice, $0.28\ \mu\text{m} \leq \lambda \leq 1.12\ \mu\text{m}$

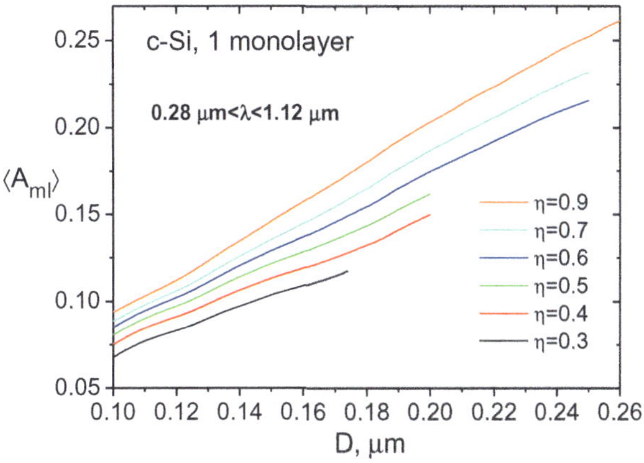


the ones for the range of $0.28\ \mu\text{m} \leq \lambda \leq 1.12\ \mu\text{m}$ (see Fig. 28). Integral absorption coefficient is an oscillating function of particle diameter. One can see (Fig. 29) that maxima and minima of these functions correlate with the ones of integral absorption efficiency factor $\langle Q_{\rm abs}\rangle$ of a single particle. In the regions of maxima, the inverse concentration dependence of $\langle A_{\rm ml}(D)\rangle$ can be observed (see, for example, the range of $0.22\ \mu\text{m} \leq D \leq 0.28\ \mu\text{m}$). Such behavior is described in Sect. 4.1.4 (Figs. 22 and 23).

Figure 30 illustrates influence of particle diameter $D$ on the integral absorption by the monolayer with two values of filling factors ($\eta = 0.5$ and $\eta = 0.9$) and shows data for the equivalent plates. As a whole, values of $\langle A_{\rm ml}\rangle$ and $A_{\rm rel}$ increase with the particle size and filling factor. In some ranges of particle sizes, the $\langle A_{\rm ml}\rangle$ decreasing can occur with concentration increasing (see region of $D$ from ~0.4 μm to ~0.55 μm in Fig. 30a). It means that larger absorption can be achieved at lesser amount of material (see the values of $\langle A_{\rm ml}\rangle$ for $D = 0.5$ μm, $D = 0.52$ μm, and $D = 0.535$ μm

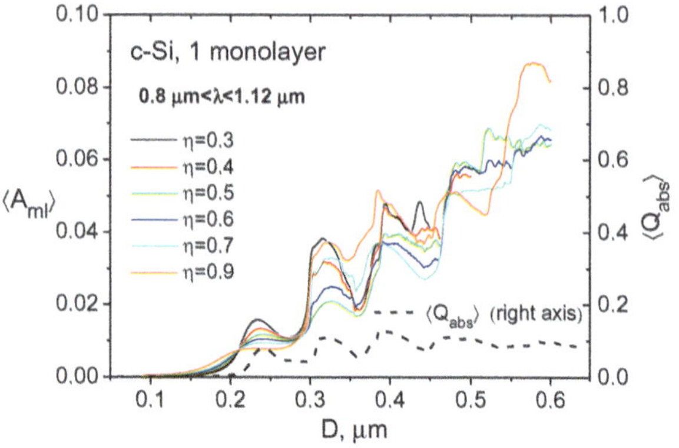


**Fig. 29** Dependences of integral absorption coefficient $\langle A_{\text{ml}}\rangle$ of monolayer with triangular lattice of c-Si particles at different monolayer filling factors and $\langle Q_{\text{abs}}\rangle$ of the c-Si particle on particle diameter. Averaging is fulfilled over the $0.8\ \mu\text{m} \leq \lambda \leq 1.12\ \mu\text{m}$ wavelengths range

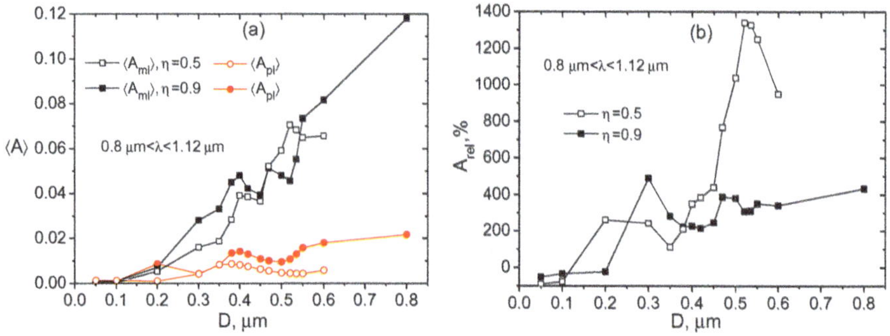


**Fig. 30** Integral absorption coefficient $\langle A_{\text{ml}}\rangle$ of monolayer, integral absorption coefficient $\langle A_{\text{pl}}\rangle$ of the equivalent plate (**a**), and relative absorption coefficient $A_{\text{rel}}$ (**b**) versus particle diameter $D$. Monolayer has triangular lattice, $\eta = 0.5$ (unfilled symbols) and $\eta = 0.9$ (filled symbols), $\sigma_0 = 0.001$, $l_c = 900$, $0.8\ \mu\text{m} \leq \lambda \leq 1.12\ \mu\text{m}$

and the data in Fig. 29). In these ranges, sharp increase of relative absorption is observed for monolayers with smaller $\eta$ as compared with monolayers having larger $\eta$ (Fig. 30b).

Figure 31 depicts the $\eta$-dependences of $\langle A_{\text{ml}}\rangle$, $\langle A_{\text{pl}}\rangle$, and $A_{\text{rel}}$ for particle diameters, at which $\langle A_{\text{ml},\eta\,=\,0.9}\rangle > \langle A_{\text{ml},\eta\,=\,0.5}\rangle$, $\langle A_{\text{ml},\eta\,=\,0.9}\rangle < \langle A_{\text{ml},\eta\,=\,0.5}\rangle$, and $\langle A_{\text{ml},\eta\,=\,0.9}\rangle \approx \langle A_{\text{ml},\eta\,=\,0.5}\rangle$ (see Fig. 30).

As seen from Fig. 31a, the dependences are nonmonotonic. In the range of small absorption index of material, the monolayer absorbs much more efficiently than the equivalent plane-parallel plate. For example, the integral absorption coefficient of monolayer with $\eta = 0.3$ is more than 20 times greater than the one of the plate (see Fig. 31b).

Analysis of the obtained results shows that, as a whole, integral absorption coefficient of the ordered monolayer of submicron crystalline silicon particles increases with particle size and the absorption efficiency is higher for monolayers of larger particles, especially in the range of small absorption index of material.

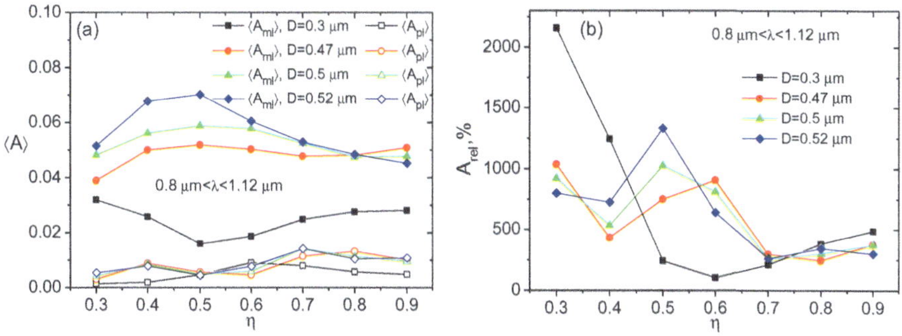


**Fig. 31** Integral absorption coefficient $\langle A_{\mathrm{ml}} \rangle$ of monolayer, integral absorption coefficient $\langle A_{\mathrm{pl}} \rangle$ of the equivalent plate (**a**), and relative absorption coefficient $A_{\mathrm{rel}}$ (**b**) versus monolayer filling factor $\eta$ at different particle diameters. Triangular lattice, $\sigma_0 = 0.001$, $l_c = 900$, $0.8\ \mu\mathrm{m} \leq \lambda \leq 1.12\ \mu\mathrm{m}$

## 4.2 *Multilayers*

### 4.2.1 Main Equations for Transmission and Reflection Coefficients

To find absorption coefficient of the three-dimensional (3D) ordered particulate structure, we consider it as a stack of monolayers (multilayer) (see Fig. 10). First, in the QCA we calculate the *amplitude* coherent transmission and reflection coefficients of individual monolayers. Second, we use these coefficients in the TMM to calculate the multilayer coherent *energy* transmission $T_{N_{\mathrm{ml}}}$ and reflection $R_{N_{\mathrm{ml}}}$ coefficients. Using these data, the absorption coefficient of multilayer is calculated next by Eq. (25). Unlike the Sect. 3.4 where large particles are considered, here we calculate the $T_{N_{\mathrm{ml}}}$ and $R_{N_{\mathrm{ml}}}$ coefficients taking into account phase relations between the waves transmitted and reflected by individual monolayers.

The $T_{N_{\mathrm{ml}}}$ and $R_{N_{\mathrm{ml}}}$ coefficients for such a case are determined by relations [44]

$$T_{N_{\mathrm{ml}}} = |t|^2 = |1/T_{11}|^2 \tag{58}$$

$$R_{N_{\mathrm{ml}}} = |r|^2 = |T_{21}/T_{11}|^2 \tag{59}$$

where $t$ and $r$ are amplitude transmission and reflection coefficients and $T_{\mathrm{ij}}$ are the elements of transfer matrix $\mathbf{T}_{0,N}$ of the multilayer. For the system of $N$ interfaces (monolayers),

$$\mathbf{T}_{0,N} = \begin{bmatrix} T_{11} & T_{12} \\ T_{21} & T_{22} \end{bmatrix} = \frac{1}{t_{0,N}} \begin{bmatrix} 1 & -r_{N,0} \\ r_{0,N} & t_{0,N} t_{N,0} - r_{0,N} r_{N,0} \end{bmatrix} \tag{60}$$

Here $t_{0,N}$ and $r_{0,N}$ ($t_{N,0}$ and $r_{N,0}$) are amplitude transmission and reflection coefficients of the layered system for the forwardly (backwardly) propagating waves.

The transfer matrix of the multilayer is found by the equation

$$\mathbf{T}_{0,N} = \left(\prod_{j=1}^{N-1} \mathbf{T}_j \mathbf{P}_j\right) \mathbf{T}_N \tag{61}$$

Here,

$$\mathbf{T}_j = \frac{1}{t_{j-1,j}} \begin{bmatrix} 1 & -r_{j,j-1} \\ r_{j-1,j} & t_{j-1,j} t_{j,j-1} - r_{j-1,j} r_{j,j-1} \end{bmatrix} \tag{62}$$

is the transfer matrix of $j$-th interface (monolayer),

$$\mathbf{P}_j = \begin{bmatrix} \exp(-ik_j s_j) & 0 \\ 0 & \exp(ik_j s_j) \end{bmatrix} \tag{63}$$

is the propagation matrix of $j$-th layer, $t_{j-1,j}$ and $r_{j-1,j}$($t_{j-1,j}$ and $r_{j-1,j}$) are amplitude transmission and reflection coefficients of $j$-th interface for the forwardly (backwardly) propagating waves, $k_j = 2\pi m_j/\lambda$, $m_j$ and $s_j$ are complex refractive index and thickness of $j$-th layer ($s$ is the distance between planes of the adjacent monolayers; see Fig. 10), and $\lambda$ is the wavelength in the incident wave.

### 4.2.2 Absorption by Multilayer Consisting of Identical Monolayers

The significant part of incident light can be transmitted or reflected by the monolayer (see, for example, Fig. 23). It means large losses in the amount of the absorbed incident light. They can be reduced by using the layered structure (multilayer) consisting of a stack of the plane-parallel monolayers of particles. The multilayers with sufficiently large (in comparison with the wavelengths) particles and intermonolayer spacings were considered in the Sect. 3.

In the present subsection, we consider the multilayers consisting of identical monolayers with particle sizes and intermonolayer spacings comparable with the wavelengths. The multiple scattering of waves in individual monolayers and multiple reflections between them are taken into account. The calculation technique is described above in Sect. 4.2.1

Figure 32 shows calculated spectral absorption coefficients $A_{N_{\mathrm{ml}}}$ of multilayer consisting of monolayers of the 0.05-μm-diameter c-Si particles at different filling factors $\eta$. The spacing $s$ between the adjacent monolayers equals to the particle diameter: $s = 0.05$ μm.

Light absorption by a multilayer increases with the number of monolayers $N_{ml}$. In the range of moderate and small absorption index of silicon, the absorption coefficient increases with $\eta$. In the range of large absorption index, it increases at transition from small to moderate values of $\eta$ (see $A_{N_{\mathrm{ml}}} = 10$ for $\eta = 0.1$ and $\eta = 0.5$) and decreases at transition from moderate to large values of $\eta$ (see $A_{N_{\mathrm{ml}}} = 10$ for

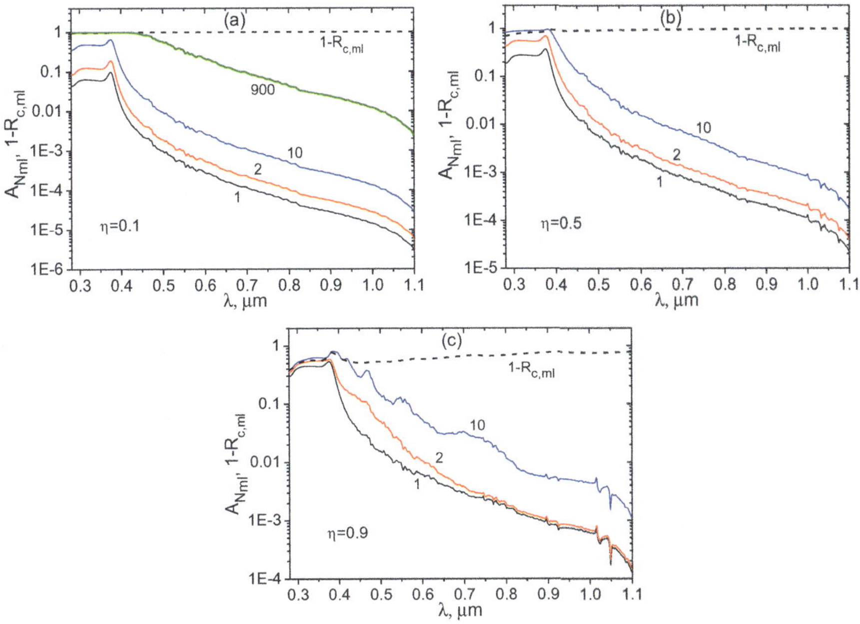


**Fig. 32** Spectral absorption coefficient $A_{N_{\mathrm{ml}}}$ of multilayer consisting of identical monolayers with triangular lattice of c-Si particles at $\eta = 0.1$ (**a**), $\eta = 0.5$ (**b**), and $\eta = 0.9$ (**c**). $D = 0.05$ μm, $\sigma_0 = 0.001$, $l_c = 900$. $s = 0.05$ μm. The dashed lines indicate the 1-$R_{c,ml}$ quantity where $R_{c,ml}$ is coherent reflection coefficient of monolayer. Numbers near the lines indicate the number of monolayers $N_{\mathrm{ml}}$

$\eta = 0.5$ and $\eta = 0.9$). The $A_{N_{\mathrm{ml}}}$ decreasing correlates with increasing the coherent reflection coefficient $R_{c,ml}$ of monolayer.

The dashed lines in Fig. 32 show spectral dependences of the $1-R_{c,ml}$ quantity, which allows one to estimate roughly maximally possible absorption coefficient of the multilayer. As can be seen from the figure, 1-$R_{c,ml}$ decreases with concentration increasing. It is remarkable that the 1-$R_{c,ml}$ quantity is practically unity for monolayers with small particle concentrations (see Fig. 32a). This result means that one can obtain the multilayer absorption coefficient close to unity at the sufficient number of sparse monolayers of small particles. Such a way is impractical in the considered scheme, because the huge number of monolayers is required. For example, although $A_{N_{\mathrm{ml}}\;=\;900} \approx 1$ in the spectral range of large absorption index of silicon, it is still remains significantly less than unity in the ranges of moderate and small absorption index (see Fig. 32a). However, it is worth noting that the results described can be useful at consideration of nano- and microcrystalline silicon solar cells.

Figure 33a shows integral absorption coefficient $\langle A_{N_{\mathrm{ml}}} \rangle$ of the layered system with various number $N_{\mathrm{ml}}$ of monolayers and the integral absorption coefficient $\langle A_{\mathrm{pl}} \rangle$ of the equivalent plates. Figure 33b shows the corresponding dependences of

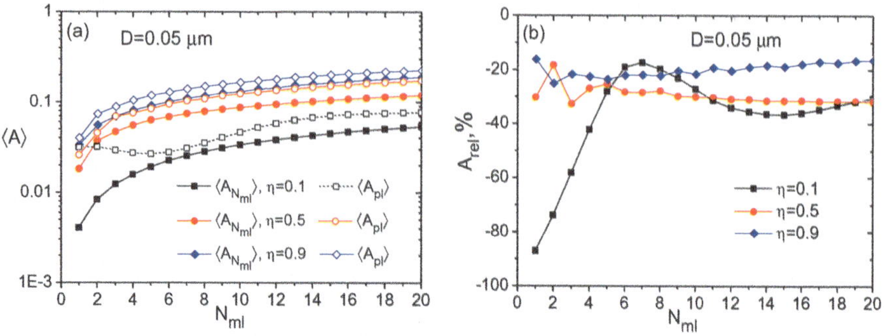


**Fig. 33** Integral absorption coefficient $\langle A_{N_{ml}} \rangle$ of multilayer, $\langle A_{pl} \rangle$ of equivalent plate (**a**), and relative absorption coefficient $A_{rel}$ (**b**) versus the number of monolayers $N_{ml}$. Individual monolayers have triangular lattice, $D = 0.05$ μm, $\sigma_0 = 0.001$, $l_c = 900$, $s = 0.05$ μm, 0.28 μm $\leq \lambda \leq$ 1.12 μm

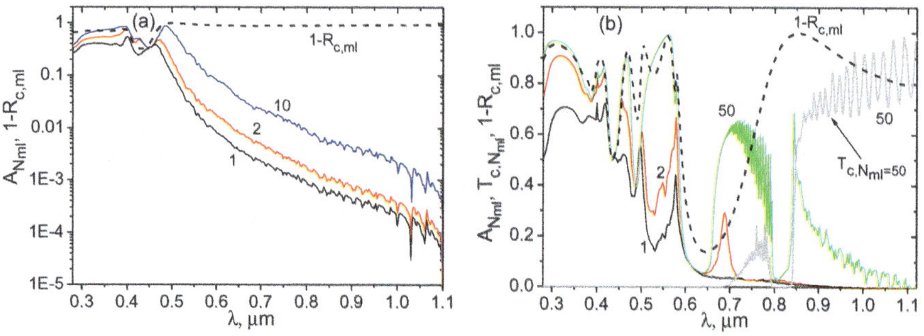


**Fig. 34** Spectral absorption coefficient $A_{N_{ml}}$ of systems consisting of identical monolayers with triangular lattice of c-Si particles at $D = 0.1$ μm, $\eta = 0.5$, $s = 0.1$ μm (**a**) and at $D = 0.2$ μm, $\eta = 0.7$, $s = 0.2$ μm (**b**). Spectral coherent transmission coefficient $T_{c,N_{ml}}$ of multilayer consisting of 50 monolayers of 0.2-μm-diameter particles, $\eta = 0.7$, $s = 0.2$ μm (b). QCA, $\sigma_0 = 0.001$, $l_c = 900$. The number of monolayers $N_{ml}$ is indicated near the lines

$A_{rel}(N_{ml})$. As follows from the simulation results, integral absorption coefficient of the systems of the 0.05-μm-diameter particles is smaller than the one of the equivalent plates.

Consider light absorption by the systems of larger particles. Figure 34 shows spectral absorption coefficient of multilayers consisting of monolayers of the 0.1-μ m-diameter and 0.2-μm-diameter particles. As in the case of 0.05-μm-diameter particle multilayer, concentration increasing can result in decreasing the absorption coefficient of the multilayer in the range of large absorption index of silicon. In the range of small absorption index of silicon, the absorption coefficient of multilayer depends on particle concentration, size, and wavelength. As a whole, $A_{N_{ml}}$ grows with particle size and concentration. In the wavelength range of the photonic bandgap (PBG), it sharply decreases (see $T_{c,N_{ml}}$ in the range

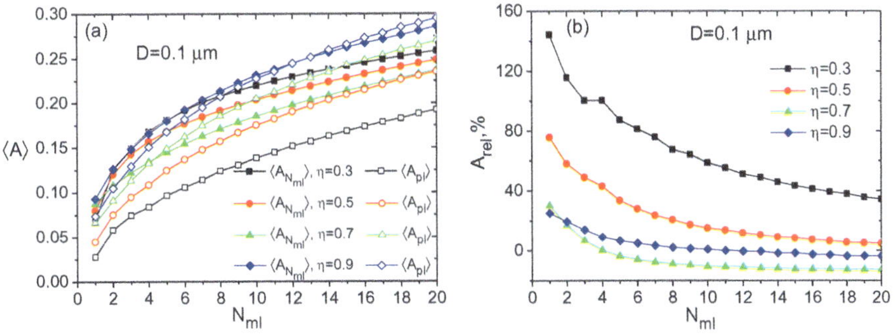


**Fig. 35** Integral absorption coefficient $\langle A_{N_{\mathrm{ml}}} \rangle$ of multilayer, $\langle A_{\mathrm{pl}} \rangle$ equivalent plate (**a**), and relative absorption coefficient $A_{rel}$ (**b**) versus the number of monolayers $N_{\mathrm{ml}}$. Individual monolayers have triangular lattice, $D = 0.1$ μm, $\sigma_0 = 0.001$, $l_c = 900$. $s = 0.1$ μm, $0.28\ \mu\mathrm{m} \leq \lambda \leq 1.12\ \mu\mathrm{m}$

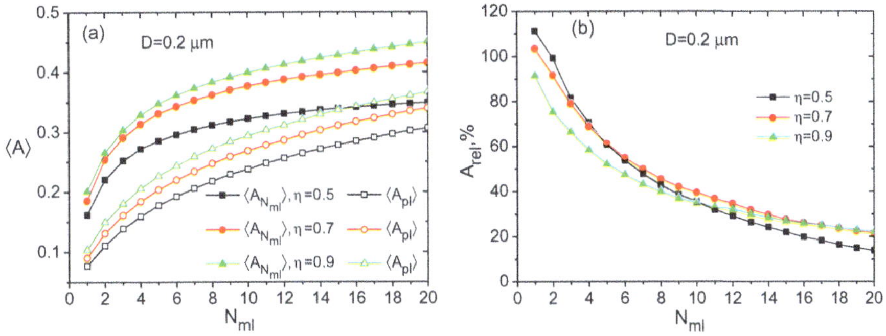


**Fig. 36** Integral absorption coefficient $\langle A_{N_{\mathrm{ml}}} \rangle$ of multilayer, $\langle A_{\mathrm{pl}} \rangle$ of equivalent plate (**a**) and relative absorption coefficient $A_{\mathrm{rel}}$ (**b**) versus the number of monolayers $N_{\mathrm{ml}}$. Individual monolayers have triangular lattice, $D = 0.2$ μm, $\sigma_0 = 0.001$, $l_c = 900$. $s = 0.2$ μm, $0.28\ \mu\mathrm{m} \leq \lambda \leq 1.12\ \mu\mathrm{m}$

~0.78 μm $\leq \lambda \leq$ ~0.84 μm in Fig. 34b). However, near the bounds of the PBG, absorption coefficient significantly grows up.

Figure 35a depicts integral absorption coefficients $\langle A_{N_{\mathrm{ml}}} \rangle$ of multilayers with monolayers of the 0.1-μm-diameter particles and integral absorption coefficients $\langle A_{\mathrm{pl}} \rangle$ of the equivalent plates. Figure 36a shows data for structures with the 0.2-μm-diameter particles. Figures 35b and 36b display the corresponding dependences of $A_{\mathrm{rel}}(N_{\mathrm{ml}})$.

As one can see from Figs. 33, 35, and 36, the integral absorption coefficient increases with particle size. The multilayers of considered particles can absorb more light than the equivalent plates. The particulate system absorbs light more efficiently (in comparison with the equivalent plate) at lower particle concentrations and smaller number of monolayers (see Fig. 35b and 36b). Maximum relative increasing in absorption for the considered cases is about 145%.

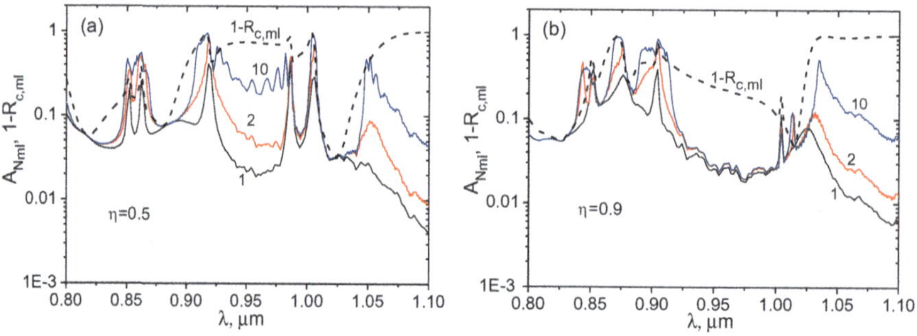


**Fig. 37** Spectral absorption coefficient $A_{N_{\text{ml}}}$ of systems consisting of identical monolayers with triangular lattice of c-Si particles with diameter $D = 0.6$ μm. $s = 0.6$ μm, $\sigma_0 = 0.001$, $l_c = 900$, $0.8\ \mu\text{m} \leq \lambda \leq 1.12\ \mu\text{m}$. $\eta = 0.5$ (**a**), $\eta = 0.9$ (**b**). The number of monolayers $N_{\text{ml}}$ is indicated near the lines

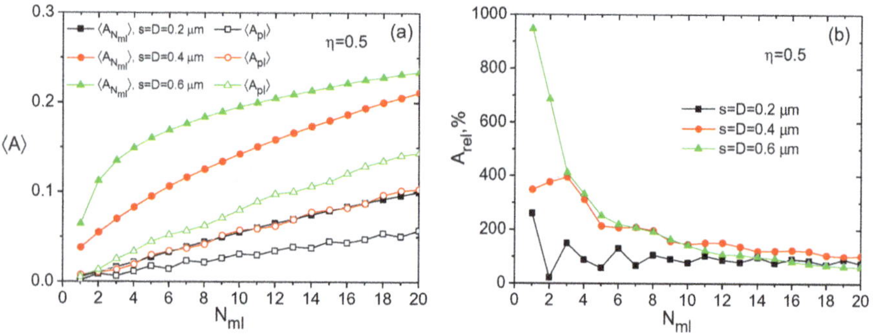


**Fig. 38** Integral absorption coefficient $\langle A_{N_{\text{ml}}} \rangle$ of multilayer, $\langle A_{\text{pl}} \rangle$ the equivalent plate (**a**), and relative absorption coefficient $A_{\text{rel}}$ (**b**) versus the number of monolayers $N_{\text{ml}}$. Individual monolayers have triangular lattice, $\eta = 0.5$, $\sigma_0 = 0.001$, $l_c = 900$, $0.8\ \mu\text{m} \leq \lambda \leq 1.12\ \mu\text{m}$

Consider absorption by multilayer in the range of small absorption index of silicon. Shown in Fig. 37 is spectral absorption coefficient of multilayers composed of monolayers of the 0.6-μm-diameter particles with filling factors $\eta = 0.5$ (a) and $\eta = 0.9$ (b). In some wavelength ranges, spectral absorption coefficient significantly increases with the number of monolayers, while in the other ones it is not practically changed. Such behavior is mainly determined by optical properties of the individual monolayers.

Shown in Fig. 38a and 39a are integral absorption coefficients $\langle A_{N_{\text{ml}}} \rangle$ of multilayers at filling factors of monolayers $\eta = 0.5$ and $\eta = 0.9$, respectively, and integral absorption coefficient $\langle A_{\text{pl}} \rangle$ of the equivalent plates in wavelength range of small absorption index of silicon. The corresponding dependences of relative absorption coefficients $A_{\text{rel}}(N_{\text{ml}})$ are shown in Fig. 38b and 39b. As follows from Figs. 38 and 39, integral absorption increases with particle size. As a whole the $A_{\text{rel}}(N_{\text{ml}})$ is larger for multilayers with the smaller number of monolayers and filling

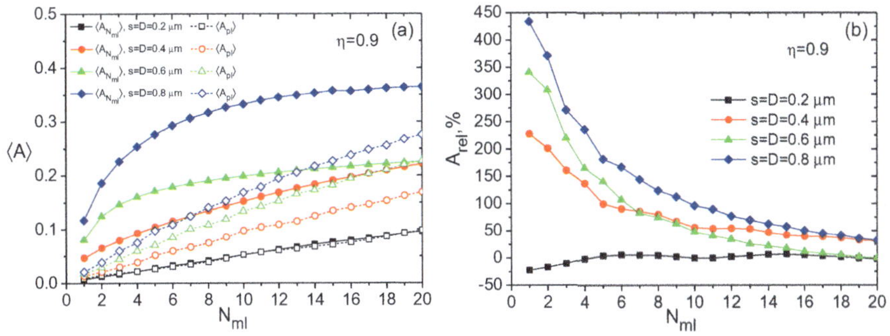


**Fig. 39** Integral absorption coefficient $\langle A_{N_{ml}} \rangle$ of multilayer, integral absorption coefficient $\langle A_{pl} \rangle$ of equivalent plate (**a**), and relative absorption coefficient $A_{rel}$ (**b**) versus the number of monolayers $N_{ml}$. Individual monolayers have triangular lattice, $\eta = 0.9$, $\sigma_0 = 0.001$, $l_c = 900$, $0.8\ \mu m \le \lambda \le 1.12\ \mu m$

**Fig. 40** Integral absorption coefficient $\langle A_{N_{ml}} \rangle$ of multilayer at different number $N_{ml}$ of monolayers versus monolayer filling factor $\eta$. Individual monolayers have triangular lattice, $D = 0.5\ \mu m$, $\sigma_0 = 0.001$, $l_c = 900$, $s = 0.5\ \mu m$. $0.8\ \mu m \le \lambda \le 1.12\ \mu m$

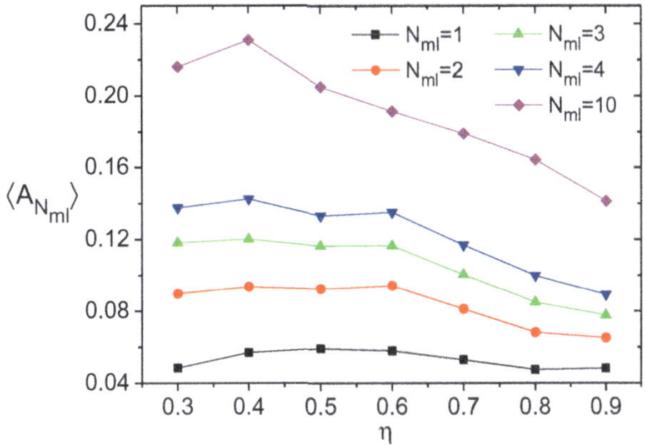


factors. As in the case of full considered wavelength range ($0.28\ \mu m \le \lambda \le 1.12\ \mu m$), integral absorption coefficient of the multilayer corresponding to the range of small absorption index of material ($0.8\ \mu m \le \lambda \le 1.12\ \mu m$) can increase or decrease with the concentration increasing. Maximum relative increasing in absorption for the considered cases is about 950%.

The simulation results for single monolayer show that dependence of the integral absorption coefficient $\langle A_{ml} \rangle$ on concentration can have maximum and/or minimum (see Fig. 31). Shown in Fig. 40 are the $\eta$-dependences of $\langle A_{N_{ml}} \rangle$ for multilayers when the maximum of $\langle A_{ml} \rangle$ has occurred (diameter of particles $D = 0.5\ \mu m$).

In the region of small filling factors, absorption grows monotonically. Note that the same behavior takes place for clusters of particles considered in [100]. A maximum of integral absorption coefficient occurs with the further $\eta$ increasing. This result correlates with the experimentally obtained data for layers of the silver halide particles in gelatin [101, 102]. The maximum shifts to lower $\eta$-value with $N_{ml}$ increasing.

### 4.2.3 Absorption Optimization by Multilayer Consisting of Different Monolayers

#### Particle Diameter and Intermonolayer Spacing Variations in Three-Monolayer System

Till now we considered multilayers composed of the identical monolayers with constant spacing between adjacent monolayers. The absorption enhancement can be achieved using the multilayers consisting of different monolayers and intermonolayer spacings.

Let us use the data of Fig. 29 to estimate absorption by the layered systems composed of three monolayers of particles with diameters corresponding to minima and maxima of the integral absorption coefficient in the spectral range of small absorption index of c-Si (0.8 μm $\leq \lambda \leq$ 1.12 μm).

Some results to illustrate absorption dependence on the arrangement of monolayers in the multilayer are presented below. The different combinations of monolayers yield different absorption coefficients of multilayer. The simulation results show that the minimum and maximum absorptions in three-monolayer systems occur when "gradient" and "inverse gradient" over the particle diameter in monolayer sequences are implemented (see Fig. 41). Other combinations of the monolayers give intermediate values of the $\langle A_{N_{\rm ml}} \rangle$. The comparison of the data of Fig. 41a, b shows that significantly larger absorption coefficient can be achieved when the particle diameters in the monolayer sequence correspond to maxima of the $\langle A_{\rm ml}(D) \rangle$ dependence as compared to the case of minima (Fig. 29). Thus, using the lesser amount of material (multilayer with monolayers of particles with diameters $D_1 = 0.394$ μm, $D_2 = 0.316$ μm, $D_3 = 0.235$ μm), one can obtain significantly larger absorption coefficient than using greater amount of material ($D_1 = 0.426$ μm, $D_2 = 0.361$ μm, $D_3 = 0.273$ μm). The $D_1$, $D_2$, and $D_3$ designate the particle diameters in first, second, and third monolayer in the multilayer.

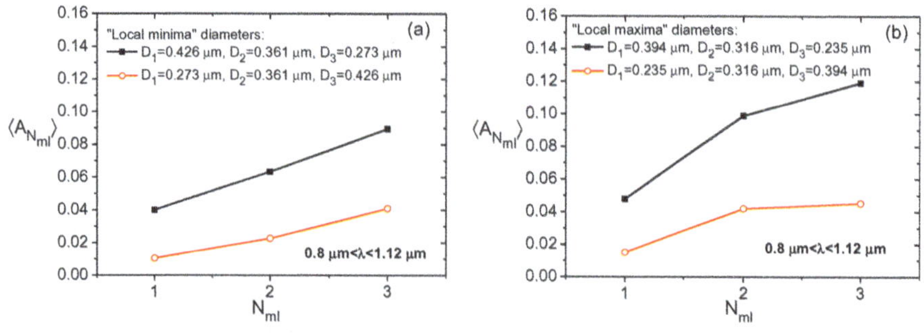


**Fig. 41** Dependences of integral absorption coefficient of multilayer on the number of monolayers with triangular lattice of c-Si particles with diameters corresponding to minima (**a**) and maxima (**b**) of the $\langle A_{\rm ml}(D) \rangle$ dependence (Fig. 29). The spacings between the adjacent monolayers are equal to the sum of particle radii in these monolayers. $\eta = 0.3$, $l_c = 115$, $\sigma(u) = 0.01u$, 0.8 μm $\leq \lambda \leq$ 1.12 μm

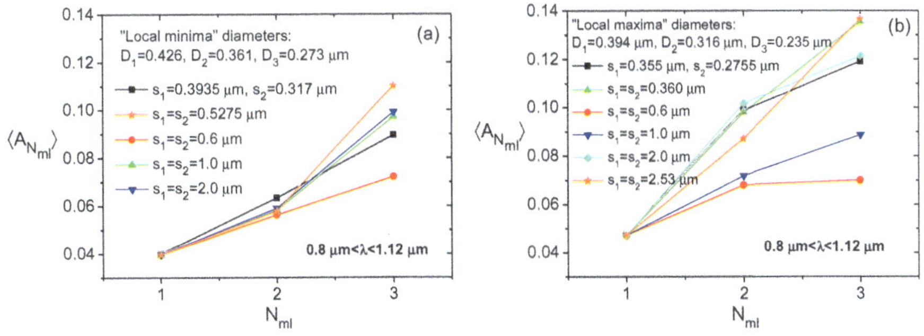


**Fig. 42** Dependences of integral absorption coefficient of multilayer on the number of monolayers with triangular lattice of c-Si particles with diameters corresponding to minima (**a**) and maxima (**b**) of the $\langle A_{\mathrm{ml}}(D)\rangle$ dependence (Fig. 29) at different spacings between the adjacent monolayers. $\eta = 0.3$, $l_c = 115$, $\sigma(u) = 0.01u$, $0.8\ \mu\mathrm{m} \leq \lambda \leq 1.12\ \mu\mathrm{m}$

Influence of spacings between adjacent monolayers on the integral absorption coefficient of multilayer is shown in Fig. 42. The $s_1$ and $s_2$ are the spacings between the first and second and the second and third monolayers (see Fig. 10) of the three-monolayer multilayer.

One can see that significant differences in the integral absorption coefficient of the three-monolayer system occur at spacing variations. In particular, the enhancement factor near the 18% ($A_{\mathrm{rel}} \approx 18\%$) for "local minima" and near the 12% ($A_{\mathrm{rel}} \approx 12\%$) for "local maxima" multilayers is achieved as compared with the multilayers in which the spacings are equal to the radii sum in the adjacent monolayers (see Fig. 41).

The results of Figs. 41 and 42 demonstrate possibilities of the absorption enhancement in the multilayer due to variations of particle diameters and the distances between the adjacent monolayers.

## Gradient Multilayers

As has shown in the "Particle Diameter and Intermonolayer Spacing Variations in Three-Monolayer System" the multilayers with different monolayers and intermonolayer spacings can be used to light harvesting optimization as compared with the multilayers composed of identical monolayers and spacings.

We have shown (Sect. 4.2.2) that increase of particle size and monolayer filling factor in the multilayers results in increase of the multilayer absorption coefficient in the range of small absorption index of silicon and can decrease it in the range of large absorption index. In this subsection, we study the gradient multilayers composed of monolayers with different filling factors and particle sizes.

Consider absorption spectra of gradient multilayers consisting of planar crystals in a wide range of monolayer filling factors and sizes of c-Si particles on the example of multilayer with seven monolayers. Figures 43a, b show spectral and integral

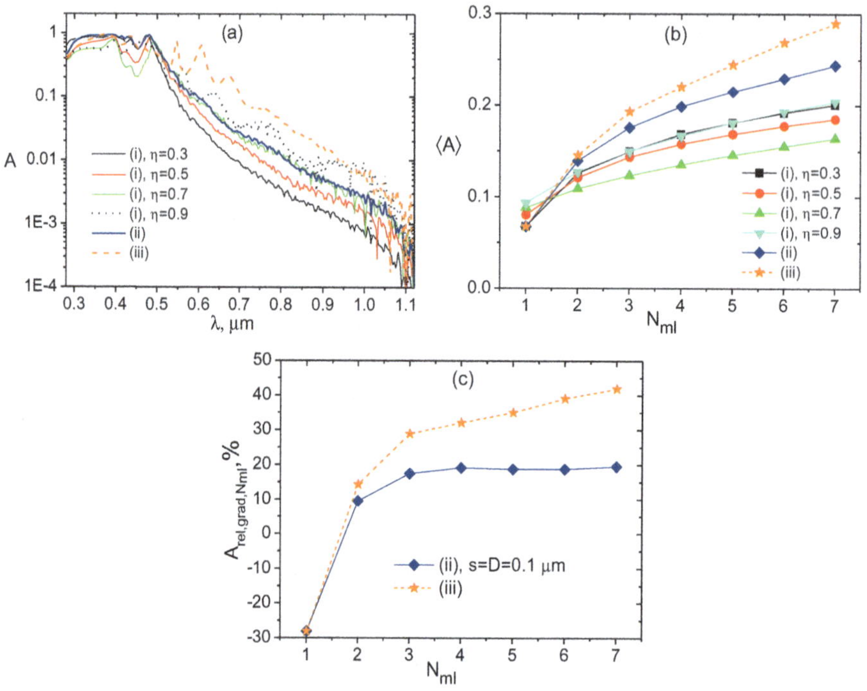


**Fig. 43** (**a**) Spectral absorption coefficient of (i) non-gradient multilayers consisting of seven identical monolayers at $\eta = 0.3$, $\eta = 0.5$, $\eta = 0.7$, $\eta = 0.9$, $D = 0.1$ μm, $s = 0.1$ μm; (ii) gradient multilayer consisting of seven monolayers with different filling factors, $\eta_1 = 0.3$, $\eta_2 = 0.4$, $\eta_3 = 0.5$, $\eta_4 = 0.6$, $\eta_5 = 0.7$, $\eta_6 = 0.8$, $\eta_7 = 0.9$, $D = 0.1$ μm, $s = 0.1$ μm; and (iii) gradient multilayer consisting of seven monolayers with different filling factors, $\eta_1 = 0.3$, $\eta_2 = 0.4$, $\eta_3 = 0.5$, $\eta_4 = 0.6$, $\eta_5 = 0.7$, $\eta_6 = 0.8$, $\eta_7 = 0.9$ and particle diameters $D_1 = 0.1$, $D_2 = 0.11$, $D_3 = 0.12$, $D_4 = 0.14$, $D_5 = 0.16$, $D_6 = 0.18$, $D_7 = 0.2$ μm; $s_1 = 0.105$, $s_2 = 0.115$, $s_3 = 0.13$, $s_4 = 0.15$, $s_5 = 0.17$, $s_6 = 0.19$ μm. The subscripts at $\eta$ and $D$ indicate the monolayer numbers; the subscripts at $s$ indicate number of intermonolayer spacing between $i$-th and $i+1$-th monolayer. (**b**) Integral absorption coefficient of multilayer versus the number of monolayers $N_{\mathrm{ml}}$. (**c**) Relative absorption coefficient $A_{\mathrm{rel,grad},N_{\mathrm{ml}}}$. Triangular lattice, $\sigma_0 = 0.001$, $l_c = 900$, $0.28\ \mu\mathrm{m} \le \lambda \le 1.12\ \mu\mathrm{m}$

absorption coefficients of multilayers consisting of monolayers with different filling factors, respectively. The results for non-gradient and gradient multilayers are presented. Figure 43c shows the relative absorption coefficient $A_{\mathrm{rel,\,grad},N_{\mathrm{ml}}}$ describing the relative absorption by gradient multilayer in comparison with the non-gradient one. It is calculated by Eq. (35), where $\langle A_{L_1} \rangle = \langle A_{\mathrm{grad},N_{\mathrm{ml}}} \rangle$ is integral absorption coefficient of the gradient multilayer and $\langle A_{L_2} \rangle = \langle A_{N_{\mathrm{ml}}} \rangle$ is integral absorption coefficient of the non-gradient multilayer consisting of identical monolayers (here calculations of $\langle A_{N_{\mathrm{ml}}} \rangle$ are made at filling factor $\eta = 0.9$).

As can be seen from Fig. 43a, spectral absorption coefficient of the $\eta$-gradient multilayer is significantly larger than the one of the non-gradient multilayer (consisting of identical monolayers with $\eta = 0.9$) in the range of large absorption index of silicon. It results in significant increasing of the integral absorption (see Fig. 43b). The integral absorption coefficient of the seven-monolayer gradient system is ~20% larger than the one of the non-gradient system (Fig. 43c). Thus, the larger absorption coefficient can be obtained at smaller volume of material.

More powerful gain can be obtained at simultaneous increase of the filling factor and diameter of particles in the monolayer sequence. This case is demonstrated in Fig. 43 by the dashed lines. At that, the multilayer has the following structure characteristics: $D_1 = 0.1$ μm, $\eta_1 = 0.3$ (first monolayer); $D_2 = 0.11$ μm, $\eta_2 = 0.4$ (second monolayer); $D_3 = 0.12$ μm, $\eta_3 = 0.5$ (third monolayer); $D_4 = 0.14$ μm, $\eta_4 = 0.6$ (fourth monolayer); $D_5 = 0.16$ μm, $\eta_5 = 0.7$ (fifth monolayer); $D_6 = 0.18$ μm, $\eta_6 = 0.8$ (sixth monolayer); and $D_7 = 0.2$ μm, $\eta_7 = 0.9$ (seventh monolayer). The spacings between the monolayers are specified as the sum of particle radii in the adjacent monolayers. As can be seen from the figure, the simultaneous $\eta$ and $D$ growing results in more pronounced increasing of the spectral and integral absorption coefficients and the value of $A_{\mathrm{rel,grad},N_{\mathrm{ml}}}$. The integral absorption coefficient of this system is more than 40% larger than the one of the non-gradient system of monolayers with $\eta = 0.9$ and $D = 0.1$ μm (see Fig. 43c).

It is useful to note that some results on the absorption by the gradient particulate structure using the radiative transfer equation are described in [103], and the influence of the particle shape on the radiation transfer is demonstrated in [104, 105].

### 4.2.4 Sketch of Solar Cell Based on Gradient Particulate Structure of Active Layer

One of the possible realizations of the solar cell based on gradient multilayer of the active layer is schematically depicted in Fig. 44 [20].

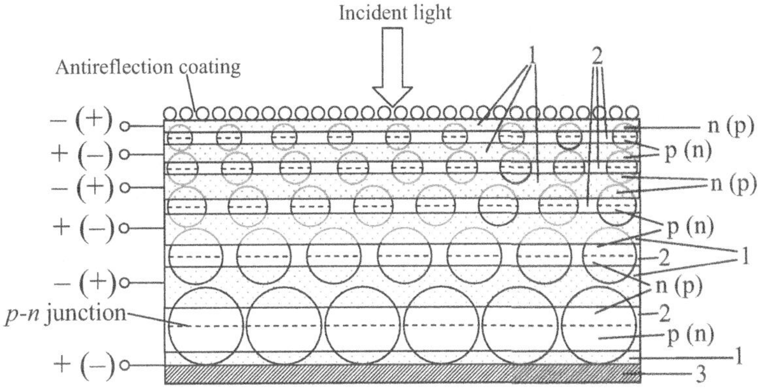


**Fig. 44** Sketch of the solar cell based on gradient layered structure of active layer (side view). The circles with the dashed lines are particles of the semiconductor material. 1, transparent electrodes; 2, transparent dielectric layers; 3, rear electrode

Here the particle size and concentration gradient multilayer are shown. The solar cell consists of active layer, antireflection coating, transparent electrodes 1, transparent dielectric layers 2, and rear electrode 3. The active layer is the stack of monolayers of particles (which generally can be of any shape) of semiconductor (it can be silicon or other material). The circles sectioned by the dashed lines are the particles with the *p–n* junctions designated by these lines. The transparent electrodes connect the parts of particles with the same conduction type (*p* or *n*). The transparent dielectric layers separate the transparent electrodes and zones of different conduction types of particles. The rear electrode 3 provides the electrical conductivity and reflection of light back into the active layer.

## 5 Conclusions

The spectral and integral absorption of light by single spherical crystalline silicon particle and by ensembles of them is theoretically examined in the wavelength range from 0.28 μm to 1.12 μm. Monolayers and multilayers of particles with diameters from 0.05 μm to 1000 μm are considered. Absorption coefficient of monolayer of large (as compared with the wavelength of incident light) particles is calculated in the single scattering approximation. Absorption coefficient of monolayers of small and wavelength-sized particles is calculated in the quasicrystalline approximation which takes into account multiple scattering of waves. Absorption by the layered systems (multilayers) consisting of monolayers is simulated using the transfer matrix method. Comparison with the data for homogeneous plane-parallel silicon plates of the equivalent volume of material (equivalent plates) is carried out.

It is shown that monolayer and multilayer consisting of silicon particles with sizes significantly smaller than radiation wavelength absorb lesser than the equivalent silicon plates. The absorption coefficient of multilayers consisting of large amount of monolayers with low concentration of small particles can be close to unity.

Absorption coefficient of the monolayer of large particles is smaller than the one of equivalent plate. Absorption by three- and more-monolayer systems of such particles is larger than the one of the equivalent plates. For six- and more-monolayer system, it can exceed 0.8 in the spectral range of large absorption index of silicon and 0.9 in the spectral range of the moderate and small values of the absorption index of silicon. The integral over the solar illumination spectrum (in the wavelength range from 0.28 μm to 1.1 μm) absorption coefficient of multilayer of large particles grows with the particle size and the number of monolayers. It can attain value of 0.9. For multilayer of seven monolayers of particles with diameters of 5 μm, the coefficient is about 1.45 times larger than the one of the equivalent plate.

Absorption coefficient of monolayer of the wavelength-sized particles can be significantly larger than the one of the equivalent plate. It is caused by strong resonance scattering by individual submicron silicon particles and by strong multiple scattering in particle arrays. Absolute absorption coefficient and relative absorption coefficient of layers of such particles as a whole grow with particle size. Relative absorption coefficient attains maximum when particle sizes and distances between them are comparable with the wavelengths of incident light. In the narrow wavelength intervals (up to 10 nm) of the resonance peaks, the spectral absorption coefficient of monolayer can be more than 100 times larger than the one of the equivalent plate. In the wavelength range from 0.8 μm to 1.12 μm, integral absorption coefficient of monolayer can be more than 20 times higher than the one of the plate. There are spectral ranges where this ratio can be essentially more.

Enhancement of light absorption due to tuning of the multilayer parameters is considered.

It was shown that more effective absorption by the three-monolayer system in the wavelength range of small absorption index of c-Si (0.8 μm $\leq \lambda \leq$1.12 μm) is achieved when multilayer consists of monolayers of particles with diameters corresponding to the local maxima of the absorption coefficients of these monolayers (and maxima of the particle absorption efficiency factor). The "inverse gradient" over the particle diameter in the monolayer sequences in such multilayers is more effective. Optimization of intermonolayer spacings in three-monolayer multilayer results in ~18% enhancement of the integral absorption coefficient as compared to the multilayer with spacings equal to the sum of particle radii of particles in the adjacent monolayers.

Influence of gradients in filling factor $\eta$ and particle diameter $D$ in the monolayer sequence on the multilayer absorption coefficient was studied. The integral absorption coefficient of $\eta$-gradient multilayer consisting of seven monolayers with 0.1-μ m-diameter particles is about 20% higher than the one of the non-gradient multilayer consisting of seven monolayers with $\eta = 0.9$ and $D = 0.1$. Integral absorption coefficient of the $\eta$- and $D$-gradient multilayer is more than 40% larger than the one of the non-gradient system of monolayers with $\eta = 0.9$ and $D = 0.1$ μm.

The sketch of the solar cell based on gradient particulate structure of active layer is presented.

All simulation results are performed for the particulate structures in air just only to simplify description of the outlined problem. The methods we have described allow one to consider specific configurations of the cell. Top, bottom, and interlayer electrodes, as well as matrix where the particles are embedded, can be included in the simulation procedure.

The approach developed is applicable not only to the solar cells based on c-Si but also for cells based on micro- and nanocrystalline silicon. It is applicable to the solution of the problem of light harvesting optimization in organic polymer solar cells, perovskite solar cells, etc.

**Acknowledgment** This work was supported in part by the Belarusian Republican Foundation for Fundamental Research (project F15IC-005).

## References


1. A. Luque, S. Hegedus (eds.), *Handbook of Photovoltaic Science and Engineering*, 2nd edn. (Wiley, Chichester, 2011)
2. S.M. Sze, K.K. Ng, *Physics of Semiconductor Devices*, 3rd edn. (Wiley Interscience, Hoboken, 2007)
3. S. Domínguez, O. García, M. Ezquer, M.J. Rodríguez, A.R. Lagunas, J. Pérez-Conde, J. Bravo, Photonics Nanostruct. Fundam. Appl. **10**, 46–53 (2012). https://doi.org/10.1016/j.photonics.2011.07.001
4. M.A. Tsai, P.C. Tseng, H.C. Chen, H.-C. Kuo, P. Yu, Opt. Express **19**, A28–A34 (2011). https://doi.org/10.1364/OE.19.000A28
5. X. Sheng, S.G. Johnson, J. Michel, L.C. Kimerling, Opt. Express **19**, A841–A850 (2011). https://doi.org/10.1364/OE.19.00A841
6. G. Kocher-Oberlehner, M. Bardosova, M. Pemble, B.S. Richards, Sol. Energy Mater. Sol. Cells **104**, 53–57 (2012). https://doi.org/10.1016/j.solmat.2012.04.018
7. C.S. Schuster, S. Moraviec, M.J. Mendes, M. Patrini, E.R. Martins, L. Lewis, I. Crupi, T.F. Krauss, Optica **2**, 194–200 (2015). https://doi.org/10.1364/OPTICA.2.000194
8. Z.R. Abrams, A. Niv, X. Zhang, J. Appl. Phys. **109**, 114905-1–114905-9 (2011). https://doi.org/10.1063/1.3592297
9. A. Deinega, I. Valuev, B. Potapkin, Y. Lozovik, J. Opt. Soc. Am. A **28**, 770–777 (2011). https://doi.org/10.1364/JOSAA.28.000770
10. R.B. Wehrspohn, J. Üpping, *3D photonic crystals for photon management in solar cells*. Paper presented at conference frontiers in optics 2012: Laser Science XXVIII, Rochester, New York, United States, 14–18 October 2012. OSA Technical Digest (online) (Optical Society of America, 2012), paper LTh3G.5. (2012). https://doi.org/10.1364/LS.2012.LTh3G.5
11. L. Ji, V.V. Varadan, J. Appl. Phys. **110**, 043114-1–043114-8 (2011). https://doi.org/10.1063/1.3626827
12. V.F. Gremenok, M.S. Tivanov, V.B. Zalesski, *Solnechnye elementy na osnove poluprovodnikovykh materialov* (*Solar Cells Based on Semiconductor Materials*). (BSU, Minsk, 2007) (in Russian)
13. L. Tsakalakos, Mater. Sci. Eng. R. Rep. **62**, 175–189 (2008). https://doi.org/10.1016/j.mser.2008.06.002
14. B.M. Kayes, H.A. Atwater, N.S. Lewis, J. Appl. Phys. Ther. **97**, 114302–114312 (2005). https://doi.org/10.1063/1.1901835
15. K. Vynck, M. Burresi, F. Riboli, D.S. Wiersma, Nat. Mater. **11**, 1017–1022 (2012). https://doi.org/10.1038/nmat3442
16. M. Saritas, H.D. McKell, Solid State Electron. **31**, 835–842 (1988). https://doi.org/10.1016/0038-1101(88)90036-6
17. J. Toušek, S. Dolhov, J. Toušková, Sol. Energy Mater. Sol. Cells **76**, 205–210 (2003). https://doi.org/10.1016/S0927-0248(02)00371-9
18. A.K. Sharma, S.N. Singh, N.S. Bisht, Z.H. Khan, Sol. Energy Mater. Sol. Cells **100**, 48–52 (2012). https://doi.org/10.1016/j.solmat.2011.04.027
19. A.A. Miskevich, V.A. Loiko, J. Quant. Spectr. Rad. Transf. **146**, 355–364 (2014). https://doi.org/10.1016/j.jqsrt.2013.12.008i
20. A.A. Miskevich, V.A. Loiko, J. Quant. Spectr. Rad. Transf. **167**, 23–39 (2015). https://doi.org/10.1016/j.jqsrt.2015.08.003
21. L. Shi, T.U. Tuzer, R. Fenollosa, F. Meseguer, Adv. Mater. **24**, 5934–5938 (2012). https://doi.org/10.1002/adma.201201987
22. L. Shi, J.T. Harris, R. Fenollosa, I. Rodriguez, X. Lu, B.A. Korgel, F. Meseguer, Nat. Commun. **4**, 1904 (2013). https://doi.org/10.1038/ncomms2934
23. I. Staude, A.E. Miroshnichenko, M. Decker, N.T. Fofang, S. Liu, E. Gonzales, J. Dominguez, T.S. Luk, D.N. Neshev, I. Brener, Y. Kivshar, ACS Nano **7**, 7824–7832 (2013). https://doi.org/10.1021/nn402736f

24. A. Bapat, C. Anderson, C.R. Perrey, C.B. Carter, S.A. Campbell, U. Kortshagen, Plasma Phys. Controlled Fusion **46**, B97–B109 (2004). https://doi.org/10.1088/0741-3335/46/12B/009
25. S. Barcikowski, A. Hahn, A. Kabashin, B.N. Chichkov, Appl. Phys. A Mater. Sci. Process. **87**, 47–55 (2007). https://doi.org/10.1007/s00339-006-3852-1
26. C.Q. Li, C.-Y. Zhang, Z.-S. Huang, X.-F. Li, Q.-F. Dai, S. Lan, S.-L. Tie, J. Phys. Chem. C **117**, 24625–24631 (2013). https://doi.org/10.1021/jp408865p
27. A. Vladimirov, S. Korovin, A. Surkov, E. Kelm, V. Pustovoy, Laser Phys. **21**, 830–835 (2011). https://doi.org/10.1134/S1054660X11080032
28. U. Zywietz, C. Reinhardt, A.B. Evlyukhin, T. Birr, B.N. Chichkov, Appl. Phys. A Mater. Sci. Process. **114**, 45–50 (2014). https://doi.org/10.1007/s00339-013-8007-6
29. U. Zywietz, A.B. Evlyukhin, C. Reinhardt, B.N. Chichkov, Nat. Commun. **5**., Article no. 3402, 1–7 (2014). https://doi.org/10.1038/ncomms4402
30. G. Mie, Ann. Phys. **25**, 377–445 (1908). https://doi.org/10.1002/andp.19083300302
31. H.C. van de Hulst, *Light Scattering by Small Particles* (Dover, New York, 1981)
32. G.J. Rosasco, H.S. Bennett, J. Opt. Soc. Am. **68**, 1242–1250 (1978). https://doi.org/10.1364/JOSA.68.001242
33. P.R. Conwell, P.W. Barber, C.K. Rushforth, J. Opt. Soc. Am. A **1**, 62–67 (1984). https://doi.org/10.1364/JOSAA.1.000062
34. C.S. Zender, J. Talamantes, J. Quant. Spectrosc. Radiat. Transfer **98**, 122–129 (2006). https://doi.org/10.1016/j.jqsrt.2005.05.084
35. A.B. Evlyukhin, C. Reinhardt, A. Seidel, B.S. Luk'yanchuk, B.N. Chichkov, Phys. Rev. B **82**, 045404-1–04540412 (2010). https://doi.org/10.1103/PhysRevB.82.045404
36. R. Bachelard, P.W. Courteille, R. Kaiser, N. Piovella, Eplasty **97**, 14004-p1–14004-p6 (2012). https://doi.org/10.1209/0295-5075/97/14004
37. Z.Y. Wang, R.J. Zhang, S.Y. Wang, M. Lu, X. Chen, Y.X. Zheng, L.Y. Chen, Z. Ye, C.Z. Wang, K.M. Ho, Sci. Rep. **5**, 7810-1–7810-6 (2015). https://doi.org/10.1038/srep07810
38. V.A. Loiko, V.P. Dick, V.I. Molochko, J. Opt. Soc. Am. A **15**, 2351–2354 (1998). https://doi.org/10.1364/JOSAA.15.002351
39. V.A. Babenko, L.G. Astafyeva, V.N. Kuzmin, *Electromagnetic Scattering in Disperse Media* (Praxis Publishing, Chichester, 2003)
40. V.A. Loiko, A.A. Miskevich, Appl. Opt. **44**, 3759–3768 (2005). https://doi.org/10.1364/AO.44.003759
41. M. Lax, Rev. Mod. Phys. **23**, 287–310 (1951). https://doi.org/10.1103/RevModPhys.23.287
42. M. Lax, Phys. Rev. **85**, 621–629 (1952). https://doi.org/10.1103/PhysRev.85.621
43. K.M. Hong, J. Opt. Soc. Am. **70**, 821–826 (1980). https://doi.org/10.1364/JOSA.70.000821
44. C.C. Katsidis, D.I. Siapkas, Appl. Opt. **41**, 3978–3987 (2002). https://doi.org/10.1364/AO.41.003978
45. E. Centurioni, Appl. Opt. **44**, 7532–7539 (2005). https://doi.org/10.1364/AO.44.007532
46. M.C. Troparevsky, A.S. Sabau, A.R. Lupini, Z. Zhang, Opt. Express **18**, 24715–24721 (2010). https://doi.org/10.1364/OE.18.024715
47. Reference Solar Spectral Irradiance: Air Mass 1.5 (American Society for Testing and Materials (ASTM) Terrestrial Reference Spectra for Photovoltaic Performance Evaluation), http://rredc.nrel.gov/solar/spectra/am1.5/ASTMG173/ASTMG173.html. Accessed 10 Oct 2016
48. E. D. Palik (ed.), *Handbook of Optical Constants of Solids* (Academic, San Diego, 1985)
49. C.F. Bohren, D.R. Huffman, *Absorption and Scattering of Light by Small Particles* (Wiley, New York, 1983)
50. M. I. Mishchenko, J. W. Hovenir, L. D. Travis (eds.), *Light Scattering by Nonspherical Particles* (Academic Press, San Diego, 2000)
51. M.I. Mishchenko, L.D. Travis, A.A. Lacis, *Scattering, Absorption, and Emission of Light by Small Particles* (University Press, Cambridge, 2002)

52. Y.H. Fu, A.I. Kuznetsov, A.E. Miroshnichenko, Y.F. Yu, B. Luk'yanchuk, Nat. Commun. **4**, 1527 (2013). https://doi.org/10.1038/ncomms2538
53. S. Person, M. Jain, Z. Lapin, J.J. Sáenz, G. Wicks, L. Novotny, Nano Lett. **13**, 1806–1809 (2013). https://doi.org/10.1021/nl4005018
54. J.M. Geffrin, B. García-Cámara, R. Gómez-Medina, P. Albella, L.S. Froufe-Pérez, C. Eyraud, A. Litman, R. Vaillon, F. González, M. Nieto-Vesperinas, J.J. Sáenz, F. Moreno, Nat. Commun. **3**, 1–8 (2012). https://doi.org/10.1038/ncomms2167
55. M. Kerker, D.-S. Wang, C.L. Giles, J. Opt. Soc. Am **73**, 765–767 (1983). https://doi.org/10.1364/JOSA.73.000765
56. A.A. Kokhanovsky, *Light Scattering Media Optics. Problems and Solutions*, 3rd edn. (Springer, Berlin, 2004)
57. A. Doicu, T. Wried, Y.A. Eremin, *Light Scattering by Systems of Particles* (Springer, Berlin, 2006)
58. S. Ishii, S. Inoue, A. Otomo, J. Opt. Soc. Am. B **31**, 218–222 (2014). https://doi.org/10.1364/JOSAB.31.000218
59. S. Ishii, R.P. Sugavaneshwar, K. Chen, T.D. Dao, T. Nagao, Opt. Mater. Exp. **6**, 640–648 (2016). https://doi.org/10.1364/OME.6.000640
60. V. Savinov, V.A. Fedotov, N.I. Zheludev, Phys. Rev. B **89**, 205112-1–20511212 (2014). https://doi.org/10.1103/PhysRevB.89.205112
61. A.E. Miroshnichenko, A.B. Evlyukhin, Y.F. Yu, R.M. Bakker, A. Chipouline, A.I. Kuznetsov, B. Luk'yanchuk, B.N. Chichkov, Y.S. Kivshar, Nat. Commun. **6**, 8069 (2015). https://doi.org/10.1038/ncomms9069
62. A.B. Evlyukhin, C. Reinhardt, B.N. Chichkov, Phys. Rev. B **84**, 235429 (2011). https://doi.org/10.1103/PhysRevB.84.235429
63. A.B. Evlyukhin, C. Reinhardt, E. Evlyukhin, B.N. Chichkov, J. Opt. Soc. Am. B **30**, 2589–2598 (2013). https://doi.org/10.1364/JOSAB.30.002589
64. M. Polyanskiy, Refractive index database. http://www.refractiveindex.info. Accessed 10 Oct 2016
65. A. Ishimaru, *Wave Propagation and Scattering in Random Media. Single Scattering and Transport Theory*, vol 1 (Academic, New York, 1978)
66. M. Born, E. Wolf, *Principles of Optics*, 7th edn. (University Press, Cambridge, 2002)
67. S. Kachan, O. Stenzel, A. Ponyavina, Appl. Phys. B Lasers Opt. **84**, 281–287 (2006). https://doi.org/10.1007/s00340-006-2252-8
68. V.A. Loiko, A.A. Miskevich, Opt. Spectrosc. **115**, 274–282 (2013). https://doi.org/10.1134/S0030400X13070096
69. A.A. Miskevich, V.A. Loiko, J. Quant. Spectrosc. Radiat. Transfer **136**, 58–70 (2014). https://doi.org/10.1016/j.jqsrt.2013.05.013
70. A.A. Miskevich, V.A. Loiko, Photocell. Republic of Belarus Patent BY 18325, 27 Feb 2012
71. A.A. Miskevich, V.A. Loiko, Photocell. Russian Federation Patent RU 2491681, 11 Mar 2012
72. A.A. Miskevich, V.A. Loiko, J. Quant. Spectrosc. Radiat. Transfer **112**, 1082–1089 (2011). https://doi.org/10.1016/j.jqsrt.2010.11.019
73. A.A. Miskevich, V.A. Loiko, J. Experiment. Theor. Phys. **113**, 1–13 (2011). https://doi.org/10.1134/S1063776111050153
74. K. Ohtaka, J. Phys. C: Solid St. Phys. **13**, 667–680 (1980). https://doi.org/10.1088/0022-3719/13/4/022
75. K. Ohtaka, M. Inoue, Phys. Rev. B **25**, 677–688 (1982). https://doi.org/10.1103/PhysRevB.25.677
76. M. Inoue, K. Ohtaka, S. Yanagawa, Phys. Rev. B **25**, 689–699 (1982). https://doi.org/10.1103/PhysRevB.25.689
77. K. Ohtaka, Y. Suda, S. Nagano, T. Ueta, A. Imada, T. Koda, J.S. Bae, K. Mizuno, S. Yano, Y. Segawa, Phys. Rev. B **61**, 5267–5279 (2000). https://doi.org/10.1103/PhysRevB.61.5267
78. A. Modinos, Phys. A **141**, 575–588 (1987). https://doi.org/10.1016/0378-4371(87)90184-1

79. N. Stefanou, A. Modinos, J. Phys. Condens. Matter **3**, 8135–8148 (1991). https://doi.org/10.1088/0953-8984/3/41/012
80. G. Gantzounis, N. Stefanou, Phys. Rev. B **73**, 035115-1–03511510 (2006). https://doi.org/10.1103/PhysRevB.73.035115
81. V. Yannopapas, J. Opt. Soc. Am. B **31**, 631–636 (2014). https://doi.org/10.1364/JOSAB.31.000631
82. A.B. Evlyukhin, C. Reinhardt, U. Zywietz, B.N. Chichkov, Phys. Rev. B **85**, 245411-1–24541112 (2012). https://doi.org/10.1103/PhysRevB.85.245411
83. J.M. Ziman, *Models of Disorder* (University Press, Cambridge, 1979)
84. Z. Fisher, *Statistical Theory of Liquids* (University Press, Chicago, 1964)
85. D.A. Varshalovich, A.N. Moskalev, V.K. Khersonskii, *Quantum Theory of Angular Momentum* (World Scientific, Singapore, 1988)
86. G.V. Arfken, H.J. Weber, F.E. Harris, *Mathematical Methods for Physicists*, 7th edn. (Academic Press, Oxford, 2012)
87. Y. Rosenfeld, Phys. Rev. A **42**, 5978 (1990.) https://doi.org/10.1103/PhysRevA.42.5978
88. V.P. Dick, V.A. Loiko, Opt. Spectrosc. **117**, 111–117 (2014). https://doi.org/10.1134/S0030400X14070066
89. V.A. Loiko, A.V. Konkolovich, Opt. Spectrosc. **85**, 563–567 (1998)
90. V.A. Loiko, A.V. Konkolovich, Opt. Spectrosc. **85**, 568–573 (1998)
91. V.K. Varadan, V.N. Bringi, V.V. Varadan, A. Ishimaru, Radio Sci. **18**, 321–327 (1983). https://doi.org/10.1029/RS018i003p00321
92. J.A. Lock, C.L. Chiu, Appl. Opt. **33**, 4663–4671 (1994). https://doi.org/10.1364/AO.33.004663
93. L. Tsang, J.A. Kong, K.-H. Ding, C.O. Ao, *Scattering of Electromagnetic Waves: Numerical Simulations* (Wiley, New York, 2001)
94. L.S. Ornstein, F. Zernike, Proc. Acad. Sci. **17**, 793–806 (1914)
95. J.K. Percus, G.J. Yevick, Phys. Rev. **110**, 1–13 (1958). https://doi.org/10.1103/PhysRev.110.1
96. A.P. Ivanov, V.A. Loiko, V.P. Dick, *Rasprostranenie sveta v plotnoupakovannykh dispersnykh sredakh* (Propagation of Light in Close-packed Disperse Media). (Nauka i Tekhnika, Minsk, 1988) (in Russian)
97. R. Rengarajan, D. Mittleman, C. Rich, V. Colvin, Phys. Rev. E **71**, 016615-1–01661511 (2005). https://doi.org/10.1103/PhysRevE.71.016615
98. T. Prasad, V.L. Colvin, D.M. Mittleman, Opt. Express **15**, 16954–16965 (2007). https://doi.org/10.1364/OE.15.016954
99. A. Ishimaru, Y. Kuga, J. Opt. Soc. Am. **72**, 1317–1320 (1982). https://doi.org/10.1364/JOSA.72.001317
100. Y. Okada, A.A. Kokhanovsky, J. Quant. Spectrosc. Radiat. Transf. **110**, 902–917 (2009). https://doi.org/10.1016/j.jqsrt.2008.12.007
101. V.A. Loiko, G.I. Ruban, Opt. Spectrosc. **88**, 756–761 (2000). https://doi.org/10.1134/1.626872
102. V.A. Loiko, G.I. Ruban, J. Quant. Spectrosc. Radiat. Transf. **89**, 271–278 (2004). https://doi.org/10.1016/j.jqsrt.2004.05.040
103. V.A. Loiko, V.V. Berdnik, The Journal of Photographic Science **48**, 12–25 (2003)
104. V.V. Berdnik, V.A. Loiko, J. Quant. Spectrosc. Radiat. Transf. **63**, 369–382 (1999). https://doi.org/10.1016/S0022-4073(99)00025-4
105. V.A. Loiko, V.V. Berdnik, Part. Part. Syst. Charact. **15**, 115–121 (1998.) 10.1002/(SICI)1521-4117(199817)15:3<115::AID-PPSC115>3.0.CO;2-N

# Modeling and Simulation of New Generation of Thin-Film Silicon Solar Cells Using Efficient Light-Trapping Structures

**R.S. Dubey and S. Saravanan**

**Abstract** Thin-film solar cells are the alternative over the crystalline solar cells in terms of manufacturing cost. The conversion efficiency of the thin-film solar cells is still less as compared to the conventional solar cells. This drawback has opened many doors of research in terms of the improvement of top antireflective coating, properties of the absorber layer, and use of light-trapping structure at the bottom. For antireflective coating, several materials such as $SiO_2$, $TiO_2$, ZnO, ITO, porous silicon, etc. have been investigated and reported the satisfactory results with respect to their electrical and optical limitations. Conversely, the light-trapping structure at the bottom is one of the crucial factors which allow the reuse of electromagnetic waves which are not being absorbed by the thin absorber layer. Accordingly, several perceptions have been reported to manipulate the electromagnetic waves with the use of an efficient light-trapping structure which reflects and guides the waves towards the thin absorber layer. With this passion, distributed Bragg reflectors, dielectric/metal nanogratings, and nanoparticles have been employed and claimed the enhanced photoconversion. In this chapter, we explore the design of thin-film silicon solar cells based on various light-trapping structures and study of their photovoltaic performance.

## 1 Introduction

Nonconventional energy sources are the demanding alternatives to overcome the problem of power scarcity worldwide. This need has generated a huge research scope in the fundamental and advanced branches of the science and technology. Severe work on silicon thin-film solar cells including design and fabrication has been reported by the scientific community. Silicon technology is a well-established,

R.S. Dubey (✉) · S. Saravanan
Advanced Research Laboratory for Nanomaterials and Devices, Department of Nanotechnology, Swarnandhra College of Engineering and Technology, Seetharampuram, Narsapur, Andhra Pradesh, India
e-mail: rag_pcw@yahoo.co.in



S. Ikhmayies (ed.), *Advances in Silicon Solar Cells*,
https://doi.org/10.1007/978-3-319-69703-1_4

safe, nontoxic, and cheaper for the fabrication of thin-film solar cells; however, weak absorption in longer wavelength is a major drawback which needs attention. To overcome this problem, a new design engineering of solar devices is spotlighted which includes an efficient light-trapping structure. For an efficient light-trapping structure, one-dimensional photonic crystal also known as distributed Bragg reflector and diffraction grating has been explored as the bottom component in the thin-film silicon solar cells. The use of distributed Bragg reflector (DBR) provides total internal reflection of longer wavelength light that passes through absorbing layer of solar cell; however, diffraction grating diffracts and scatters the light. These ideas are found to be effective because of scattering the incident light and coupling it into the fundamental material. Recently, metal nanoparticles or nanogratings have also been demanded in solar cells for the light-trapping application. The metal-induced surface plasmons are found to be promising for the harvesting of solar light by the way of guiding and manipulation of photons through the mechanism of collective oscillation of electrons at the surface of the metal nanostructures. Surface plasmon resonance can be induced between the propagation path of light and metal surface which can ultimately enhance the absorption of light to an optimal level. In brief, surface plasmon is nothing but the gathering of electrons at the metal surface which makes a propagation path along it, while surface plasmonic energy is getting concentrated at the tip of the metal nanostructures.

## 2 Solar Cells Using Dielectric Diffraction Grating and Distributed Bragg Reflector

Thin-film-based solar cells have weak absorption in longer wavelength spectrum due to indirect bandgap of silicon which limits their overall efficiency. Hence, light-trapping mechanism is essential for the enhancement of incident light absorption. This requirement can be fulfilled by two mechanisms: first is diffraction or scattering which can change the direction of incident photons so that as much as photons can be propagated at higher angles with the prolonged path length within the cell, while second is the coupling of incident photons provided with the guided mode in the active region. In other words, once the incident photons entered into the device, their mean residing time in active region must be long enough, and they are absorbed before escaping the device. Several literatures have been reported on the design and fabrication of light-trapping structure for the improvement of light absorption in silicon solar cells. Yablonovitch et al. have first reported the light trapping concept using analytical solutions for light path enhancement in bulk solar cell with an ideal Lambertian light trapping theory, and later M.A. Green has extended the calculations for any degree of absorption in the active material [1, 2]. The main light-trapping mechanisms are the top antireflective coating film to reduce the surface reflections, the back reflector to reuse the unabsorbed light, and the diffraction gratings to couple the light. The direct incident of light on silicon yields optical loss around 35% so an antireflection coating (ARC) layer at the top of solar cells is significant which helps to reduce surface reflections. The one-dimensional photonic crystal (1DPC)/distributed Bragg reflector (DBR) is used to increase the total internal reflection of light on the backside of solar cell. The diffraction grating is used to bend the light waves at a titled angle for the coupling. Numerous research papers have been reported on the design and fabrication of solar cells for the better performance with the combination of antireflection coating layer, the diffraction grating, and the one-dimensional photonic crystal. L. Zeng et al. have presented an experimental application of a textured photonic crystal as backside reflector in thin-film silicon solar cells [3]. Light absorption was found to be enhanced due to the high reflection and large-angle diffraction. They have experimentally demonstrated a 5-μm thin-film silicon solar cells and found an increment in short-circuit current density by 19% as compared to a theoretical prediction of 28%. Xianqin Meng et al. have proposed the thin-film-based solar cell design with the combined front and back 1D and 2D diffraction gratings of irregular periods [4]. The absorption was found to be increased with 750-nm long-period back grating; however, reflection of the incident light was observed to be decreased by using 250-nm short-period front grating. The simulated results showed an increment in short-circuit current up to 30.3 $mA/cm^2$ as compared to 18.4 $mA/cm^2$ of the reference cell. Xing Sheng et al. have explored the mechanism for an efficient light trapping in thin-film silicon solar cell structure by using a distributed Bragg reflector (DBR) and the periodic gratings. They have reported that the light can be scattered into the DBR by gratings with an unusual way of light trapping from the metal reflectors and photonic crystals [5]. Alongkarn Chutinan et al. have reported a designing of solar cell with light-trapping concept and theoretically demonstrated a significant enhancement of efficiency through thin crystalline silicon solar cells by using photonic crystal [6]. They have observed the relative increment in conversion efficiency about 11.15% and 3.87% for 2 μm and 10 μm thick solar cells respectively. L. Zhao et al. have proposed a design of solar cell with an indium tin oxide diffraction grating, a-Si:H/ITO DBR and Ag reflector [7]. With the use of metal reflector, they have observed 69% and 72% weighted absorptance by the solar cells designed with four and eight pairs of a-Si:H/ITO DBR, respectively. The use of metal reflector was helpful to trap light in a better way with the reduced number of DBR pairs, and, hence, it makes easy manufacturing. Ning Feng et al. have presented the optimization of an efficient light-trapping structure for the crystalline solar cells with the combination of antireflection coating layer, grating, and DBR [8]. They have observed an improvement in efficiency with optimized parameters and achieved up to 18.88% for 100 μm solar cell. Here, we present a complete design of solar cells based on DBR and diffraction grating which showed enhanced performance in the red and infrared wavelength regions.

For the design of solar cells, commercial available Rsoft package was used. Figure 1 shows a design of 5-μm-thick solar cell structure with DBR and diffraction grating as back reflector. The solar cell comprises of an antireflection coating layer of silicon nitride, crystalline silicon layer as active region, combination of a

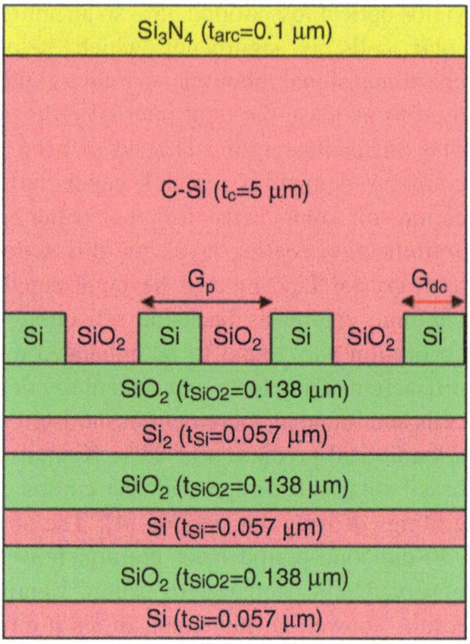


**Fig. 1** Thin-film silicon solar cell structure with DBR and diffraction grating

distributed Bragg reflector (DBR), and the diffraction grating as bottom region. The DBR is composed of periodic layers of silicon (Si) and silicon dioxide ($SiO_2$) with their refractive indices 1.46 and 3.5 and thicknesses 0.057 μm and 0.138 μm, respectively, at center wavelength 0.8 μm, while $SiO_2$ grating was embedded into active silicon region. Using FDTD method, periodic boundary conditions were applied in x- and y-directions where perfectly matched layer boundary condition was applied in z-direction [9].

## 2.1 *Optimization of Various Parameters*

For the optimal design of proposed solar cell design, various design parameters such as refractive index and thickness of antireflective coating layer, grating width, center wavelength, number of DBR pairs, grating period, grating thickness, and the duty cycle are explored.

### 2.1.1 Refractive Index of ARC Layer

Refractive index of ARC layer is a significant parameter in order to limit the surface reflections. According to antireflection coating theory, ARC layer has zero reflection with refractive index $n_{arc} = \sqrt{n_c}$ and thickness $t_{arc} = \lambda_c/4n_{arc}$ where $n_c$ and $\lambda_c$ are the refractive index of active crystalline silicon region and center wavelength,

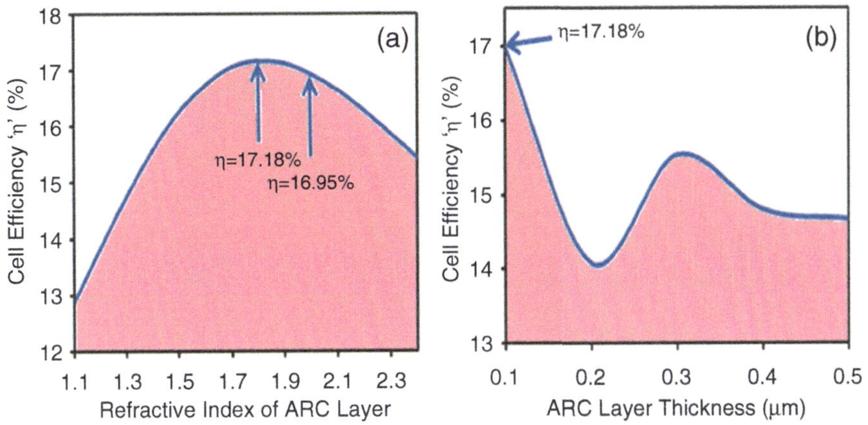


**Fig. 2** Solar cell efficiency as a function of refractive index (**a**) and thickness (**b**) of antireflection coating, respectively (Dubey et al. [9]. AIP: With permission)

respectively. Figure 2a shows the solar cell efficiency (η) as a function of refractive index of antireflection coating ($n_{arc}$). The simulated result showed the cell efficiency dependent on the refractive index of ARC coating, and maximum efficiency was observed to be at $n_{arc} = 1.8$. Analytically, refractive index can be calculated as $n_{arc} = \sqrt{n_c}$, and the simulated result was found similar.

### 2.1.2 Thickness of ARC Layer

To observe the effect of ARC layer thickness, we have plotted Fig. 2b by keeping $t_{arc} = 0.1$ μm and $n_{arc} = 1.8$. Here, we have observed that the cell efficiency is decreased as the ARC layer thickness is increased, and up to a maximum 17.18% was obtained for 5-μm cell thickness. According to ray theory, ARC layer thickness should be $t_{arc} = \lambda_c/4n_{arc} = 0.1$ μm, and our simulated result was found similar with the analytical value and thus validated.

### 2.1.3 Grating Width

For the next level of optimization, we kept $n_{arc} = 1.8$ and $t_{arc} = 0.1$ μm, while other parameters were constant. Figure 3a depicts the variation in cell efficiency in accordance with grating width "$G_w$." An increase in efficiency, 17.52%, can be observed at $G_w = 0.3$ μm rather than 17.18% at $G_w = 0.4$ μm (assumed one). This result indicates that the choice of $G_w = 0.3$ μm was worthy.

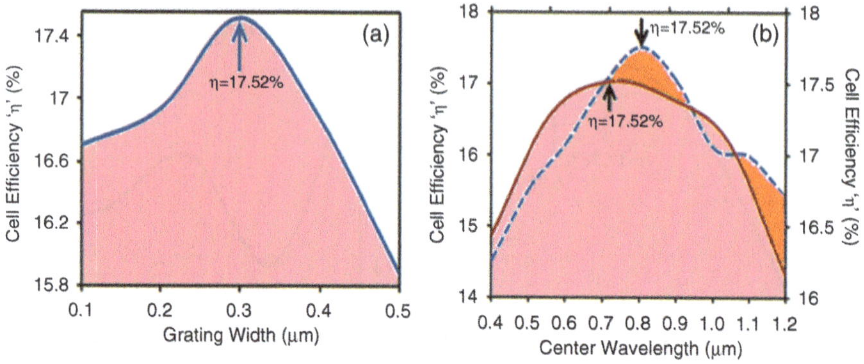


**Fig. 3** Solar cell efficiency as a function of grating width (**a**) and center wavelength and no. of DBR (**b**) (Dubey et al. [9]. AIP: With permission)

### 2.1.4 Center Wavelength and Number of Pairs of DBR

Distributed Bragg reflector (DBR) is used in solar cells as a back reflector part and helpful to redirect the light transmitted through the thin active region. To possess this required property of the DBR, a quarter-wavelength thickness of the constituent layers ($t_{Si} = 0.057\ \mu m$ and $t_{SiO_2} = 0.138\ \mu m$) were considered. If a DBR is properly designed, then the twice optical path length can be assured according to the ray theory. In this context, center wavelength is a significant parameter which corresponds to maximum (~100%) reflection of the incident light with the sufficient number of DBR pairs. To observe the effect of the center wavelength and the number of DBR pairs on overall efficiency of solar cell, we have plotted Fig. 3b. The designed device shows maximum efficiency at center wavelength $\lambda_c = 0.8$ μm (dotted line). Till now in our designing, $\lambda_c$ was the same value which shows the validation of the simulated result. The selection of center wavelength is mainly important for thin solar cells; however for thicker cells, the maximum photons get absorbed in a single pass, and the trapping of light is needed only for the wavelength near to the bandgap. Solid curve shows solar cell efficiency as a function of number of DBR and can be observed that the maximum efficiency was obtained with the use of three DBR pairs only. The performance of DBR was saturated and decayed after three DBR pairs which has been attributed to the evanescent wave decayed into the DBR structure.

### 2.1.5 Grating Period, Grating Thickness, and Duty Cycle

Diffraction grating is an important component in solar cells in addition to the DBR as part of back reflector and depending upon the design it diffracts the light at various angles. The design of diffraction grating has mainly three considerations such as grating period ($G_p$), grating thickness ($G_t$), and grating duty cycle ($G_{dc}$).

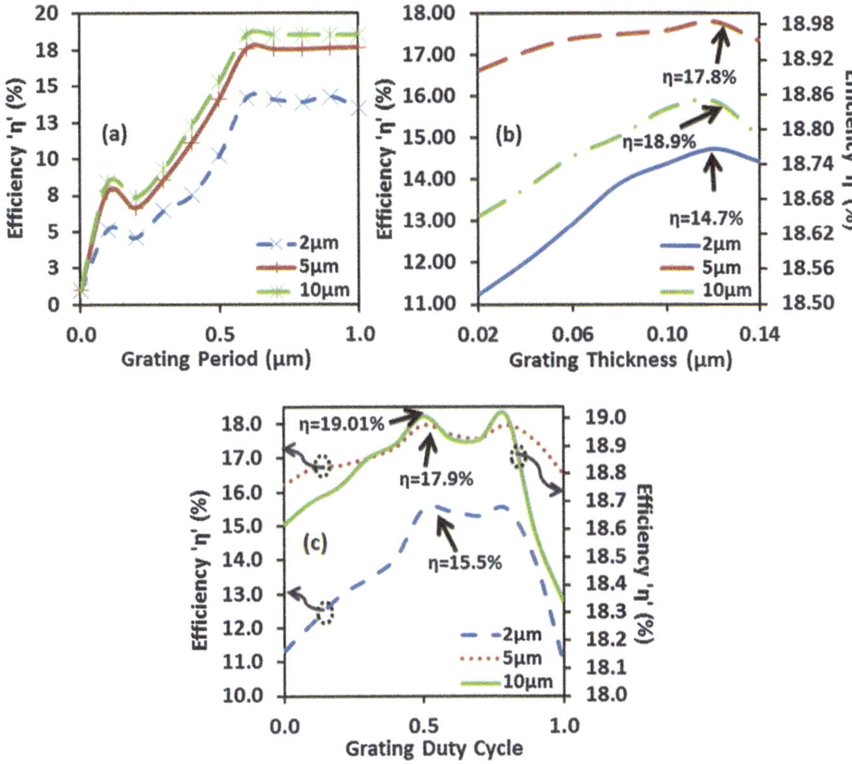


**Fig. 4** Solar cell efficiency variation in accordance with grating period (**a**), thickness (**b**), and duty cycle (**c**) for 2-, 5-, and 10-μm cell thicknesses (Dubey et al. [9]. AIP: With permission)

Figure 4a shows the effect of grating period on the cell efficiency for various cell thicknesses. As can be seen in the figure, there are two maxima points which has been attributed to the first- and second-order diffractions of the grating. An improvement in cell efficiency can be observed for all the thickness of cells at grating period $G_p = 0.6$ μm with maximum efficiency $\eta = 18.8\%$ for 10 μm solar device. Figure 4b shows a variation of grating thickness ($G_t$) for various thicknesses of the solar cells. The grating thickness 0.1 μm was fixed and simulated from 0.02 to 0.14 μm. At $G_t = 0.12$ μm, maximum cell efficiency was observed, i.e., 14.7%, 17.8%, and 18.9% corresponding to the cell thicknesses 2, 5, and 10 μm, respectively. A significant variation in the cell efficiency can be observed due to change in grating thickness in case of thinner cells. Further, $G_t = 0.12$ μm was used, and simulation was performed to study the effect of duty cycle on the device performance. The effect of grating duty cycle ($G_{dc}$) on the solar cell efficiency is plotted in Fig. 4c for various thicknesses of solar cells. The maximum efficiency has been observed at $G_{dc} = 0.5$ for all cell thicknesses, and, further, it is observed to be reduced after 0.9 duty cycle which indicates a significant role of grating duty cycle in solar cells.

### 2.1.6 Performance Analysis After Optimization

Further, we have extended the designing of device with 5-μm cell thickness by embedding three bilayers of grating ($Si/SiO_2$) within silicon active region while keeping same parametrical values of grating thickness, duty cycle, and period (see inset of Fig. 4). As it was expected, we have noticed the better performance of this design as compared to a previous one with a single-layer grating. Similar design of solar device has been reported where they had used three-layer-based antireflection coating [10]. Figure 5a shows the absorption and the quantum efficiency of the bilayer grating-based solar cell. An enhancement in absorption can be clearly seen in the wavelength range of 450–1100 nm. In addition, an overlapping of absorption curve with incident solar spectrum is noticed in a short-wavelength range of 650–780 nm. The quantum efficiency curve (solid line) reveals the better performance of the device particularly in the wavelength range 580–850 nm.

Figure 5b shows J–V characteristics of the single-layer and the bilayer grating-based solar cells of 5, 10, 20, and 30 μm thicknesses. J–V curves demonstrate an improvement in short-circuit current with the use of three bilayer grating-based devices. The maximum obtained short-circuit current densities are 29.6, 32.9, 34.6, and 36.05 mA/cm$^2$ for the 5-, 10-, 20-, and 30-μm-thick solar cells, respectively. An inset figure shows a comparison of the bilayer and the single-layer grating-based solar cells of 5, 10, 20, and 30 μm thicknesses. In comparison to the reference solar cell with 14% efficiency, we have achieved 42.8% relative enhanced efficiency for the three bilayer grating, while it was 28.5% for the single-layer grating-based solar cell of 5-μm thickness. Remarkably, the modified design of the bilayer grating embedded in active silicon region has competed the previous design by giving 24% cell efficiency for 30 μm cell thickness.

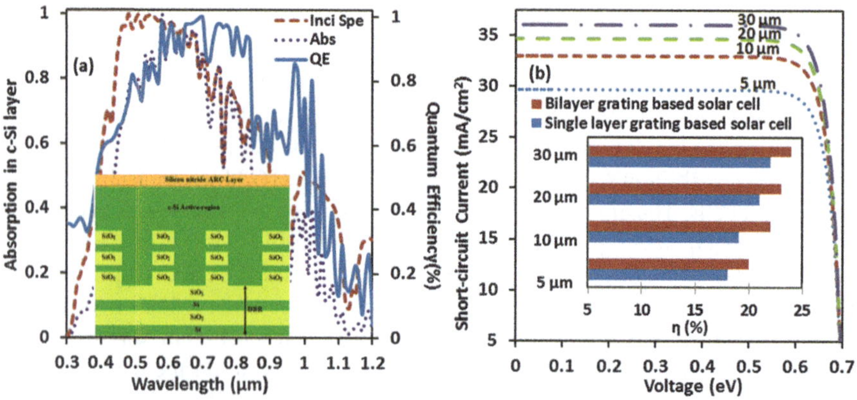


**Fig. 5** Absorption and quantum efficiency of three bilayer grating-based solar cell (**a**) and short-circuit current (**b**) with cell efficiency chart (inset figure) (Dubey et al. [9]. AIP: With permission)

## 3 Ultrathin-Film Silicon Solar Cells Using Metal Grating and Dielectric Distributed Bragg Reflector

In conventional solar cells, the recombination of charge carriers may yield phonons or photons, and, therefore, the performance is degraded. However, ultrathin-film solar cells are having advantages such as reduced recombination rate and low-manufacturing cost. Plasmonic solar cells are another class of solar cell which showed an improvement in the photovoltaic performance due to the induction of surface plasmons associated with the metal nanostructures (nanogratings/nanoparticles) aligned in adjacent to the ultrathin absorber layer. Metal nanostructures have the light-scattering property which increases the local field intensity. However, this effect has significant impact of the particle size, shape, and refractive index of the surrounding medium. In addition to these, the plasmonic interparticle distance is an important parameter which sustains the optical resonances. Accordingly, plasmon coupling is affected by the near field which strongly depends on the oscillating electrons near the neighboring particles. In ultrathin-film solar cells, the induced surface plasmons cause a high electric field buildup, and further the reduced electron path length between the electrodes is responsible for the enhancement of photocurrent. In simple words, the induced surface plasmons increase the near field via the scattering of light from the metal surface as a result the direct generation of the charge carriers take place.

Reduced cell thickness is a critical issue in silicon solar cells, when it is less than 2 μm (c-Si) and below 300 nm (a-Si:H). Therefore, for the ultrathin solar cells, the light trapping is essential, whereas plasmonic solar cells are found to be more promising to overcome this problem. A design of ultrathin-film silicon solar cells by using one-dimensional plasmonic nanogratings at the bottom of the solar cell has been demonstrated by Xiao et al. [11]. They have observed 90% enhancement of photocurrent in the considered wavelength range from 200 nm thick crystalline silicon solar cell. The analysis of obtained result was suggested for the realization of low-cost and high-efficiency thin-film solar cells. He et al. have proposed a design of an ultrathin silicon solar cell by placing a periodic array of silver strips on a metallic nanograting substrate [12]. The designed structure could give 170% enhancement of light absorption as compared to the bare silicon thin film. This enhancement has been attributed to the excited multiple resonant and waveguiding modes within the silicon layer, localized surface plasmon resonance, and surface plasmon polaritons. Yan et al. have presented three modeling methods for a-Si:H solar cells and to observe the light absorption; parabolaconical nanoarrays were introduced in the ultrathin a-Si:H solar cells [13]. They have observed an optimal absorption enhancement about 53.9% when height/radius ratio was 1, and, further, it was increased to 61.9% for the case when height/radius ratio was 3. This enhancement was due to the graded refractive index of silicon and waveguide mode. Wang et al. have proposed a planar ultrathin absorber concept by exploiting plasmonic resonance absorption enhancement and noticed the enhanced absorption about 89.8% through 5 nm thin-film absorber which showed the single-pass absorption of only 1.7% for the case of TM polarization [14]. The absorption enhancement was broadband and angle independent. Furthermore, this concept was suggested for the two-dimensional periodic grating geometries to achieve a strong angle and polarization-independent absorption analysis. Juan et al. have reported a design of solar cell with the influence of relative position of silver metallic nanoparticles which was embedded in 100 nm thick antireflection coating layer [15]. The plasmonic antireflection coating layer could reduce the reflections as compared to the SiOx antireflection coating layer, but the addition of silver nanoparticles in front-surface geometry has generated the poor interferences due to which the efficiency of cell was found to be degraded. Sheng et al. have analytically investigated the light-trapping mechanism in plasmonic silicon solar cells [16]. This designing was explored by considering the absorption enhancement for surface plasmon polaritons (SPPs) at planar silicon-metal interfaces and localized surface plasmon resonances (LSPRs) for the metallic spheres in a silicon matrix. They have observed that the absorption enhancement factor was not bound to Lambertian limit, and the localized plasmonic resonances can be used as efficient light-trapping schemes for ultrathin silicon solar cells. Chriki et al. have proposed an ultrathin solar cell design by incorporating the two periodic layers of metallic and dielectric gratings [17]. Both layers could able to couple the incident light to photonic and plasmonic modes, and, hence, enhanced absorption of light was achieved. A relative position between the two gratings was analyzed and observed the significant effect. The proposed design was compared with a reference solar cell of a single layer of metallic and dielectric nanostructures, respectively, and found to be satisfactory in term of high absorption for the dual-grating-based design. Plasmonic solar cells are promising to produce the high efficiency due to their high carrier collection and less bulk recombination. Spinelli et al. have presented two possible ways of integrating metal nanoparticles in a solar cell: the first one was a coating of silver nanoparticles which acted as antireflective surface, and the second one was based on regular and random arrays of metal nanostructures which could couple the light in the waveguide modes [18]. By employing a relative inexpensive nanoimprint technique, the fabrication of solar cell was attempted which showed an improvement in the cell efficiency. In plasmonic solar cells, photons are trapped into localized surface plasmon (LSP) as a result, it induces the surface plasmon (SP) which propagates transversely into active layer. Chao et al. have proposed a plasmonic multilayer structure (PMS) for the application in ultrathin solar cell with 30 nm thick amorphous silicon ($\alpha$-Si) as active layer. With the use of plasmonic multilayer structure, they have observed enhanced absorption (~ 28.7%) as compared to the indium tin oxide (ITO)/$\alpha$-Si/Ag structure for the normal incident case of transverse magnetic (TM) polarization [19]. Lee et al. have numerically presented the design of amorphous silicon (a-Si) thin-film solar cell by employing ultrathin top grating into a-Si active layer. They have observed the enhanced absorption with a wide range of incident angle for TM polarization through the solar cell with 30 nm thickness. The overall absorption for TM polarization was improved about 25% as compared to a solar cell of thicker metal grating; however, 2.5 times improvement was observed for the TE polarization [20].

Figure 6 shows a schematic design of crystalline silicon (c-Si) thin-film solar cell with an ultrathin active layer (50 nm). It is comprised of an ARC layer ($Si_3N_4$), a DBR structure (a-Si/$SiO_2$), and the metal grating structure (Al). The DBR pairs was composed of alternate layers of a-Si and $SiO_2$ with their refractive indices 3.6 and 1.45 and thicknesses 41($t_1$) and 103 nm ($t_2$), respectively, whereas the center wavelength of DBR was 600 nm. The assumed thicknesses of ARC layer ($t_5$) and Al grating ($t_3$) were 70 and 50 nm, respectively. In this simulation, the refractive index of the active region "$n_c$=3.5" and its thickness "$t_4$ = 50 nm" were used.

The aluminum (Al) gratings help to diffract the incident light at oblique angles, whereas distributed Bragg reflector (DBR) resists the metal diffusion and sustains the mechanism of light trapping. The back reflector with the combination of DBR and metal grating is supposed to be as an efficient light-trapping structure which utilizes the longer wavelength light by enforcing those in the active region of the solar cell. Table 1 displays the initial parametrical values considered for the design of ultrathin solar cell.

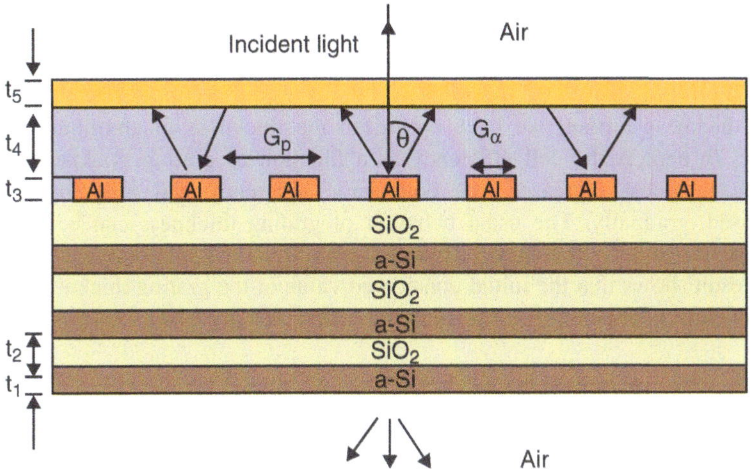


**Fig. 6** Schematic diagram of ultrathin silicon solar cell

**Table 1** Design parameters of ultrathin silicon solar cell

| Parameters | Values |
|---|---|
| Grating period ($G_p$) | 600 nm |
| Grating duty cycle ($G_{dc}$) | 0.5 |
| Grating thickness ($t_3$) | 50 nm |
| Grating tooth width ($G_w$) | 300 nm |
| No. of DBR pairs | 5 |
| Antireflection coating thickness ($t_5$) | 70 nm |
| DBR center wavelength ($\lambda_c$) | 600 nm |
| Active layer thickness ($t_4$) | 50 nm |
| DBR first-layer a-Si thickness ($t_1$) | 41 nm |
| DBR second-layer $SiO_2$ thickness ($t_2$) | 103 nm |

For the simulation, finite-difference time-domain method (FDTD) was employed with the periodic boundary conditions which were applied in x- and y-directions, and the perfectly matched layer boundary condition was performed in z-direction [21].

## 3.1 Optimization of Various Parameters

Various design parameters such as thickness of antireflective coating layer, grating thickness, active layer thickness, grating period, and duty cycle are discussed.

### 3.1.1 ARC and Grating Thickness

Figure 7a depicts the cell efficiency in accordance with ARC thickness. Reasonably, an enhancement in cell efficiency from 10.5% to 11.6% is observed with ARC layer thickness ($t_5$) of 65 nm instead of 70 nm as used, previously. This result is a supplementary evidence of importance of ARC layer in term of its thickness which can be tuned to have optimal transmission of solar light into silicon active region.

The diffraction grating as a part of the back reflector plays vital role for the light trapping in longer wavelength region. By replacing the optimized value of ARC layer thickness further, we have optimized the thickness of aluminum grating. Figure 7b depicts the cell efficiency as a function of grating thickness. As the grating thickness increased, the conversion efficiency was also found to be increased, gradually. The usual behavior of grating thickness can be observed, whereas maximum efficiency 11.6% (same as previous) was obtained at 50 nm. This result shows that the initial considered value of the grating thickness was the correct choice. The obtained short-circuit current density was 17.67 mA/cm$^2$.

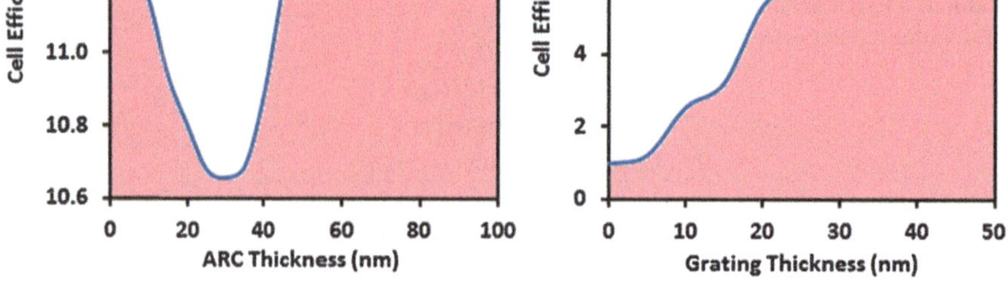


**Fig. 7** Solar cell efficiency as a function of ARC thickness (**a**) and grating thickness (**b**) (Saravanan et al. [21]. AIP: With permission)

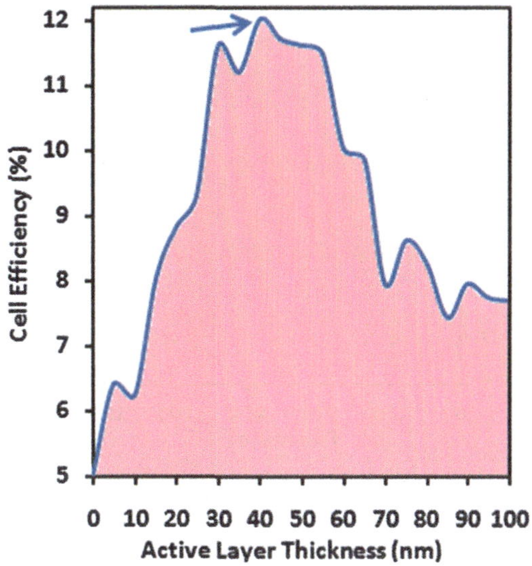


**Fig. 8** Solar cell efficiency in accordance with active layer thickness (Saravanan et al. [21]. AIP: With permission)

### 3.1.2 Active Layer Thickness

We have explored our analysis to study the effect of active layer thickness by keeping all parameters constant except active layer thickness ($t_4$). Figure 8 shows the variation of the active layer thickness against the cell efficiency. An exponential enhancement in cell efficiency is observed, and further a decrease in cell efficiency is observed.

This reduction of cell efficiency indicates no longer use of back reflector due to increased cell thickness. This result is validated with reported work in which effective role of the back reflector was found below 10-μm cell thickness and beyond it; the contribution by the back reflector was rapidly decreased [8]. An enhancement in cell efficiency (12%) is observed with cell thickness 40 nm, while short-circuit current density was 18.28 mA/cm$^2$. For further simulation, 50-nm cell thickness was replaced with an optimized one, i.e., 40 nm.

### 3.1.3 Grating Period and Duty Cycle

As the grating period and duty cycle are the important parameters for the diffraction of light in terms of diffraction angle, for example, a larger diffraction angle represents the long optical path length of photons. Accordingly, Fig. 9 shows the effect of the grating duty cycle and period.

As the duty cycle was increased, the cell efficiency was found to be increased exponentially, and it was maximum at 0.8 as depicted in Fig. 9a. Further, the cell efficiency was reduced beyond 0.8 which shows the importance of the duty cycle as a sensitive parameter for the performance of solar cell. The obtained optimal cell

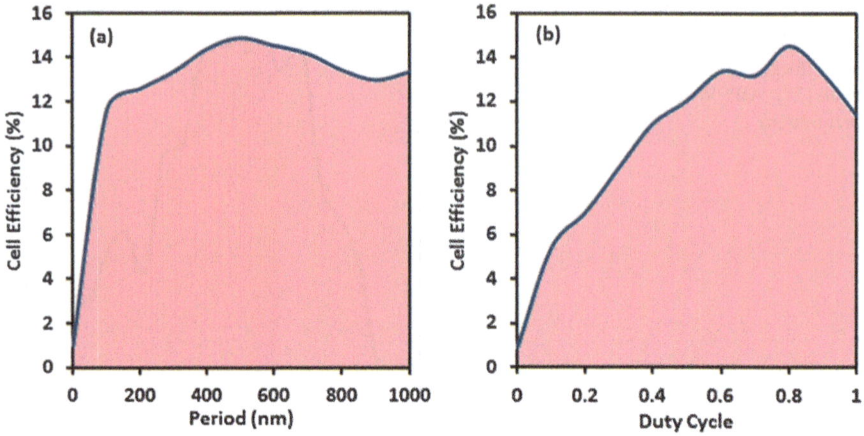


**Fig. 9** Solar cell efficiency in accordance with grating duty cycle (**a**) and grating period (**b**) (Saravanan et al. [21]. AIP: With permission)

efficiency is 14.55% with current density 22.13 mA/cm$^2$ at 0.8 duty cycle. For the next level of simulation, all parameters were kept constant, while duty cycle value was replaced with its optimal one. Figure 9b depicts a variation in cell efficiency as a function of grating period. It shows optimal cell efficiency (14.9%) at 500 nm period which would be attributed to the high-order diffraction angle. Our previous value of grating period was 600 nm whereas the optimized one is 500 nm.

### 3.1.4 Performance Analysis After Optimization

After all the optimization, we could able to obtain ~15% cell efficiency with 40 nm cell thickness. To conclude the simulation results after above discussed optimization, we have plotted the absorption curves in Fig. 10. If we observe the absorption curve of grating thickness (t3), an enhanced absorption can be observed in infrared wavelength region with a peak centered at 680 nm. The curves of grating thickness ($t_3$) and ARC layer thickness ($t_5$) are superimposed as the assumption of grating thickness value was coincided with the optimized one. The curve of duty cycle ($G_{dc}$) shows high absorption in visible range with intense absorption peaks centered at wavelength 540, 640, and 680 nm. However, the absorption curve of ($G_p$) revealed the extraordinary wider absorption peaks in visible and infrared region centered at 500, 600, and 800 nm. In case of $t_3$, $t_4$, and $t_5$, various absorption peaks were observed from 400 to 1020 nm. However, for the case of $G_{dc}$ and $G_p$, absorption is observed to be enhanced in far infrared region too. It indicates that the enhanced light absorption was not only in visible but also infrared part of the solar spectrum. Due to this, our designed solar cell showed an amazing enhancement in cell efficiency with an ultrathin active layer thickness "40 nm" which has not been reported yet.

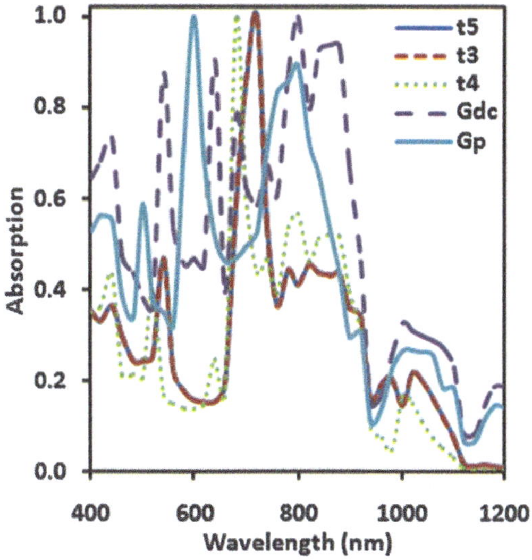


**Fig. 10** Light absorption behavior in silicon active layer of each optimization (Saravanan et al. [21]. AIP: With permission)

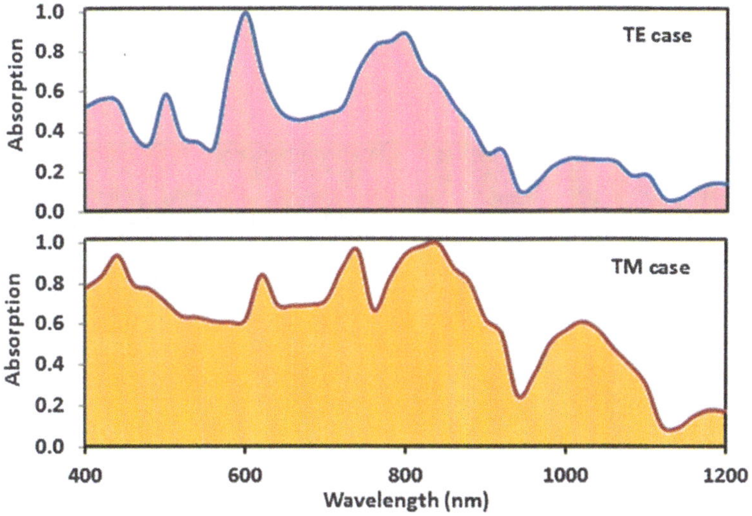


**Fig. 11** Light absorption behavior in silicon active layer for TE and TM cases (Saravanan et al. [21]. AIP: With permission)

This enhancement in absorption was due to the use of aluminum grating and ultrathin absorbing layer. An ultrathin active layer can induce flow of generated electrons and holes towards the electrodes with the less recombination chance, whereas localized surface plasmon excited by aluminum grating can generate surface plasmon.

To observe the effect of metal grating for the TE and TM polarization cases, Fig. 11 is plotted. The absorption in active silicon region is considerably enhanced for the case of TM as comparison to TE polarization. For TE case, strong absorption peaks were centered at 600 and 800 nm, whereas for TM case it was at 741 and 841 nm. Remarkably, the absorption peaks were found to be shifted to infrared part and has become wider for the magnetic transverse case.

For the validation of plasmonic effect, electric and magnetic field profiles of optimized solar cells are plotted in Fig. 12.

Two high absorption peak values 600 and 800 nm corresponding to electric field and 741 and 841 nm corresponding to magnetic field were selected for the analysis. For transverse electric case, at wavelength 600 nm Fig. 12a we can observe strong electric field within the grating and waveguiding mode in the silicon region; however, at wavelength 800 nm, DBR has supported the light reflection back in active region which would be a cause of getting wider absorption peak, and hence strong field in active region can be seen in Fig. 12b. For transverse magnetic case, localized plasmon is visible at the top of aluminum gratings which could give enhanced absorption in active region as shown in Fig. 12c at wavelength 741, and for wavelength 841 nm, surface plasmon is clearly visible as can be observed in Fig. 12d due to which enhanced absorption can be seen in Fig. 6 (dashed line). With this analysis, the enhanced efficiency of the designed solar cell has been attributed to the plasmonic effect. We have compared the performance of different solar cell structures for both TE and TM polarization modes which is shown in Table 2.

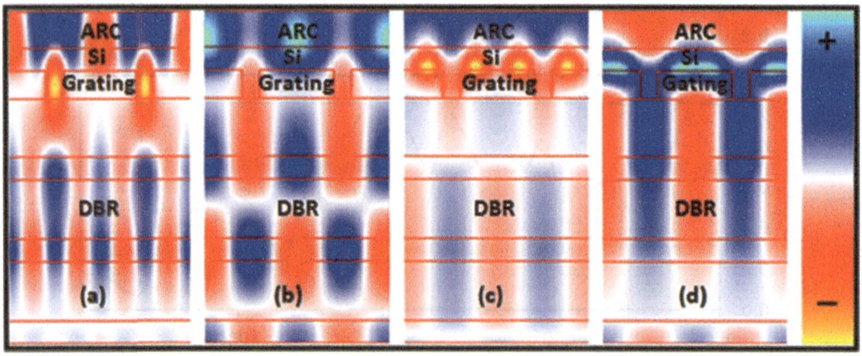


**Fig. 12** Electric field profile at 600 and 800 nm (**a**, **b**) and magnetic field profile at 741 and 841 nm (**c**, **d**) (Saravanan et al. [21]. AIP: With permission)

**Table 2** Performance comparison of various designed silicon solar cells

| | TE | | TM | |
|---|---|---|---|---|
| Structure | C. E (%) | $J_{SC}$ (mA/cm$^2$) | C. E (%) | $J_{SC}$ (mA/cm$^2$) |
| Reference (only ARC) | 0.71 | 1.09 | 0.72 | 1.1 |
| C1 (3DBR+ARC) | 0.89 | 1.35 | 0.92 | 1.4 |
| C2 (GRA+ARC) | 9.79 | 14.89 | 10.92 | 11.16 |
| C3 (3DBR+GRA+ARC) | 14.90 | 22.66 | 14.93 | 22.71 |

An extraordinary enhancement in performance can be observed from the solar cells C2 and C3 which is due to the use of back reflector of metal grating (GRA) and DBR. Compared to all solar cells, cell C3 is found to be best performed with cell efficiency ~15%; however, for TM case a small increment in cell efficiency is observed.

## 4 Ultrathin-Film Silicon Solar Cells Using Top Dielectric and Bottom Metal Gratings

Metal nanostructures have been applied in thin-film solar cells as an efficient light-trapping component supported by the plasmonic effect which not only guides light but also manipulates photons through the collective oscillation of electrons at the metal surface. Further, the integration of dielectric (photonic mode) and metal (plasmonic mode) gratings can facilitate the use of shorter as well as the longer wavelength of light. Several research articles have been reported on the performance enhancement of the solar cells due to the combined effect of photonic and plasmonic structures. Chang et al. have proposed a design of a plasmonic multilayer structure for the silicon (a-Si) solar cell of 30 nm thickness by using rigorous coupled-wave analysis (RCWA) and finite-difference time-domain (FDTD) methods. The plasmonic multilayer structure showed the trapping of incident light into localized surface plasmon (LSP) as a result of propagation of surface plasmon transversely within the active layer. Comparatively, the use of plasmonic multilayer structure could yield ~ 29% enhancement in light absorption at normal incidence transverse magnetic (TM) polarization condition [19]. Battal et al. have presented an enhancement of light absorption through silicon thin-film solar cell with the use of triangular corrugations at the bottom. They could observe ~ 22% overall absorptivity through 100-nm-thick solar cell as compared to a reference solar cell. This increased absorption was found to be insensitive to the polarization conditions and incident angle [22]. Abass et al. have numerically presented a design of silicon solar cell based on dual-interface grating systems which could reasonably enhance the light absorption. The proposed dual-grating structure with different combinations of shapes and sizes was emphasized on the coupling efficiency with respect to photonic and plasmonic modes. The combination of blazed dual-grating periodicity structures has shown a strong coupling efficiency, and as a consequence enhanced light absorption was achieved. The optimization of top dielectric and bottom metal gratings was observed to be promising for the better harvesting of light within active layer [23]. Chriki et al. have studied an ultrathin solar cell architecture design based on two periodic layers of metal and dielectric gratings. They could observe the optimal light absorption due to well-supported photonic as well as plasmonic modes. The coupling of incident light to the dark photonic modes with a lateral shift between the two periodic layers might be one cause of mode interactions. After optimization, the proposed design showed enhanced cell efficiency over a broad range of incident light [17]. Shi et al. have introduced a hybrid light-trapping structure by using front grating of indium tin oxide and silver nanoparticles at the backside. The front grating was served as antireflection purpose, whereas silver nanoparticles at the backside were supported as light-trapping mechanism. The proposed design of 1-μm-thick solar cell could yield short-circuit current ~ 29.7 mA/cm$^2$, and further this design was discussed for its low-cost fabrication using sophisticated tools [24]. Liu al. have presented a theoretical design of dual-grating-based thin-film silicon solar cells and studied the influence of geometrical parameters of gratings. The optimized design could reduce the reflections at the front grating, while enhanced light absorption in longer wavelength was noticed due to the use of the back grating. With comparison to the reference solar cell, nearly 17% enhanced short-circuit current was obtained from 1-μm-thick silicon solar cell [25]. Saravanan et al. have presented a theoretical study of an ultrathin crystalline silicon solar cell which could enhance the performance due to the absorption of light in the shorter as well as longer wavelength part of the solar spectrum. Optimization of various parameters could result the better harvesting of light through 40 nm thick ultrathin-film silicon solar cell with ~ 15% cell efficiency and ~ 23 mA/cm$^2$ short-circuit current density due to well-supported photonic as well as plasmonic modes [21].

The proposed design of an ultrathin-film silicon solar cell is depicted in Fig. 13. Two-dimensional simulations were performed for both TE and TM polarizations under AM1.5G solar radiation by using finite-difference time-domain method (FDTD) and rigorous coupled-wave analysis (RCWA). In FDTD simulation, the periodic boundary conditions (PBC) were applied in x- and y-directions, while the perfect match layer (PML) was performed in the z-direction. The use of perfect match layer eliminates the reflection of light from the propagation direction. RCWA simulation was done to analyze the field distribution within solar cell device which is also imperative to realize the effect of embedded grating structure [26].

The proposed structure consists of a $Si_3N_4$ antireflection coating layer of *107 nm* *($t_5$)*, an ultrathin silicon absorber layer of *40 nm ($t_3$)*, a top *30 nm ($t_4$)* ITO grating (GRA), a bottom Al grating of *30 nm ($t_2$)*, and *103 nm ($t_1$)* thick Al back reflector (BR). In this simulation, parameters of ARC layer such as refractive index and thickness were calculated using expressions $n_{ARC}=\sqrt{n_c}$ and $t_5 = \lambda_c / \sqrt{4 n_{ARC}}$ where $\lambda_c$ is the center wavelength. The numerical values of the ARC layer such as

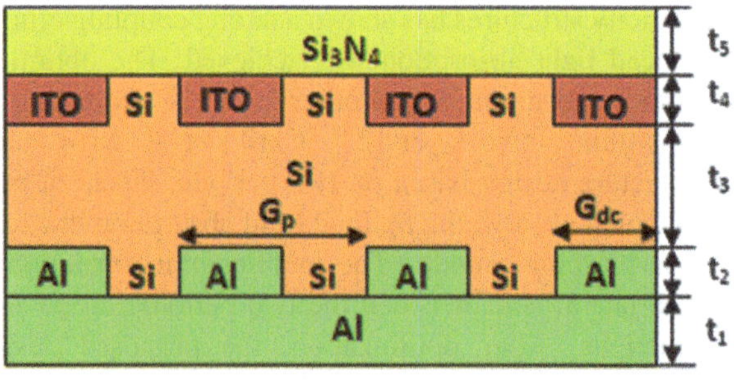


**Fig. 13** Schematic diagram of ultrathin-film silicon solar cell based on dual grating

refractive index *1.87* and thickness *107 nm* were calculated at center wavelength *800 nm*. The simulation was performed in the ideal conditions such as zero shadow and 100% collection efficiency.

### *4.1 Transverse Electric and Magnetic Field Behavior in Solar Cell Structure*

The electric (TE) and magnetic (TM) field distribution in the proposed ultrathin-film solar cell is plotted in Figs. 14 and 15. For TE case, a sharp absorption peak centered at *610 nm* (near red region) can be observed in Fig. 14a and beyond that, a reduction in the absorption can also be noticed. Electric field profiles plotted in

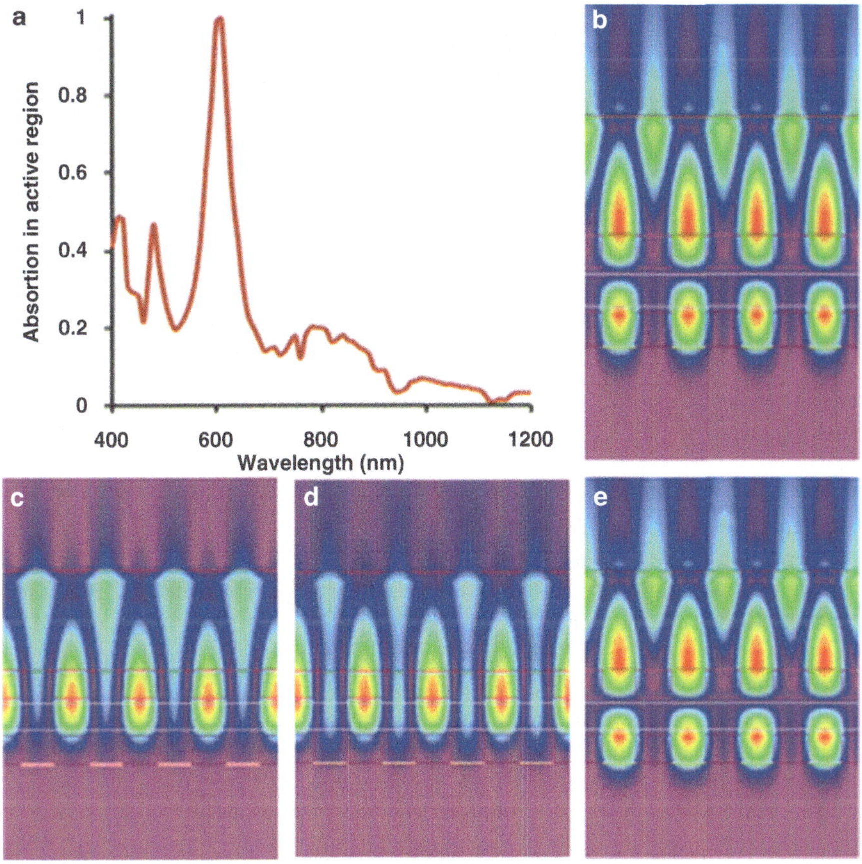


**Fig. 14** Absorption spectra and field profile in ultrathin-film solar cell for transverse electric mode (Saravanan et al. [26]. Elsevier: With permission)

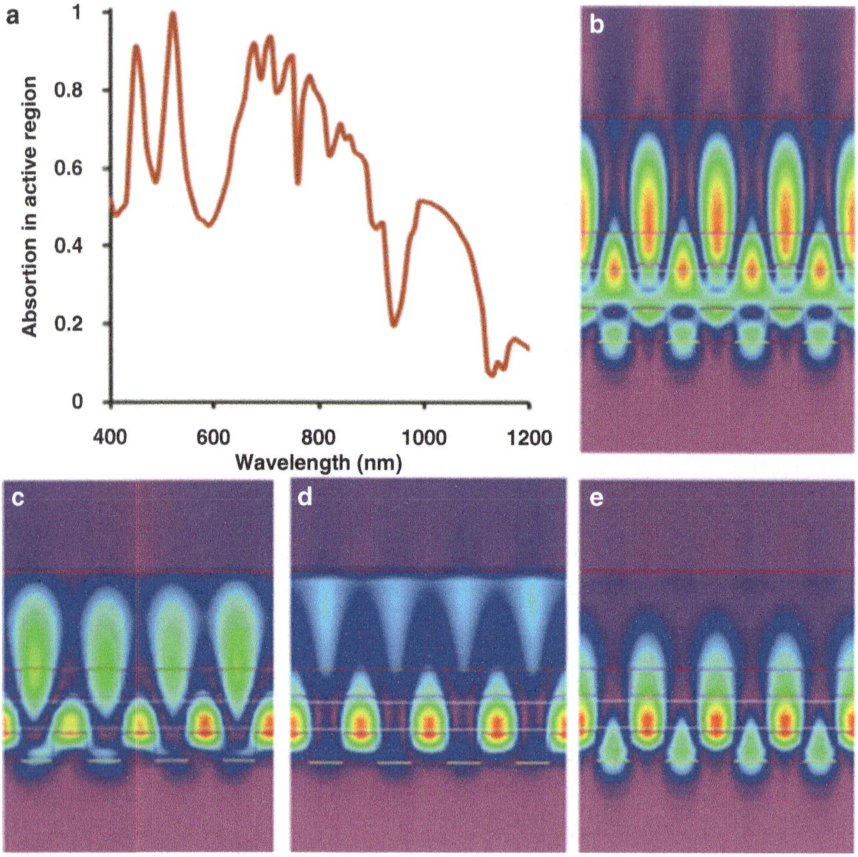


**Fig. 15** Absorption spectra and field profile in ultrathin-film solar cell for transverse magnetic mode (Saravanan et al. [26]. Elsevier: With permission)

Fig. 14b–d show dark and bright modes corresponding to the symmetric and asymmetric electric field distributions.

Figure 14b plotted at wavelength *480 nm* shows the anomalous effect of shorter wavelength with Fabry-Perot (FP) resonances between the top gratings. A distinct field profile within the active region can be observed in Fig. 14c–e plotted at wavelength *600*, *610*, and *780 nm*, respectively. However, a strong field within active region can be observed at wavelength *610 nm* as depicted in Fig. 14c. This result coincides with sharp absorption peak depicted in Fig. 14a. In general, the incident light preferentially scatters or diffracts into the active region because of dual-grating-based structure.

The metal-based solar cells are interesting due to their strong localized field in the active region supported by plasmonic effect, and, hence, we have extended our analysis for magnetic field as well. Figure 15f depicts absorption curve of the ultrathin-film solar cell for the transverse magnetic polarization with several absorption peaks as compared to the transverse electric case.

The surface plasmon polariton (SPP) can be observed between the dielectric and metallic interface at wavelength *400 nm* as shown in Fig. 15g. The nanostructured gratings were noticed to be played a pivotal role for the plasmonic effect. Figure 15h shows a strong field at the interface of metal and silicon with surface plasmon polariton modes at wavelength *710 nm* which is attributed to the effect of complex refractive index interface between ITO and Al gratings. Further, Fig. 15i–j plotted at wavelength *840* and *1000* nm, respectively, shows localized surface plasmons (LSP) leading to an enhancement of light absorption within the active semiconductor region. The magnetic field resonances appeared in shorter and longer wavelengths indicate most excited (bright) states arising due to the coupling of incident or scattered light at metal and silicon interface. The higher-order diffraction mechanism is able to guide the excited photons amplitude oscillations parallel to the resonance surfaces.

## 4.2 Light Absorption Through Various Solar Cell Structures

To compare the performance of the proposed solar cell structure, we have designed various solar cell structures and hereafter named as solar cell C1, ARC+ITO GRA+BR; C2, ARC+Al GRA+BR; C3, ARC+Dual GRA; and C4, ARC+Dual GRA+BR. The absorption spectra of various designed solar cells for TE and TM polarizations are shown in Fig. 16. Referring to Fig. 16a an improvement in light absorption can be observed for solar cells C2, C3, and C4. The solar cell C1 composed of ARC, top ITO grating (GRA), and Al back reflector (BR) shows absorption peaks from *600–780 nm*. Solar cells C2 and C3 show absorption peaks

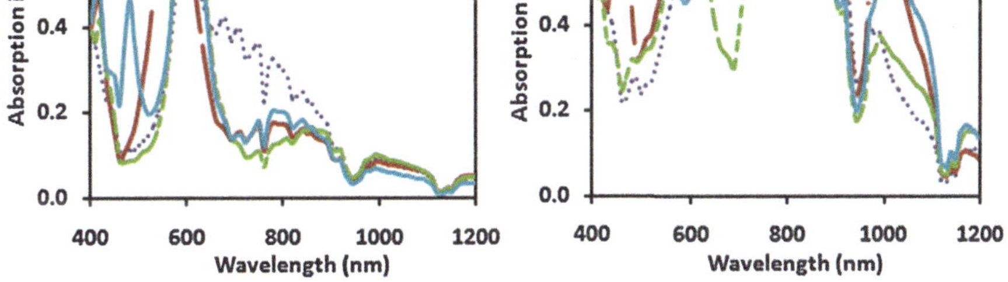


**Fig. 16** Absorption spectra for TE (**a**) and TM (**b**) polarizations of solar cells C1, C2, C3, and C4 (Saravanan et al. [26]. Elsevier: With permission)

from *420–810* nm and *600–860* nm, respectively. The dual-grating-based solar cell C4 reveals improved light absorption from wavelength *410–810 nm*.

Referring to Fig. 16b several absorption peaks can be observed for various designed solar cells. The solar cells C1, C2, and C3 yield absorption peaks at *640–970 nm*, *460–990 nm*, and *580–990 nm*, respectively, in addition to longer wavelength. However, dual-grating-based solar cell C4 reveals enhanced light absorption with sharp and broad peaks in the wavelength range from *450–1020 nm*. In case of TM polarization mode, remarkably enhanced light absorption is observed, not only in shorter wavelength but also in longer wavelength. The enhanced light absorption for TM polarization is associated to the metal grating which could yield strong field localization at the interface of silicon metal.

## *4.3 Optimization of Grating Parameters*

Further, we have optimized the design of solar cell C4 by varying top ITO and bottom Al gratings, grating period, and the duty cycle which are plotted in Fig. 17.

An enhancement in short-circuit current density from *12.56* to *12.7* $mA/cm^2$ (TE case) was observed with *35 nm* ITO grating thickness as referring to a solid line in Fig. 17a. A similar trend of improvement in short-circuit current density (dashed line) with respect to Al grating thickness can also be observed, i.e., from *12.56* to *12.76* $mA/cm^2$. In addition to the grating thickness, grating duty period and duty cycle are the important parameters as these involves the diffraction of light waves and, hence, responsible for the enhanced light absorption. The influence of grating

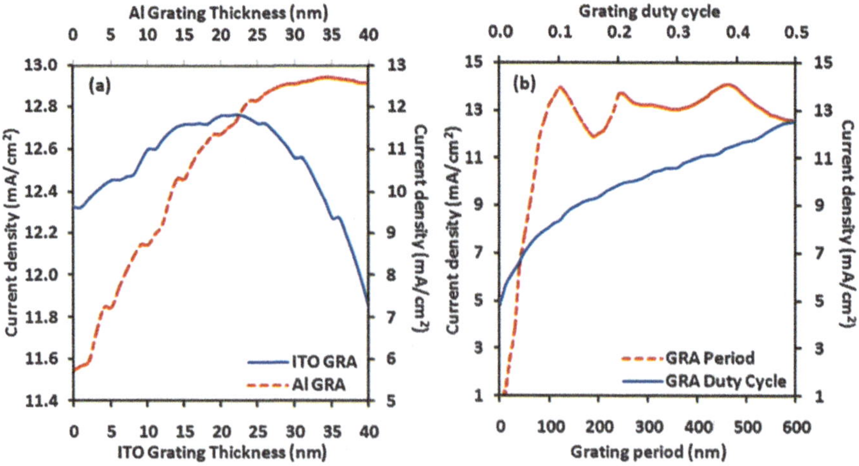


**Fig. 17** Short-circuit current density in accordance to top ITO grating and bottom Al grating thicknesses (**a**) and grating period with duty cycle (**b**) (Saravanan et al. [26]. Elsevier: With permission)

period and duty cycle on short-circuit current density is depicted in Fig. 17b. A significant enhancement in short-circuit current, i.e., from *12.56* to *14.1 mA/cm*$^2$ is observed (dashed line) with grating period *460 nm* instead *600 nm* (as the initial value). Further, an increase in grating period could reduce short-circuit current due to low diffraction angle. The grating duty cycle is defined as the ratio between the rectangular grating tooth width and period of the grating which is plotted (solid line) in Fig. 17b. The maximum obtained short-circuit current density is *12.56 mA/cm*$^2$ at duty value 0.5, i.e., same result as initial and validates our simulation.

## 4.4 Performance Analysis After Optimization

The optimized values of grating parameters were replaced with initial values for the further simulation, and the optimized design of solar cell was found to be well performed. Figure 18 shows the absorption and short-circuit current density of optimized solar cell for both TE and TM polarizations. As shown in Fig. 18a, enhanced absorption in the whole wavelength range from *400* to *1200 nm* can be observed for both TE and TM waves.

Figure 18b shows short-circuit current density for TE and TM waves. The maximum obtained short-circuit current density is *14.41* and *33.86 mA/cm*$^2$ corresponding to transverse electric and magnetic fields. The optimized ultrathin-film silicon solar cell with *40 nm* absorbing layer could produce cell efficiency as much as ~ *10%* (TE) and *22%* (TM) as compared to un-optimized solar cell C4. Chriki et al. have reported an enhanced photocurrent density up to *1.86* from *100 nm* thick crystalline silicon solar cell based on dual-grating plasmonic structure [17]. Shi et al. have achieved enhanced short-circuit current density up

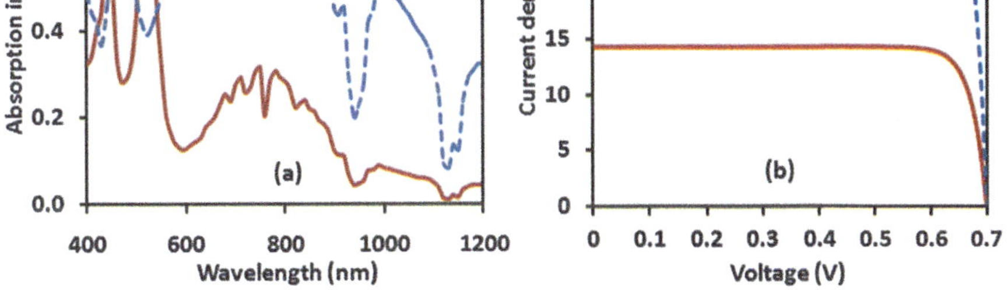


**Fig. 18** Absorption spectra (**a**) and short-circuit current density (**b**) of optimized ultrathin-film solar cell (Saravanan et al. [26]. Elsevier: With permission)

to *29.7 mA/cm²* in a *1-μm*-thick silicon solar cell by using optimized gratings and metal nanoparticles [24]. Liu al. have reported ~ 17% enhanced short-circuit current density from *1-μm*-thick silicon solar cell in comparison to reference solar cell [25]. Guo et al. have reported enhanced optical absorption from 200 nm thick crystalline silicon solar cell based on the dual grating with *38.71 mA/cm²* short-circuit current density [27]. Xiao et al. have demonstrated a solar cell design by using the front and backside metallic gratings and reported the enhanced photocurrent up to a factor *1.9* from a crystalline silicon solar cell of *200 nm* thick [11]. As compared to these investigations, our result shows the superior performance of ultrathin-film solar cell with a relative enhancement in short-circuit density *14.7%* (TE) and *36.7%* (TM).

## 5 Ultrathin-Film Silicon Solar Cells Using Metal Nanoparticles Decorated Metal Nanograting

An improvement in light absorption has been reported with the use of metal nanoparticles decorated on the planer metal and nanograting reflectors. Here, we propose an ultrathin-film solar cell design using silver nanoparticles decorated on aluminum nanograting which is shown in Fig. 19a. The silver nanoparticles of diameter 20 nm were embedded within the 40-nm silicon absorber region, while 70 nm distance between the nanoparticles was maintained.

A silicon nitride ($Si_3N_4$) thin layer of 107 nm was used as an antireflection coating (ARC). The aluminum (Al) surface layer was employed as a back reflector with its thickness 103 nm onto which the metal grating (Al) with its height 30 nm was preferred. This numerical study was carried out by using rigorous coupled-wave analysis (RCWA) method which is employed for the periodic media.

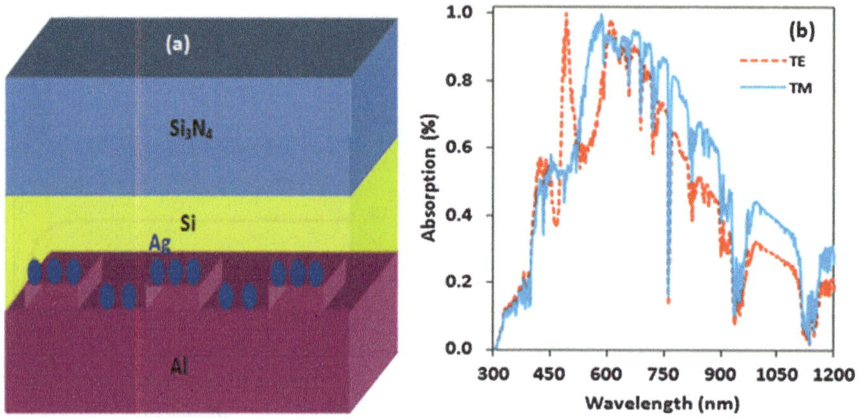


**Fig. 19** Schematic diagram of ultrathin-film silicon solar cell (**a**) and light absorption in absorber region (**b**)

## 5.1 Light Absorption and Field Analyses

Figure 19b shows the absorption of ultrathin-film silicon solar cell with the use of silver nanoparticles decorated on aluminum grating as back reflector. As comparison to TE case, an enhanced light absorption can be noticed for the TM waves. The use of silver (Ag) nanoparticles shows the increased light absorption in longer wavelength (IR) region due the strong plasmonic effect.

Figure 20a–d depicts the electric field distribution for the TE mode at wavelengths 490, 607, 994, and 1176 nm. The combined effect of nanoparticles and nanogratings shows the strong light scattering as can be seen Fabry-Perot resonance within the absorber region at wavelength 490 and 607 nm referring to Fig. 20a, b. Similarly, the occurrence of Fabry-Perot resonance is visible in Fig. 20c, d along with the guided modes in-between the dielectric and metal interface.

Figure 20 depicts the magnetic field distribution at wavelengths 583, 670, 995, and 1176 nm. A strong field excitation between the semiconductor and metal nanostructure (red color) can be observed at wavelength 583 nm as depicted in Fig. 20e. However, the localized surface plasmon resonance (LSPR) near the surface of silver (Ag) nanoparticles is observable Fig. 20f–h. As a result of this, an enhanced field is built-up which makes significant contribution for the direct generation of electrons, and this mechanism has been demanded in analytical

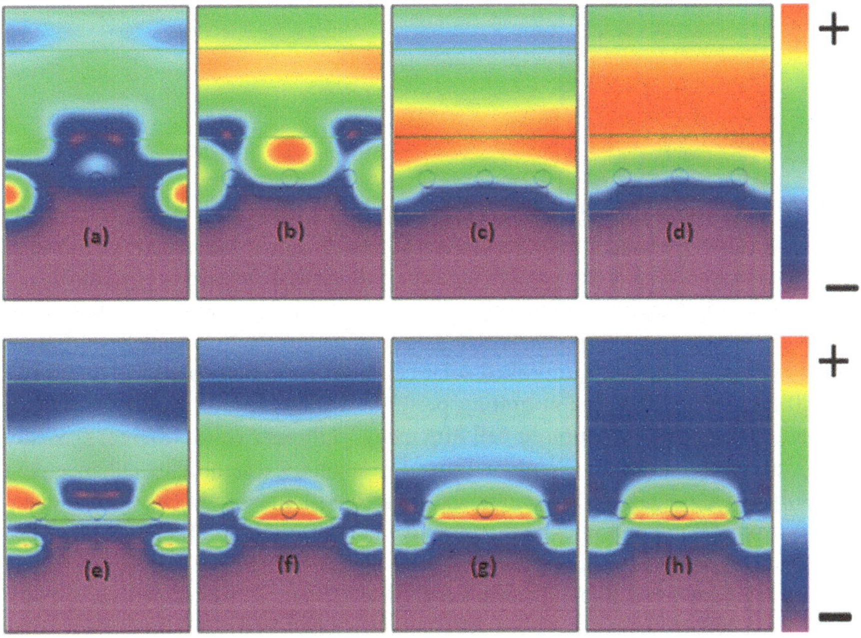


**Fig. 20** Electric field distribution at wavelengths 490 nm (**a**), 607 nm (**b**), 994 nm (**c**), and 1176 (**d**) nm and magnetic field distribution at wavelengths 583 nm (**e**), 670 nm (**f**), 995 nm (**g**), and 1176 nm (**h**)

sciences. Generally, the metal nanostructures have the capability to scatter maximum light with respect to the size and shape of the particles which affects the surface plasmonic excitations as well. This design showed the enhanced optical absorption due to the surface plasmons, guided modes, and the localized field. The metal nanoparticles decorated back reflector showed the short-circuit current about 27 and 35 mA/cm$^2$ for the TE and TM modes, respectively.

This chapter is limited to the modeling and simulation of thin-film solar cells; however, such solar cell structures can be fabricated by using lift-off technique. The epitaxial lift-off technique has been explored for the fabrication of cost-effective solar cells due to easy separation of III/V device layers from the GaAs substrate and, hence, enables substrate reuse [28, 29]. Using this technique, the first step is the fabrication of a porous silicon layer on the silicon substrate by employing electrochemical etching method. By using an epitaxial technique, a crystalline silicon thin film can be grown on the previous prepared porous silicon. Then diffraction gratings can be fabricated on the epitaxially grown layer by sputtering/evaporation with suitable dry/wet etching. During the process, the fabricated structure can adhere to the desired substrate such as glass/stainless steel/plastic film, and further prepared solar cell can be separated from the host silicon substrate either by electrochemical etching or mechanical force [30, 31]. After detaching completely from porous silicon layer and doing the desired level of doping through diffusion/implantation technique, an antireflection coating can be deposited by using plasma-enhanced chemical vapor deposition method, and at last electrical contacts are made.

## 6 Conclusions

In this chapter, four design approaches of thin-film silicon solar cells using FDTD and RCWA methods have been presented. In first design approach, improved solar cell performance was observed with the use of three DBR pairs designed at 0.8 μm center wavelength and 0.3 μm grating width. By optimizing grating period, we have observed the two maxima points which are attributed to the first- and second-order diffractions of the grating. Improved solar cell efficiency was observed for 5-, 10-, and 20-μm cell thicknesses at grating period 0.6 μm. Grating thickness analysis showed a significant variation in cell efficiency in thinner cells, whereas 0.5 duty cycle was the optimal value. With optimized parameters, a significant enhancement in absorption and quantum efficiency was noticed with maximum cell efficiency up to 24% from 50-μm cell thickness. A relative enhancement in efficiency of 42.8% was observed for 5-μm cell thickness as compared to a reference cell. Remarkably, modified designing of bilayer grating embedded in active silicon region is competed our previous design by giving 24% cell efficiency from the 30-μm thick solar cell. Finally, efficient light-trapping structures with three bilayer grating were designed and achieved enhanced performance as comparison to single-layer grating-based solar cells. In the second design approach of ultrathin silicon solar cell, strong absorption peaks centered at 600 and 800 nm (TE case), whereas 741 and 841 nm (TM case) were observed. Remarkably, the absorption peaks were found to be shifted to infrared part and has become wider for magnetic transverse case. The optimized solar cell of 40 nm active region could show the best performance with extraordinary enhancement in cell efficiency (~15%) and current density (~ 23 mA/cm2). In the third design approach of another ultrathin-film silicon solar cell assisted by photonic and plasmonic modes, due to the effect of complex refractive index interface between ITO and Al gratings, a strong magnetic field (at wavelength *710 nm*) near the metal-silicon interface with surface plasmon polariton mode is observed. Further, the localized surface plasmons leading to an enhanced light absorption were observed at *840* and *1000* nm. The optimal performance of the solar cell has been attributed to the localized surface plasmon modes, surface plasmon polariton, and optical resonances. For normal incident transverse magnetic polarization, the optimized ultrathin-film silicon solar cell could produce enhanced absorption in a wide range of solar spectrum due to the strong plasmonic effect. Further, the optimization of gratings period, and the duty cycle has improved the absorption of light in the shorter as well as longer wavelength with the cell efficiency ~ *25%* for normal transverse magnetic polarization. In fourth design approach, silver nanoparticles decorated on aluminum grating as a back reflector has been explored, and the increased absorption was noticed assisted by the plasmonic modes. In summary, the presented design efforts would be helpful to realize the new generation of thin-film solar cells with the fabrication challenges as well.

**Acknowledgments** The financial support provided by Defence Research and Development Organisation (DRDO), New Delhi, India, is highly acknowledged.

## References

1. E. Yablonovitch, G.D. Cody, Intensity enhancement in textured optical sheets for solar cells. IEEE Trans. Electron. Dev. **29**(2), 300–305 (1982)
2. M.A. Green, Lambertian light trapping in textured solar cells and light-emitting diodes: analytical solutions. Progr. Photovolt: Res. Appl. **10**, 235–241 (2002)
3. L. Zeng, P. Bermel, Y. Yi, B.A. Alamariu, K.A. Broderick, J. Liu, C. Hong, X. Duan, J. Joannopoulos, L.C. Kimerling, Demonstration of enhanced absorption in thin film Si solar cells with textured photonic crystal back reflector. Appl. Phys. Lett. **93**, 221105 (2008)
4. X. Meng, E. Drouard, G. Gomard, R. Peretti, A. Fave, C. Seassal, Combined front and back diffraction gratings for broad band light trapping in thin film solar cell. Opt. Exp. **20**(S5), A560–A571 (2012)
5. X. Sheng, G. Steven, L.J.Z. Broderick, J. Michel, L.C. Kimerling, Integrated photonic structures for light trapping in thin-film Si solar cells. Appl. Phys. Lett. **100**, 111110 (2012)
6. A. Chutinan, N.P. Kherani, S. Zukotynski, High-efficiency photonic crystal solar cell architecture. Opt. Exp. **17**(11), 8871–8878 (2009)
7. L. Zhao, Y.H. Zuo, C.L. Zhou, H.L. Li, H.W. Diao, W.J. Wang, A highly efficient light-trapping structure for thin-film silicon solar cells. Sol. Energy **84**, 110–115 (2010)
8. N.-N. Feng, J. Michel, L. Zeng, J. Liu, C.-Y. Hong, L.C. Kimerling, X. Duan, Design of highly efficient light-trapping structures for thin-film crystalline silicon solar cells. IEEE Trans. Electron Dev. **54**(8), 1926–1932 (2007)

9. R.S. Dubey, S. Saravanan, S. Kalainathan, Performance enhancement of thin film silicon solar cells based on distributed Bragg reflector & diffraction grating. AIP Adv. **4**, 127121-1-6 (2014)
10. G. James, S.S. Mutitu, C. Chen, T. Crezzo, A. Barnett, C. Honsberg, D.W. Prather, Thin film silicon solar cell design based on photonic crystal and diffraction grating structure. Opt. Exp. **16**(19), 15238–15248 (2008)
11. S. Xiao, E. Stassen, N. Asger Mortensen, Ultrathin silicon solar cells with enhanced photocurrents assisted by plasmonic nanostructures. J. Nanophoton. **6**, 061503-1-061503-7 (2012)
12. Jinna He, Chunzhen Fan, Junqiao Wang, Yongguang Cheng, Pei Ding, and Erjun Liang, Plasmonic nanostructure for enhanced light absorption in ultrathin silicon solar cells, Adv. OptoElectron., Article ID 592754, doi:https://doi.org/10.1155/2012/592754, 8 (2012).
13. W. Yan, G. Min, Comparative study of light absorption enhancement in ultrathin a-Si:H solar cells with conformal parabolaconical nanoarrays. J. Opt. **16**, 045003–045006 (2014)
14. Z. Wang, T.P. White, K.R. Catchpole, IEEE Photon. J. **5**(5), 8400608–8400608 (2013)
15. F. Cortes-Juan, C. Chaverri Ramos, J.P. Connolly, C. David, F.J. García de Abajo, J. Hurtado, V.D. Mihailetchi, S. Ponce-Alcantara, J. Guillermo Sanchez, Renew. Sustain. Energy **5**, 033116–033112 (2013)
16. X. Sheng, H. Juejun, J. Michel, L.C. Kimerling, Light trapping limits in plasmonic solar cells: an analytical investigation. Opt. Exp. **20**(S4), A496–A501 (2012)
17. R. Chriki, A. Yanai, J. Shappir, U. Levy, Enhanced efficiency of thin film solar cells using a shifted dual grating plasmonic structure. Opt. Exp. **21**(S3), A382–A391 (2013)
18. P. Spinelli, V.E. Ferry, J. van de Groep, M. van Lare, M.A. Verschuuren, R.E.I. Schropp, H.A. Atwater, A. Polman, Plasmonic light trapping in thin-film Si solar cells. J. Opt. **14**, 024002–024011 (2012)
19. C.-C. Chao, C.-M. Wang, Y.-C. Chang, J.-Y. Chang, Plasmonic multilayer structure for ultrathin amorphous silicon film photovoltaic cell. Opt. Rev. **16**(3), 343–346 (2009)
20. S. Lee, S. Kim, Optical Absorption Characteristic in Thin a-Si Film Embedded Between an Ultrathin Metal Grating and a Metal Reflector. IEEE Photon. J. **5**(5), 4800610–4800611 (2013)
21. S. Saravanan, R.S. Dubey, S. Kalainathan, M.A. More, D.K. Gautam, Design and optimization of ultrathin crystalline silicon solar cells using an efficient back reflector. AIP Adv. **5**, 057160-1-9 (2015)
22. E. Battal, T.A. Yogurt, L.E. Aygun, A.K. Okyay, Triangular metallic gratings for large absorption enhancement in thin film Si solar cells. Opt. Exp. **20**(8), 9458–9464 (2012)
23. A. Abass, K.Q. Le, A. Alu, M. Burgelman, B. Maes, Dual-interface gratings for broadband absorption enhancement in thin-film solar cells. Physical Review B **85**, 115449-1-8 (2012)
24. Y. Shi, X. Wang, W. Liu, T. Yang, F. Yang, Hybrid light trapping structures in thin-film silicon solar cells. J. Opt. **16**, 075706-1-17 (2014)
25. R. Liu, Z. Xia, W. Yonggang, H. Jiao, Z. Liang, J. Zhou, Light trapping enhancement in thin film silicon solar cells with different front and back grating periodicities. Chinese Opt. Lett. **11** (12), 120501-1-3 (2013)
26. Saravanana, R.S. Dubey, Optical absorption enhancement in 40 nm ultrathin film silicon solar. Optics Commun. **377**, 65–69 (2016)
27. X. Guo, J. Liu, S. Zhang, Design of light trapping structures for ultrathin solar cells. Photon. Optoelectron. **3**, 66–68 (2014)
28. J.J. Schermer, P. Mulder, G.J. Bauhuis, M.M.A.J. Voncken, J. van Deelen, E. Haverkamp, P.K. Larsen, Epitaxial Lift-Off for large area thin film III/V devices. Phys. Stat. Sol **202**, 501–508 (2005)
29. H. Taguchi, T. Soga, T. Jimbo, Epitaxial lift-off process for GaAs solar cell on Si substrate. Solar Energy Mater. Solar Cells **85**(1), 85–89 (2005)
30. H.J. Kim, V. Depauw, G. Agostinelli, G. Beaucarne, J. Poortmans, Progress in thin film free-standing monocrystalline silicon solar cells. Thin Solid Films **511–512**, 411–414 (2006)
31. M. Reuter, W. Brendle, O. Tobail, J.H. Werner, 50 μm thin solar cells with 17.0% efficiency. Solar Energy Mate. Solar Cells **93**, 704–706 (2009)

# Optical Anisotropy and Compositional Ratio of Conductive Polymer PEDOT:PSS and Their Effect on Photovoltaic Performance of Crystalline Silicon/Organic Heterojunction Solar Cells

**Hajime Shirai, Qiming Liu, Tatsuya Ohki, Ryo Ishikawa, and Keiji Ueno**

**Abstract** We demonstrate the optical anisotropy of a transparent conductive polymer poly(3,4-ethylenedioxythiophene):poly(styrene sulfonate) (PEDOT:PSS) and its effect on the photovoltaic performance of n-type crystalline silicon (n-Si)/ PEDOT:PSS heterojunction solar cells. The depth profile of PEDOT/PSS compositional ratio and optical anisotropy depends on the type of polar solvent and/or external DC bias supplied to n-Si substrate during film deposition by spin coating (SC) and chemical mist deposition (CMD). N-Si/PEDOT:PSS heterojunction solar cells with higher PEDOT/PSS compositional ratio near the film surface exhibited better power conversion efficiency $\eta$ of 12.5% (@2 $\times$ 2 cm$^2$) without any light harvesting techniques. In this chapter, the correlation among the optical anisotropy, the depth profile of PEDOT/PSS compositional ratio in conductive polymer PEDOT:PSS, and the photovoltaic performance of n-Si/PEDOT:PSS heterojunction solar cells is demonstrated.

## 1 Introduction

N-type crystalline silicon (n-Si) solar cells require a hole-conducting layer to form the electron-hole separating junction; state-of-art junctions are realized by technologically activated methods, such as high-temperature boron diffusion [1] or plasma deposition of boron-doped hydrogenated amorphous silicon (a-Si:H) [2]. Today, crystalline silicon (c-Si) heterojunction solar cells with intrinsic a-Si:H thin films (HIT) have been also extensively studied for high performance devices with a power conversion efficiency $\eta$ of 22–23% [3, 4]. In 2014, a power conversion efficiency $\eta$ of 25.6% has been also demonstrated using back-contact and back-junction concept [5]. The a-Si:H and transparent conductive oxide (TCO) layer, in

H. Shirai (✉) · Q. Liu · T. Ohki · R. Ishikawa · K. Ueno
Graduate School of Science and Engineering, Saitama University, Saitama, Japan
e-mail: shirai@fms.saitama-u.ac.jp



S. Ikhmayies (ed.), *Advances in Silicon Solar Cells*,
https://doi.org/10.1007/978-3-319-69703-1_5

general, are fabricated by plasma-enhanced chemical vapor deposition (PE-CVD) and sputtering, respectively. Furthermore, a-Si:H causes parasitic light absorption, which can be reduced by implementing the junction on the rear surface of the solar cell [6].

Recently, the promising combination of inorganic and organic materials has led to the emerging research field of hybrid optoelectronic devices. An alternate approach is technologically much less demanding approach using an organic p-type layer, such as poly(3,4-ethylenedioxythiophene):poly(styrene sulfonate) (PEDOT:PSS) on the n-Si base [7–10]. PEDOT:PSS is a transparent hole-conducting polymer, which has been extensively studied for organic thin-film solar cells and electroluminescent devices [11–13]. Conductivities over 1500–2000 S/cm and work functions in the range between 4.8 and 5.2 eV can be realized with this organic material. PEDOT:PSS can be easily deposited by spin, spray, dip coating, and so on as low as 100 C. In addition, a single junction of n-Si/PEDOT:PSS works as a photovoltaic device without using a p-n junction and transparent conductive oxide (TCO) such as Indium tin oxide (ITO). They show a $\eta$ of 10–13% by adjusting polar solvent without using any light-harvesting technique such as texturing and antireflection (AR) coating layer. More recently, a $\eta$ of 20% has been established in n-Si/PEDOT:PSS junction device combining with well-established c-Si semiconductor technology [14]. Thus, we named this type of solar cell as a c-Si heterojunction with organic thin-layer "HOT" solar cell. Recently, several mechanisms of carrier transport in c-Si/PEDOT:PSS heterojunction have been proposed [15, 16]. However, the effect of the optical anisotropy, the molecular orientation, the depth profile of PEDOT/PSS compositional ratio, and the carrier transport at the n-Si/PEDOT:PSS interface is still not understood.

In this chapter, we demonstrate the optical anisotropy of transparent conductive polymer PEDOT:PSS, the depth profile of PEDOT/PSS compositional ratio, and its effect on the photovoltaic performance of n-Si/PEDOT:PSS heterojunction solar cells.

## 2 PEDOT:PSS

So far, PEDOT:PSS has been extensively studied as a hole-transporting layer and hole-injection layer, which has been applied for organic thin-film solar cells and electroluminescent devices. Figure 1a shows the molecular structure of PEDOT: PSS, which consists of the hydrophobic conductive PEDOT polymer coordinated with hydrophilic insulating PSS coil by ionic chemical bond. A schematic of the c-Si/PEDOT:PSS heterojunction solar cell structure used in this study is also shown in Fig. 1b. These films show a relatively high electrical conductivity of 1000–1500 S/cm by adding polar solvent such as methanol (MeOH) and ethylene glycol (EG) solvent alone or MeOH/EG cosolvent [17–19]. Furthermore, several reports also revealed that conductivity increased up to 2000~3000 S/cm by posttreatment using sulfonic acid, which was almost identical to that of ITO.

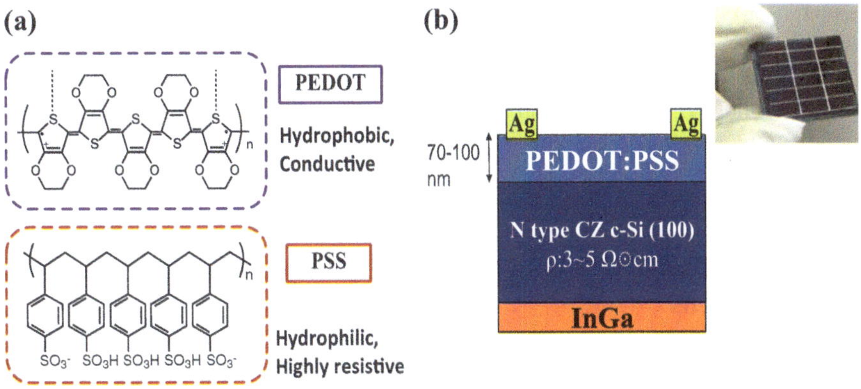


**Fig. 1** (**a**) Molecular structure of PEDOT:PSS, (**b**) device structure of n-S/PEDOT:PSS heterojunction solar cells

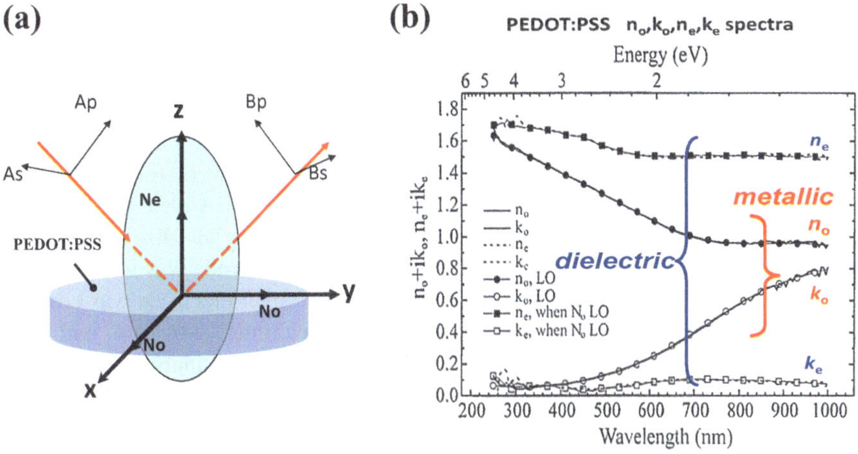


**Fig. 2** (**a**) Optical anisotropy of PEDOT:PSS, (**b**) complex refractive index spectra ($N = n + ik$) determined from the spectroscopic ellipsometry [20]

## 3 Optical Anisotropy of PEDOT:PSS

Optical point of view, PEDOT:PSS is optically anisotropic and of uniaxial character with the optical axis normal to the film surface as shown in Fig. 2a. In Fig. 2b, the complex index of refraction ($N = n + ik$) is shown determined from the spectroscopic ellipsometry (SE) by L. A. Pettersson in 1997 [20, 21]. The uniaxial optical anisotropy of PEDOT films reveals an entire different appearance for the ordinary and extraordinary indices of refraction. Major difference is in infrared region corresponding to free carrier absorption. The former show a metallic character and the dielectric character in the latter, although its degree depends on the preparation condition and solvent. Thus, PEDOT:PSS shows a metallic character in the in-plane with a dielectric in the normal to the film surface. These features are one of origins that a single PEDOT:PSS/c-Si junction works as a photovoltaic device without p-n junction and TCO layer, although the resistivity of n-Si and passivation quality also influence on the photovoltaic performance.

## 4 Fabrication of c-Si/PEDOT:PSS Heterojunction Solar Cells

Commercialized PEDOT:PSS (Clevios™ PH1000) was used as a source material with a 2.5 wt% PEDOT/PSS. One-side-polished N-type CZ c-Si(100) (n-Si) with a resistivity of 3–5 Ω·cm and a thickness of 300 μm was used as a substrate. The c-Si wafers were RCA cleaned using SC-1 and SC-2 treatments and then dipped in 5% hydrofluoric (HF) acid to remove any native oxide. 5 wt% dimethyl sulfoxide (DMSO) was added to filter (0.45 μm pore size) purified PEDOT:PSS to render it conductive. In general, soluble PEDOT:PSS is not coated uniformly on hydrophobic c-Si. The usage of a 0.1% Zonly-treated soluble PEDOT:PSS composite improved the adhesion of precursor solution on hydrophobic c-Si wafer without any oxidation process [21]. The blend solution was made using MeOH and EG as a cosolvent with different weight ratios. PEDOT:PSS films were formed by spin coating (SC) and chemical mist deposition (CMD) on RCA-cleaned c-Si(100) wafers at 1000 rpm for 60 s, followed by thermal annealing at 140°C for 30 min to remove residual solvent. Finally, the top silver grid electrode was formed using screen printing method Fig. 1b. The outer area of the silver electrode was covered with an opaque mask to block the incident light. Electronic, structural, and morphological characterizations of the PEDOT:PSS films were analyzed using SE, atomic force microscopy (AFM: SII nanotechnology, SPA-300/SPI-3800), and X-ray photoelectron spectroscopy (XPS: ULVAC-PH1). The solar cell performance was characterized by current density-voltage (*J*-V) measurements in the dark and under AM1.5G illumination, 100 mW/cm$^2$ of simulated solar light (Bunkoukeiki CEP-25BX). The optical properties of several PEDOT:PSS films with different solvent treatment was determined from SE measurements at an angle of incidence at $\theta_i = 70°$, in the energy region of 0.8–5.0 eV.

## 5 Optical Anisotropy of PEDOT:PSS by SC Using MeOH/EG Cosolvent

Figure 3a shows measured *n* and extinction coefficient *k* spectra denoted by $<n>$ and $<k>$ of ~100-nm-thick pristine, MeOH-modified, and MEOH/EG cosolvent-modified PEDOT:PSS films on n-Si at a probe light incidence angle of $\theta_i = 70°$. No

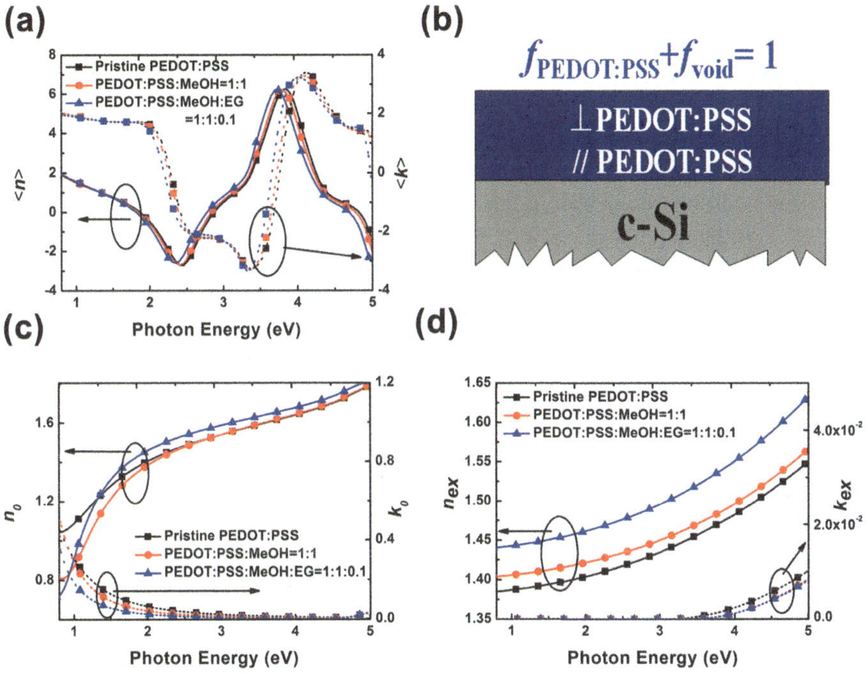


**Fig. 3** (**a**) The <n> and <k> spectra of n-Si/PEDOT:PSS films: pristine, MeOH-modified, and MeOH/EG-modified PEDOT:PSS films on n-Si at a $\theta_i$ of 70°. (**b**) Optical model used. (**c**) $n_o$ and $k_o$ spectral components and (**d**) $n_{ex}$ and $k_{ex}$ spectral components for the corresponding cosolvent-modified PEDOT:PSS films

marked differences in the <n> and <k> spectra were observed, suggesting no in-plane optical anisotropy. The <n> and <k> spectra were analyzed using a Tauc-Lorentz (TL) oscillator and a Drude model taking the optical anisotropy into consideration. The optical model used for the spectra analysis is shown in Fig. 3b, which consists of a single layer with uniaxial optical anisotropy: ||PEDOT:PSS (ordinary, ($n_o$,$k_O$)) and ⊥PEDOT:PSS (extraordinary, ($n_{ex}$,$k_{ex}$)). The probable structure was determined by minimizing the mean squared error between the measured and calculated values of the ellipsometric parameter $\chi^2$ using a linear regression method:

$$\chi^2 = \frac{1}{2N - M} \sum_{i=1}^{N} \left[ \left( \tan \Psi_i^c - \tan \Psi_i^m \right)^2 - \left( \cos \Delta_i^c - \cos \Delta_i^m \right)^2 \right], \qquad (1)$$

where the superscripts $c$ and $m$ represent calculated and measured values, and $N$ and $M$ are the numbers of the measured and calculated wavelength, respectively. The best fitted ($n_O$, $k_O$) and ($n_{ex}$, $k_{ex}$) spectra for the bulk component in pristine, MeOH-modified, and cosolvent-modified PEDOT:PSS films are also shown in Fig. 3c, d, respectively. The $n_o$ and $k_o$ spectra showed metallic behavior, whereas dielectric characteristics were more dominant for the $n_{ex}$ and $k_{ex}$ spectra of all samples. The $n_o$ and $k_o$ spectra components showing metallic character were nearly independent of the MeOH/EG concentration ratio. On the other hand, change in the $n_{ex}$ and $k_{ex}$ spectra components was prominent. In particular, the $k_o$ spectra in the infrared region of 0.8–1.0 eV corresponding to free carrier absorption were slightly changed [22]. Figure 3c implies that the free carrier concentration $N$ was almost same by cosolvent addition. In contrast, $n_{ex}$ was significantly increased by adding cosolvent, with no significant changes in $k_{ex}$. The standard optical transmittance/reflectance spectra of PEDOT:PSS films were nearly independent of EG concentration.

The XRD peak feature around $2\theta = 26^{\circ}$ assigned to (020) plane of an orthorhombic unit cell of PEDOT [23] enhanced, and the film crystallization promoted for MeOH-modified and MeOH/EG cosolvent-modified PEDOT:PSS rather than that of pristine PEDOT:PSS despite of thinner film thickness (Fig. 4a). Figure 4b shows the XPS spectra of S (2p) core energy region of PEDOT:PSS fabricated with and without using cosolvent addition. Here, high-energy peak appeared at 169 eV is due to S atom in PSS, and low-energy peak appeared at 164 eV is PSS. The spectra are shown referred to the high-energy peak at 169 eV in PSS. The spectra revealed that compositional ratio PEDOT/PSS was higher near the surface region in cosolvent-modified films. It is reported that polar organic solvent with high dielectric constant can suppress the electrostatic interaction between PEDOT and PSS by screening effect and promote the interaction among PEDOT molecules to form the crystalline structure [24]. These findings imply that the cosolvent addition enhances the phase separation of PEDOT-PSS polymer chains and improves the crystallinity of PEDOT molecules; thus, promoted densification of the PEDOT:PSS network may be achieved which can affect the optical anisotropy of corresponding PEDOT:PSS thin films.

In Fig. 4c, the conductivity measured by $\sigma = 1/R_{sh}d$, in which the sheet resistance $R_{sh}$ was determined by four-point probe measurement and film thickness $d$ was observed from SEM characteristic, is summarized for pristine, MeOH-modified, and MeOH/EG cosolvent-modified PEDOT:PSS films of three different samples. The conductivity increased slightly for solvent-modified PEDOT:PSS, whereas single EG addition was not sufficient. Wang et al. reported that the electrical conductivity of PEDOT:PSS films was increased by EG mixing and reached a maximum at an EG concentration of 10–15 wt% [25]. Thus, there exists an appropriate MeOH:EG concentration ratio that minimizes the sheet resistance.

The carrier transport parameters, i.e., the free carrier concentration $N$ and carrier mobility $\mu$, were determined from plasma frequency $\omega_p(=Ne^2/\varepsilon_0 m^*)^{0.5}$ and $\mu = e/m^*\Gamma_d$ relations, where $\Gamma_d$ is a damping factor determined by the dielectric constant $\varepsilon(E) = \varepsilon_\infty\text{-}[A_d/(E^2 + \Gamma_d^2)]$. Here, $\mu$ represents the average hole mobility within an average-sized grain [26, 27]. Thus, $\mu$ proportional to $1/\Gamma_d$ does not include the influence of scattering at the grain boundary. $\varepsilon_\infty$ and $A_d$ are the dielectric constant $\varepsilon$ in the high-frequency region and a constant, respectively. Here, an effective mass $m^*$ of $0.82m_0$ was used, where $m_0$ is the free electron mass [28].

$\omega_p$, $N$, and $\mu$ were 1.41 eV, $1.18 \times 10^{21}$ cm$^{-3}$, and 1.23 cm$^2$/Vs for the pristine film; 1.38 eV, $1.13 \times 10^{21}$ cm$^{-3}$, and 2.14 cm$^2$/Vs for the MeOH modified film; and

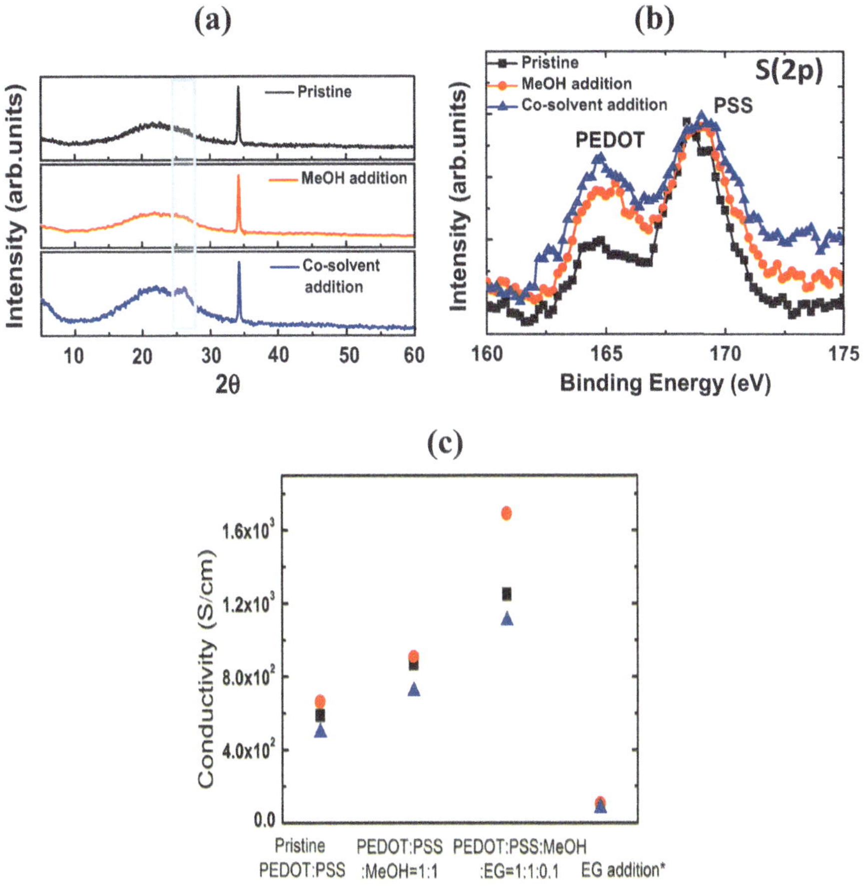


**Fig. 4** (**a**) XRD pattern, (**b**) XPS spectra of S(2p) core energy region in PEDOT and PSS, (**c**) conductivities of pristine, MeOH-modified, and cosolvent-modified PEDOT:PSS films. Conductivity was plotted in three different samples

1.44 eV, $1.23 \times 10^{21}$ $cm^{-3}$, and 3.92 $cm^2/Vs$ for the MeOH/EG cosolvent film, respectively. These $\mu$ values determined from the SE spectra agreed well with those based on conductivity ($\sigma = Ne\mu$) using the results shown in Fig. 4c. Therefore, the SE analysis supports the optical constant as well as the carrier transport parameter.

Figure 5a, b shows $J$-V characteristics observed in the dark and under AM1.5G illumination of 100 mW/$cm^2$ simulated solar light, as well as external quantum efficiency (EQE) spectra from c-Si heterojunction solar cells based on pristine, MeOH-modified, and MeOH/EG cosolvent-modified PEDOT:PSS. The photovoltaic parameters, i.e., open-circuit voltage $V_{oc}$, short-circuit current density $J_{sc}$, fill factor FF, and $\eta$ are summarized in Table 1 for devices treated with MeOH/EG cosolvents containing different amounts of EG (by wt%). The dark current density $J_d$ was significantly suppressed by MeOH mixing Fig. 5a. Therefore, the reduction

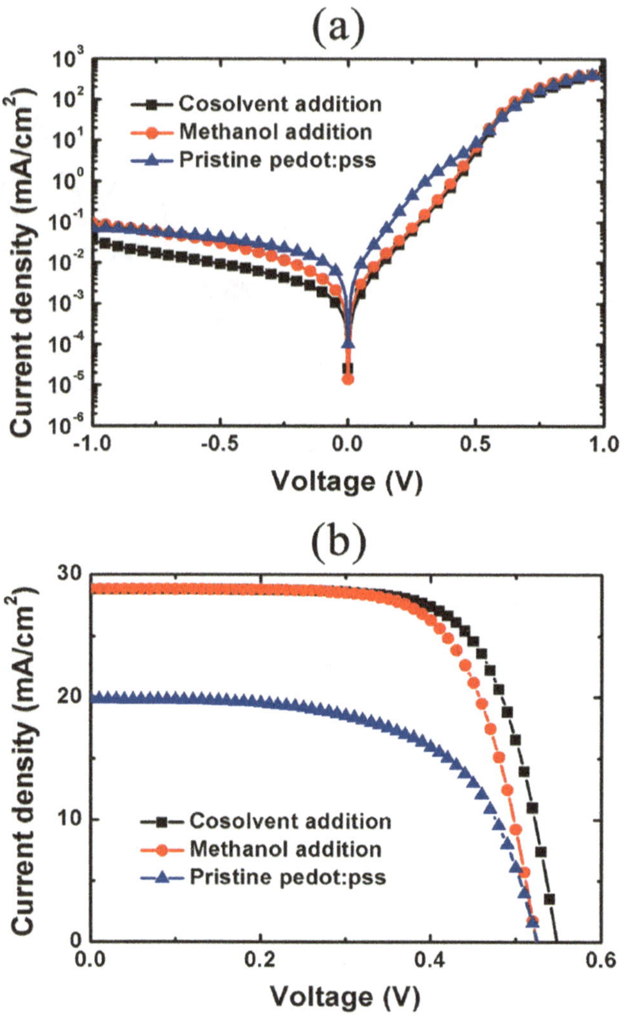


**Fig. 5** *J*-V curves for n-Si/PEDOT:PSS heterojunction solar cells with pristine, MeOH-modified, and cosolvent-modified buffer layers (**a**) in the dark and (**b**) under simulated AM1.5G illumination

in the forward and reverse dark currents originated from the MeOH solvent. Furthermore, the reduction of dark current is a clear evidence of the suppression of electron recombination. These results imply that the n-Si/PEDOT:PSS interface property can be improved by using MeOH as a solvent without creating additional defects.

The devices with a pristine PEDOT:PSS layer exhibited an average $\eta$ of 6.38%, whereas an MeOH/EG-modified device demonstrated an improved $\eta$ of 11.23% with a $J_{sc}$ of 28.79 mA/cm$^2$, a $V_{oc}$ of 0.548 V, and an FF of 0.711 Fig. 5b. Obviously, the improvements in device performance can be attributed to the use

**Table 1** Performance parameters ($J_{sc}$, $V_{oc}$, FF, $\eta$, $R_s$, and $R_{sh}$) for n-Si/PEDOT:PSS heterojunction solar cells with different MeOH and EG concentrations. The concentration is shown as the wt% ratio of (PEDOT:PSS):(MeOH):(EG)

| Conc. | $J_{sc}$(mA/cm$^2$) | $V_{oc}$(V) | *F.F.* | $\eta$ (%) | $R_s$ (Ω cm$^2$) | $R_{sh}$ (kΩ cm$^2$) |
|---|---|---|---|---|---|---|
| 1:0:0 | 19.85 | 0.525 | 0.612 | 6.38 | 0.72 | 13.0 |
| 1:1:0 | 29.02 | 0.527 | 0.669 | 10.23 | 0.77 | 16.7 |
| 1:1:0.1 | 28.79 | 0.548 | 0.711 | 11.23 | 0.64 | 56.3 |
| 1:0:0.1 | 4.71 | 0.355 | 0.453 | 0.76 | 7.06 | 3.9 |
| 1:1:0.3 | 13.88 | 0.459 | 0.612 | 4.20 | 1.12 | 10.2 |
| 1:1:0.5 | 3.45 | 0.480 | 0.644 | 1.07 | 1.38 | 10.6 |

of the MeOH/EG cosolvent. As a result, the photovoltaic performance of n-Si/PEDOT:PSS heterojunction solar cells was improved for the PEDOT:PSS layer showing higher electrical conductivity Fig. 4c.

On the other hand, the conductivity showed a maximum at a weight ratio of (PEDOT:PSS):(MeOH):(EG) of 1:1:0.1. The n-Si heterojunction with cosolvent-modified PEDOT:PSS that demonstrated the maximum conductivity exhibited a relatively high power conversion efficiency of 11.23%. These findings suggest that the increased extraordinary index of refraction contributes to the enhanced hole mobility of PEDOT:PSS, resulting in improved photovoltaic performance.

## 6 Depth Profile of PEDOT/PSS Compositional Ratio of Spin-Coated PEDOT:PSS by Substrate DC Bias Supply

The optical anisotropy and the depth profile of the PEDOT/PSS compositional ratio could be also controlled by external DC bias $V_s$ supplied to c-Si substrate during the film growth using SC and chemical mist deposition (CMD) methods [29].

Figure 6 shows the depth profile of PEDOT/PSS compositional ratio determined from the XPS S (2p) core level spectra for negative and positive DC-biased PEOT:PSS films fabricated at different $V_s$s. Here, the DC bias was supplied to n-Si substrate between −5 kV and +10 kV referred to mesh ground electrode. The depth profile of compositional ratio of conductive PEDOT to insulating PSS was almost constant. On the other hand, the compositional ratio was higher near the film surface for negative DC-biased film compared to that of the SC film without $V_s$, whereas it was lower for positively DC-biased films. Thus, the depth profile of compositional ratio PEDOT/PSS could be controlled by adjusting $V_s$ and its polarity during SC.

Figure 7a, b shows the *J*-V characteristics in the dark and under illumination of AM1.5G 100 mW/cm$^2$ simulated solar light for corresponding SC n-Si/PEDOT:PSS hybrid devices synthesized at three different $V_s$ supplies. The photovoltaic parameters, i.e., $V_{oc}$, $J_{sc}$, FF, and $\eta$ are summarized for the corresponding devices in

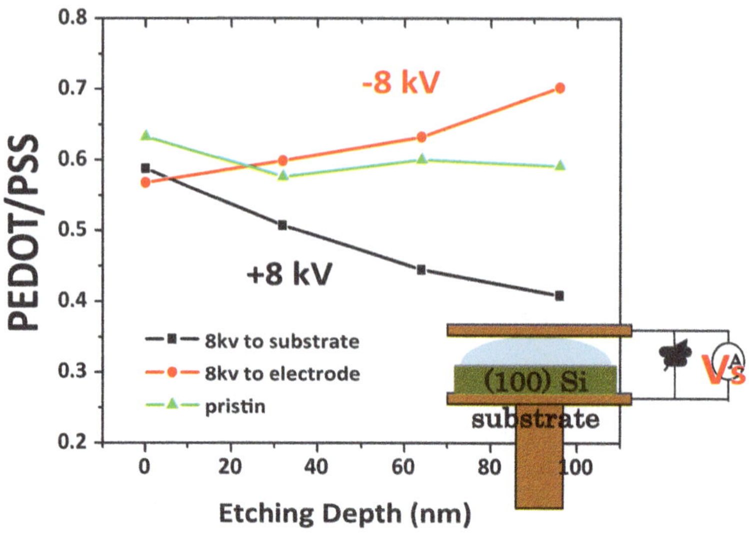


**Fig. 6** Depth profile of PEDOT/PSS compositional ratio determined from XPS characterization for SC PEDOT:PSS films fabricated at different $V_s$ of 0, +8 kV and −8 kV. The $V_s$ was supplied to n-Si substrate referred to mesh electrode set a 2 cm apart from c-Si substrate

Table 2. It is seen that dark current density $J_d$ in the reverse bias region is suppressed to be lower for the positive DC bias supplied device (Fig. 7a). The calculated series resistance $R_s$ showed a minimum of 0.75 Ω·cm$^2$ with a highest shunt resistance $R_{sh}$ of $1.2 \times 10^4$ Ω·cm$^2$ for the +8 kV supplied SC PEDOT:PSS diode. The reduction in the dark current is a clear evidence for reduced electron current and the suppression of electron recombination. These imply that the junction property is improved by the +8 kV supply during SC without creating additional defects at the junction interface compared to those of pristine and −8 kV devices. The RMS value was 1.53 nm in 10 × 10 μm$^2$ area at a $V_s$ of +8 kV, whereas it increased to 3.04 nm at a $V_s$ of −8 kV, suggesting that the film densification promoted by supplying positive DC bias.

Better photovoltaic performance was also obtained at positive $V_s$ supply to n-Si substrate rather than those of pristine and negatively biased devices (Fig. 7b). Thus, there exists the correlation between the depth profile of PEDOT/PSS compositional ratio and the photovoltaic performance. As a consequence, PEDOT:PSS films with higher PEDOT/PSS compositional ratio near the film surface exhibited higher photovoltaic performance. These findings are explained as follows. It is well known that PSS is one of good passivation layers of c-Si, which exhibit a minority carrier lifetime of 3–5 ms, which is almost same order of a-Si:H. On the other hand, XPS revealed that the composition ratio of PSS/PEDOT was higher near the film surface for SC film. Thus, the positive DC bias supply $V_s$ in the formation of SC PEDOT:PSS promoted the preferential sticking of PSS at the n-Si surface, which

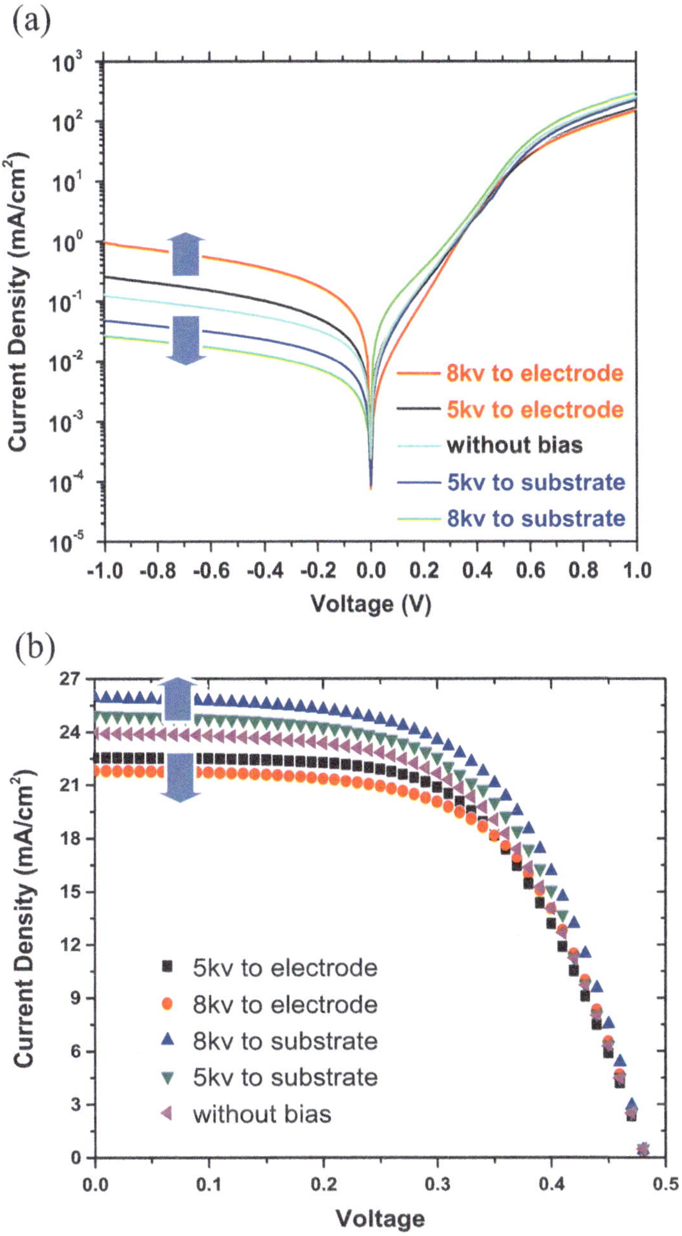


**Fig. 7** *J*-V curves (**a**) in the dark and (**b**) under AM1.5G simulated solar light exposure for n-Si/PEDOT:PSS heterojunction solar cells fabricated at different $V_s$ supplies during the SC

**Table 2** Performance parameters ($J_{sc}$, $V_{oc}$, and $\eta$) for n-Si/PEDOT:PSS heterojunction solar cell devices with three different $V_s$ values of 0, +8, and −8 kV. RMS value for corresponding films is also shown

| $V_s$ | $J_{sc}$ (mA/cm$^2$) | $V_{oc}$ (V) | $\eta$ (%) | RMS (nm) |
|---|---|---|---|---|
| Without bias | 23.97 | 0.474 | 6.75 | 2.19 |
| − 8 kV | 22.55 | 0.482 | 5.84 | 3.04 |
| + 8 kV | 25.98 | 0.478 | 8.84 | 1.53 |

contributed to better passivation quality of n-Si. Thus, conductive PEDOT:PSS acts as passivation layer as well as hole conductive layer.

## 7 Chemical Mist Deposition (CMD) of PEDOT:PSS with Negatively Charged Mist Precursor

For further increase of the photovoltaic performance, efficient light management using antireflection (AR) layer and/or textured Si substrate by solution process is required to suppress the optical loss. However, the SC PEDOT:PSS on textured n-Si substrate results in inhomogeneous film deposition. To this aim, several attempts have been made for the uniform deposition of PEDOT:PSS on textured n-Si substrate using several deposition techniques such as electrospray deposition (ESD) [30–33] and chemical mist deposition (CMD) [34–41]. Among the several deposition techniques, the gas phase deposition is a possible candidate for the uniform deposition of organic on textured Si substrate. We attempted the uniform deposition of PEDOT:PSS on textured c-Si(100) substrate using the CMD with negatively charged mist precursor for efficient PEDOT:PSS/n-Si heterojunction solar cells.

A schematic of the experimental setup for the CMD is shown in Fig. 8a. It consists of ultrasonic atomizer, transportation tube (glass tube), and substrate stage. The atomization frequency $f$ used was 2.4 or 3 MHz, which produces mist precursor, whereby transportation tube with an inner tube diameter of 10 mm passes the mist to the substrate stage using $N_2$ as a carrier gas at ambient. Aqueous solution of 5 wt% dimethyl sulfoxide (DMSO) added conductive PEDOT:PSS solution with DI water ($\varepsilon = 78$, $T_b = 100°C$), and ethylene glycol (EG) ($\varepsilon = 38.6$, $T_b = 226°C$) (10 wt%) mixture was sealed in the reactor on an ultrasonic vibrator. A mist of PEDOT:PSS transported through the glass tube using nitrogen (200–600 sccm) as a carrier gas and passed through the mesh electrode which was placed 2 cm away from the textured c-Si substrate.

Figure 9 shows the Mie scattering image of PEDOT:PSS mist using DI water/EG (10%) cosolvent at different $N_2$ gas flow, substrate temperature $T_s$, and $V_s$ conditions. The increase of $N_2$ flow promoted the traveling of mist to n-Si substrate up to 400 sccm. Further increases of $N_2$ flow, however, suppressed the sticking of mist precursor at the substrate because of its reflection. The mist flux impinging to n-Si

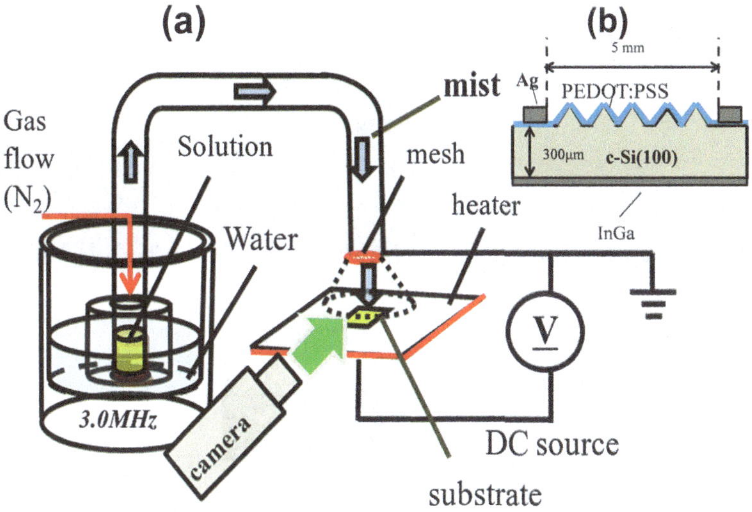


**Fig. 8** (**a**) Schematic of chemical mist deposition (CMD) apparatus, (**b**) device structure of PEDOT:PSS/textured n-Si heterojunction solar cells [29]

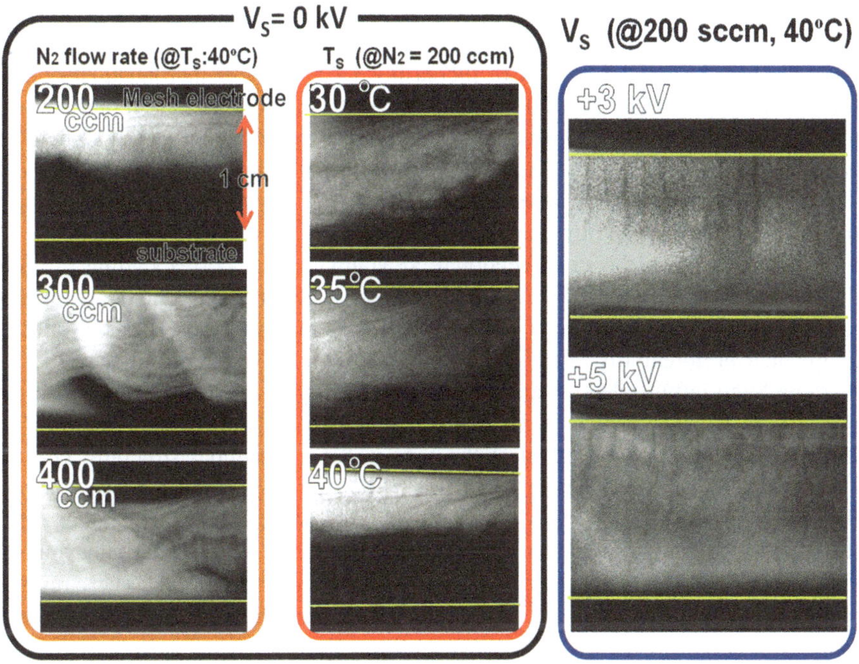


**Fig. 9** Mie scattering images of PEDOT:PSS mist with cosolvent DI water and 10 wt% EG at different $N_2$ flow, $T_s$, and $V_s$ conditions [29]

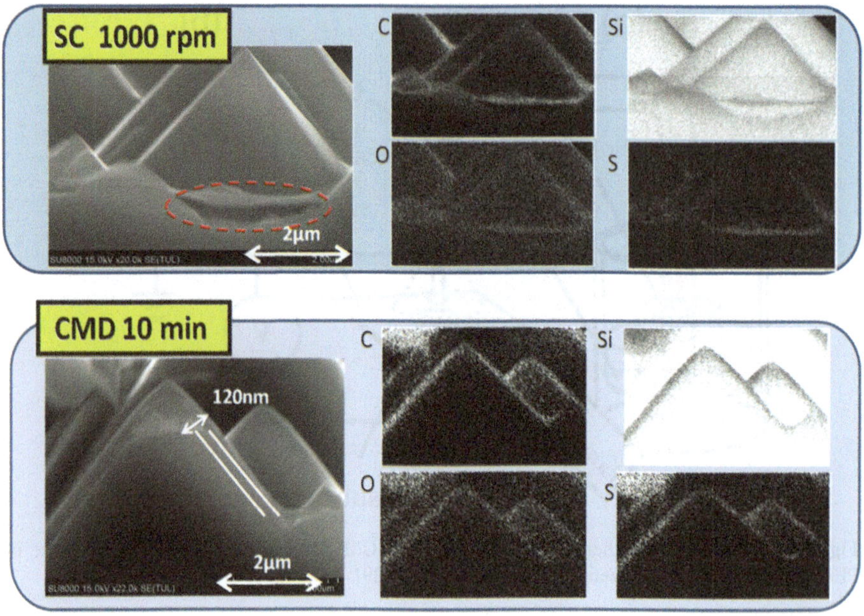


**Fig. 10** SEM image and EDS map, C, Si, O, and S elements of SC and CMD PEDOT:PSS films deposited on textured n-Si substrate. The thickness of PEDOT:PSS was ~100 nm [29]

substrate was also suppressed with increasing $T_s$ because of thermal floating. In fact, no film deposition occurred at a $T_s$ above 50°C. In addition, the average size of mist particle decreased, and mist flux and speed increased significantly, when $V_s$ was impressed at positive to c-Si substrate. Whereas no film deposition occurred at $V_s$ impressed at negative. These findings imply that the negatively charged mist mainly contribute to the film deposition.

Figure 10 shows the cross-sectional SEM image and corresponding energy dispersive spectroscopy (EDS) mapping image of C, Si, O, and S elements for SC and CMD PEDOT:PSS films on textured c-Si substrate. In the SC films, products aggregated preferentially at the valley region of each pyramidal structure. They were preferentially peeled off after thermal annealing at 140°C for 30 min, which was due to weak adhesion of PEDOT:PSS on c-Si. Inhomogeneous coating of PEDOT:PSS was also observed at the side wall of pyramidal structure, i.e., Si (111) surface. On the other hand, in the CMD, the uniform and homogeneous deposition of PEDOT:PSS was observed up to a ~100-nm thickness rather than the SC films. A scratch test using Kempton tape also showed a better adhesion of CMD films on flat and textured c-Si substrates rather than those of SC. These results imply that the adhesion of PEDOT:PSS on n-Si enhanced using negatively charged mist precursor.

Figure 11a shows the time evolution of the $<n>$ and $<k>$ spectra collected during deposition of a PEDOT:PSS film on a flat c-Si wafer. Deposition was carried out for a total of 60 s with a $V_s$ of 3 kV and a $T_s$ of 40 C. At $t = 0$ s, the spectra are for

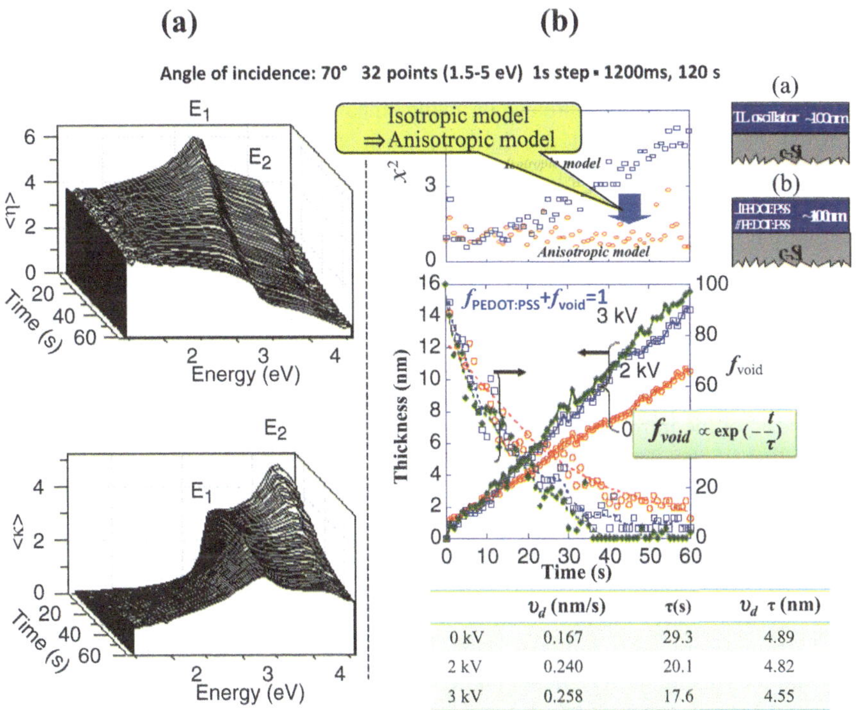


| | $v_d$ (nm/s) | τ(s) | $v_d$ τ (nm) |
|---|---|---|---|
| 0 kV | 0.167 | 29.3 | 4.89 |
| 2 kV | 0.240 | 20.1 | 4.82 |
| 3 kV | 0.258 | 17.6 | 4.55 |

**Fig. 11** (**a**) Time evolution of the <*n*> and <*k*> spectra during the PEDOT:PSS film growth on n-Si for the first 60s. The CMD of PEDOT:PSS was performed for D of 2 cm and $V_m$ of 3 kV. The $N_2$ flow rate was 500 sccm. (**b**) Time evolutions of $d$, $f^{o(e)}_{PEDOT:PSS}$, $f_{void}$, and $\chi^2$ values of PEDOT: PSS determined from spectral fitting using the isotropic and anisotropic optical models [19, 20]. The optical models used in this study are shown on the right. The $\chi^2$ values determined from the isotropic optical model are also shown for comparison [42]

the c-Si substrate and exhibit fine structures at 3.3 and 4.2 eV, attributed to $E_1$ and $E_2$ optical interband transitions, respectively. These fine structures became weaker as the deposition time increases. In the UV region, the magnitude of <*n*> and <*k*> at the UV energy regions decrease more rapidly with increasing deposition time, because they are more sensitive to film surface roughness rather than the bulk. The <*n*> and <*k*> spectra were analyzed using a TL models with and without the presence of anisotropy.

Figure 11b shows the deposition time dependence of the film thickness $d$ and the volume fraction of void $f_{void}$ determined by fitting the spectra in Fig. 11a using $V_s$ values, 0, 2, and 3 kV. The upper graph shows the $\chi^2$ values obtained using the isotropic and anisotropic TL models with $V_s$ = 3 kV. During the first 10–15 s, corresponding to a film thickness of about 5 nm, similar $\chi^2$ values of about unity were obtained using both models. However, for thicker films, large differences were found between the results obtained using the two models. From lower graph, $d$ is seen to increase roughly linearly with deposition time. As $V_s$ is varied from 0 to

3 kV, the deposition rate $v_d$ changes from 0.19 to 0.29 nm/s, when $V_s$ was varied from 0 to 3 kV. The $f_{void}$ values are initially large but decreased to zero following 30–40 s of deposition, corresponding to a film thickness of 5–10 nm.

The fits to the experimental data using linear and exponential functions were carried out to understand the initial growth stage at different $V_s$ conditions. Attempts have been made to correlate distortion and stress relaxation in polymer of the two [43, 44]. As seen in Fig. 11b, the best fits were obtained when the time evolution of the void fraction was expressed using the exponential function, $f_{void} \propto \exp(-t/\tau)$, where $t$ is the deposition time, and $\tau$ represents the time constant for the introduction of bulk anisotropy. In the initial stages of film growth, the PEDOT:PSS chains have random orientations due to island growth and the presence of substrate roughness. The results obtained by the exponential fits in Fig. 11b are summarized in table in Fig. 11b. The threshold thickness $d_t$, which is the critical thickness for the introduction of anisotropy and is equal to $v_d \cdot \tau$, was 4.68–5.39 nm, depending on $V_s$. Therefore, bulk anisotropy appears for film thicknesses above ~5 nm. Based on the results of X-ray scattering, the most likely structure of PEDOT:PSS has been reported to be PEDOT crystallites with average dimensions of about 5 nm surrounded by PSS [24]. This is similar to the value of the critical thickness determination in the present study.

To further clarify the introduction of optical anisotropy, films with thicknesses of 2–16 nm were subjected to variable angle spectroscopic ellipsometry between 1.5 and 5 eV using the TL model taking optical anisotropy into consideration [42]. Figure 12 shows the bulk ($n_o$, $k_o$) and ($n_e$, $k_e$) spectra for a 5-nm-thick film. Upper figure shows the $\chi^2$ values calculated using isotropic and anisotropic models at different angles of incidence $\theta_i$s. The $\chi^2$ values were almost the same for isotropic and anisotropic models at a $\theta_i$ value of 70°, but their differences were prominent with decreasing $\theta_i$. Lower $\chi^2$ values were obtained for anisotropic model, although both ($n_o$, $k_o$) and ($n_e$, $k_e$) spectra were almost unchanged for $\theta_i$ values of 55–65°, but they changed at a $\theta_i$>70°, suggesting less degree of anisotropy, i.e., randomly oriented polymer network and/or island growth.

Figure 13a, b shows the bulk ($n_o$, $k_o$) and ($n_e$, $k_e$) spectra for 9- and 16-nm-thick films, respectively. $\chi^2$ values calculated using isotropic and anisotropic models at different $\theta_i$ values are summarized for corresponding films on the top, showing a significant difference with decreasing $\theta_i$. The $\chi^2$ values were almost the same for isotropic and anisotropic models at a $\theta_i$ value of 70°, but their differences were more prominent with decreasing $\theta_i$. The $n_o$ and $k_o$ spectra were almost independent of $\theta_i$ for 9- and 16-nm thicknesses, whereas $n_e$ and $k_e$ values increased with increasing $\theta_i$. Here, only the $n_o$ and $k_o$ spectra at a $\theta_i$ of 70° are illustrated for 9- and 16-nm-thick films. The $n_e$ and $k_e$ values clearly increase with $\theta_i$. In contrast, $n_o$ and $k_o$ values are almost independent of $\theta_i$, so only the spectra at $\theta_i = 70°$ are shown. The $\chi^2$ value calculated using the isotropic and anisotropic optical models for different angels of incidence $\theta_i$ are shown in Fig. 6b, and it can be seen that whereas the two models give similar values for $\theta_i = 70°$, the difference between them becomes significant with decreasing $\theta_i$, with the anisotropic model giving lower values. Thus, these results revealed that optical anisotropy is introduced

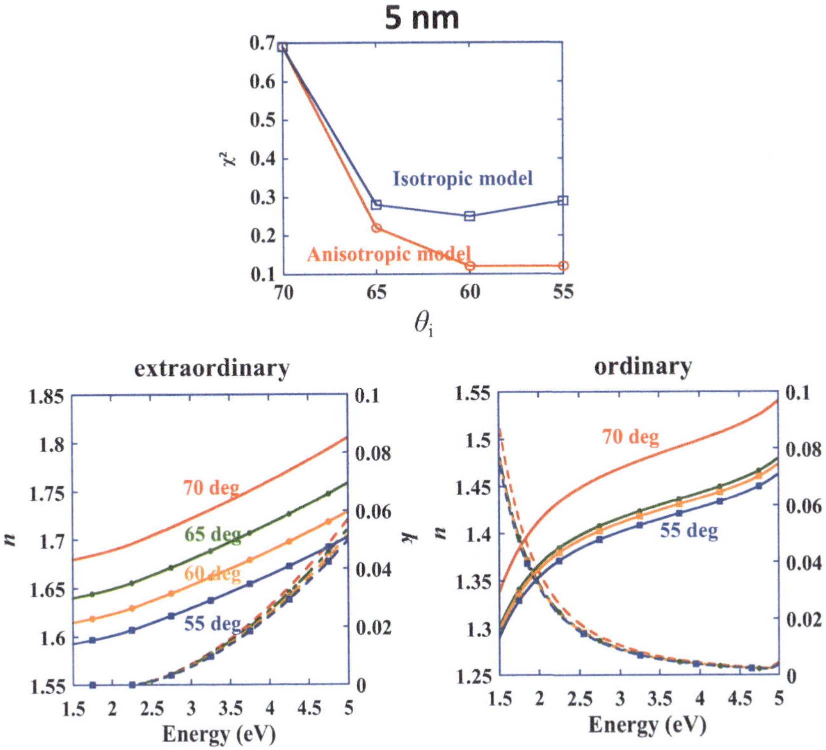


**Fig. 12** (Top) $\chi^2$ values plotted as a function of $\theta_i$ for 5-nm-thick PEDOT:PSS film using isotropic and anisotropic optical models. (Bottom) ($n_o$, $k_o$) and ($n_e$, $k_e$) spectra for PEDOT:PSS bulk components determined at different $\theta_i$ values [42]

when the PEDOT:PSS polymer units become aligned in the film plane, which begins to occur for a film thickness of about 5 nm.

Figure 14 shows the best fits to the ($n_o$, $k_o$) and ($n_e$, $k_e$) spectra of 30-nm-thick films. For all $V_s$ values, the $n_o$ and $k_o$ spectra indicate metal-like behavior and are almost independent of $V_s$ except low-energy region where free carrier absorption occurs. This is particularly the case for the $k_o$ spectra. On the other hand, dielectric-like behavior is indicated by the $n_e$ and $k_e$ spectra. In contrast to $k_e$, $n_e$ is seen to increase significantly with $V_s$. Thus, films fabricated at higher $V_s$ are more anisotropic, which implies that the degree of anisotropy can be controlled by adjusting $V_s$. In fact, in the X-ray diffraction (XRD) patterns for the samples, the intensity of a peak is at around $2\theta = 26°$, which was associated with the (020) planes in the orthorhombic unit cell of PEDOT [26]. XPS also indicated that the depth profile of PEDOT-PSS depended on $V_s$, and the PEDOT/PSS concentration ratio was higher near the surface at a higher $V_s$ of 3 kV, which was identical to that of Fig. 4.

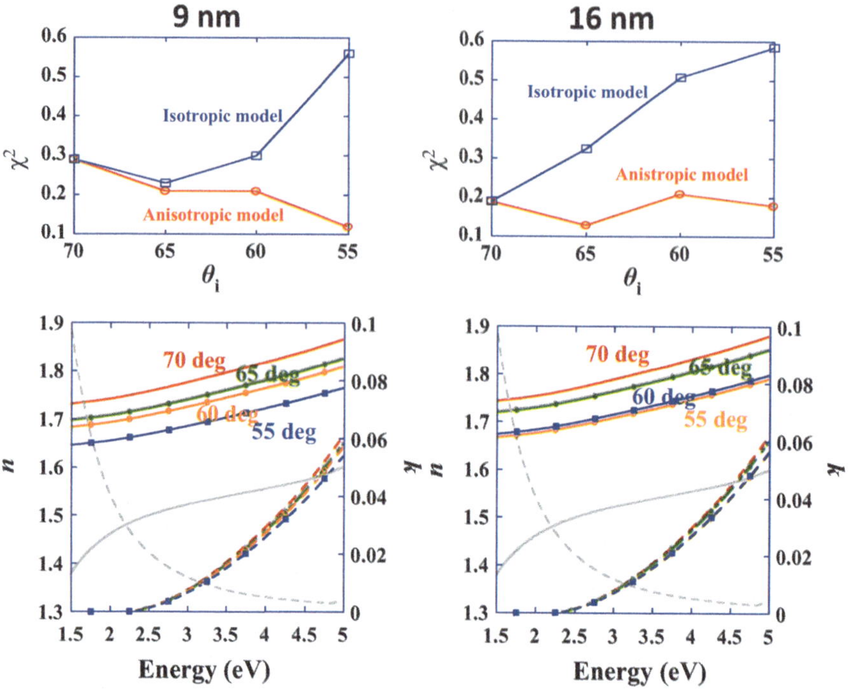


**Fig. 13** (Top) $\chi^2$ values plotted as a function of incident angle, $\theta_i$ for 9- and 16-nm-thick PEDOT:PSS films using isotropic and anisotropic optical models. (Bottom) The ($n_o$, $k_o$) and ($n_e$, $k_e$) spectra a for 9- and 16-nm-thick films determined at different $\theta_i$ values

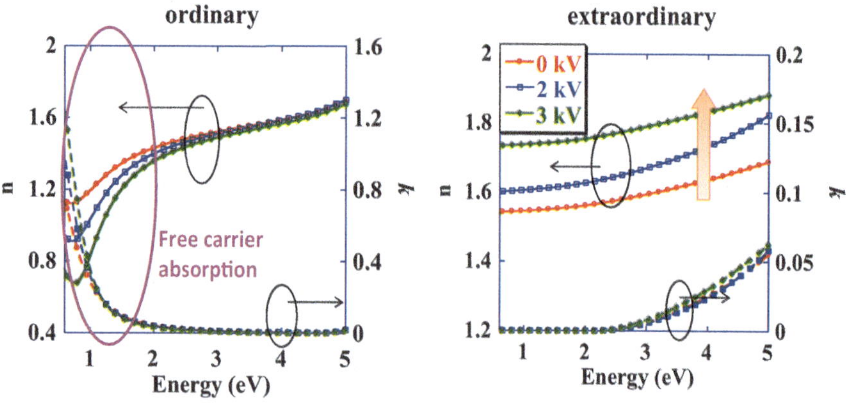


**Fig. 14** The ($n_o$, $k_o$) and ($n_e$, $k_e$) spectra for PEDOT:PSS bulk components for different $V_s$ values

Figure 15a shows the *J*-V curves for the CMD and SC PEDOT:PSS (100 nm thickness)/n-Si solar cells fabricated on flat c-Si(100) substrate under exposure of AM1.5G, 100 mW/cm$^2$ simulated solar light. The CMD device was fabricated at a $V_s$ of +5 kV supply and a $T_s$ of 40 C. A $\eta$ of 10.4% was obtained for both SC and

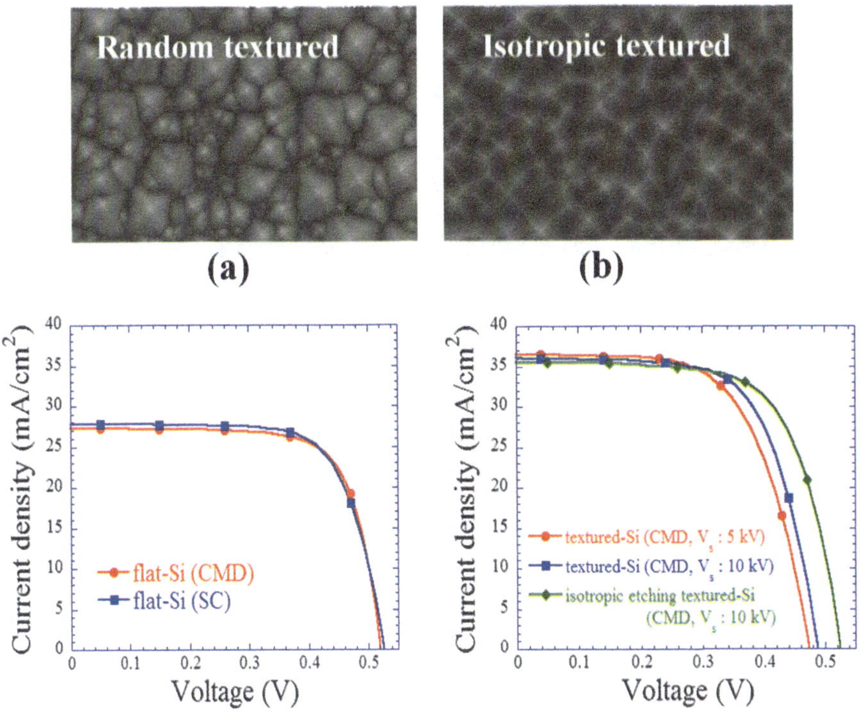


**Fig. 15** (**a**) *J*-V curve of the SC and CMD PEDOT:PSS/n-Si device formed on flat Si(100) substrate using DI water/EG cosolvent at a $T_s$ of 40 C under exposure of AM1.5G, 100 mW/cm$^2$ simulated solar light. (**b**) *J*-V curve of the CMD PEDOT:PSS/n-Si devices with two different $V_s$ values of 5 and 10 kV on random textured n-Si (shown on top). The *J*-V curve of the CMD device at a $V_s$ of 10 kV on isotropic-textured c-Si with an AR coating layer of a 20-nm-thick $MoO_x$ is also included. The random and isotropic textured c-Si substrates are shown on the top

CMD devices on flat n-Si substrate with a $J_{sc}$ of 27.3–27.8 mA/cm$^2$, a $V_{oc}$ of 0.52–0.53 V, and a FF of 0.71–0.73 by adjusting the deposition parameters. In Fig. 16b, the *J*-V curves for the CMD PEDOT:PSS/randomly textured n-Si/PEDOT:PSS heterojunction solar cells are shown at two different $V_s$ values of 5 and 10 kV. The SEM images of randomly and isotropic textured n-Si are also shown on the top. The photovoltaic parameters are summarized in Table 3. A $\eta$ of 11.0% was obtained with a $J_{sc}$ of 36.5 mA/cm$^2$, a $V_{oc}$ of 0.47 V, and a FF of 0.63 for a $V_s$ of 5 kV. A $\eta$ also increased to 11.6% with a $J_{sc}$ of 36.1 mA/cm$^2$, a $V_{oc}$ of 0.49 V, and a FF of 0.66, when a $V_s$ increased from 5 to 10 kV. In addition, $V_{oc}$ increased to 0.52 V without decreasing $J_{sc}$ for the isotropic etched textured c-Si device. The increased $V_{oc}$ and FF for the higher $V_s$ supplied device originate from the increased adhesion of PEDOT:PSS on c-Si. Furthermore, a $\eta$ increased up to 12.5% (2 × 2 cm$^2$) with a $J_{sc}$ of 35.6 mA/cm$^2$, a $V_{oc}$ of 0.53 V, and a FF of 0.67 with an antireflection (AR) coating layer of 20-nm-thick CMD molybdenum oxide $MoO_x$ ($n$ = 2.1) using negatively charged mist of a 0.1 wt% 12 Molybdo

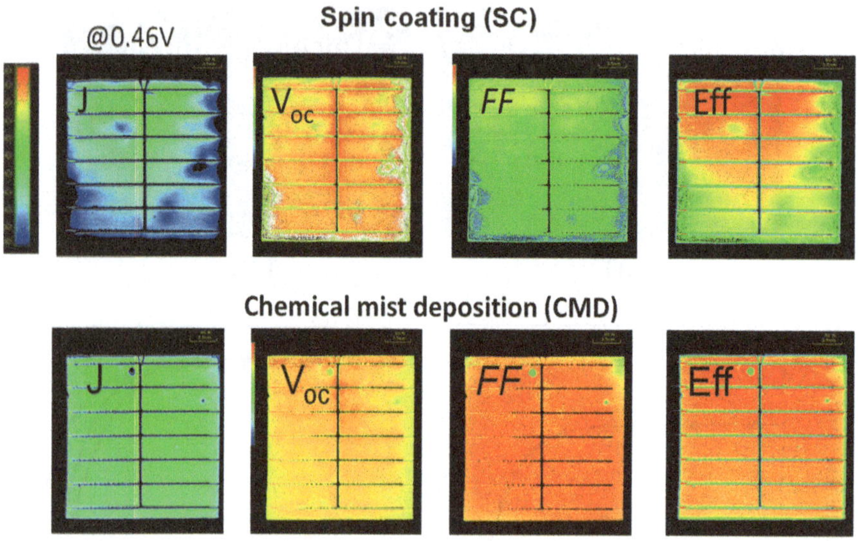


**Fig. 16** 2-dimensional map ($2 \times 2$ $cm^2$) of solar cell parameters: photocurrent density $J$ at 0.46 V, $V_{oc}$, FF, and $\eta$ for SC and +5 kV supplied CMD PEDOT:PSS on random textured n-Si solar cells. The relative peak heights of the $J$, $V_{oc}$, FF, and $\eta$ are shown as a legend next to $J$ map in the SC.

**Table 3** Photovoltaic performance of CMD n-Si/PEDOT:PSS heterojunction solar cell devices on flat and textured c-Si substrates

| | $J_{sc}$ (mA/$cm^2$) | $V_{oc}$ (V) | FF | $\eta$ (%) |
|---|---|---|---|---|
| Flat-Si (CMD) | 27.3 | 0.52 | 0.73 | 10.4 |
| Flat-Si (SC) | 27.8 | 0.53 | 0.71 | 10.4 |
| Randomly textured-Si ($V_s$: 5 kV) | 36.5 | 0.47 | 0.63 | 11.0 |
| Randomly textured-Si ($V_s$: 10 kV) | 36.1 | 0.49 | 0.66 | 11.6 |
| Isotropic textured-Si ($V_s$: 10 kV) | 35.6 | 0.53 | 0.67 | 12.5 |

(VI) phosphoric acid n-Hydrate $H_3(PMo_{12}O_{40}) \cdot nH_2O$ in methanol. These findings imply that the CMD with negatively charged mist has a great potential for the uniform deposition of PEDOT:PSS/$MoO_x$ heterostructure on textured n-Si.

Figure 16 shows the 2-dimensional map ($2 \times 2$ $cm^2$) of solar cell parameters: current density $J$ at 0.46 V, $V_{oc}$, FF, and $\eta$ for SC and CMD 100-nm-thick PEDOT:PSS devices formed on textured c-Si substrate. Apparently, the 2D map of all solar cell parameters, $J$ at 0.46 V, $V_{oc}$, FF, and $\eta$ showed better uniformity in the CMD device compared to those in SC. In particular, the PV performance was extremely poor around edge region rather than the central region in the SC device due to the inhomogeneous film thickness by the SC. In addition, scratch test of CMD films showed increased adhesion compared to the SC. Thus, the increased photovoltaic performance originate from the increased homogeneity and adhesion of PEDOT:PSS on textured c-Si substrate by the CMD using negatively charged mist precursor.

For further increasing $\eta$, in particular, to increase $V_{oc}$, the back surface field (BSF) or back-contact and back-junction-type device structure using thinner-thick Si wafer is a possible candidate. Improved stability for air storage and light soaking in the c-Si/PEDOT:PSS heterojunction solar cell device is also required for high reliability. To these problems, several approaches have been attempted and under progress.

## 8 Concluding Remarks

We demonstrated the effect of optical anisotropy and the depth profile of PEDOT/PSS compositional ratio of conductive polymer PEDOTPSS on the photovoltaic performance of n-Si/PEDOT:PSS heterojunction solar cells. The optical anisotropy and the depth profile of PEDOT/PSS compositional ratio can be controlled using MeOH/EG cosolvent and external DC bias impressed to c-Si substrate during the film growth using SC and CMD. The higher compositional ratio of PEDOT/PSS near the film surface exhibited better photovoltaic performance of c-Si/PEDOT: PSS heterojunction solar cells. The MeOH/EG cosolvent and positive $V_s$ supply during the film growth provide the densification of PEDOT:PSS network with increased $\mu$. Real-time monitoring of optical anisotropy during chemical mist deposition (CMD) of PEDOT:PSS using negatively charged mist precursor was carried out using spectroscopic ellipsometry. The substrate bias voltage $V_s$ was found to increase the film deposition rate and densify the PEDOT:PSS polymer network. Bulk uniaxial anisotropy began to appear for film thicknesses above 5 nm, which corresponds to the PEDOT/PSS composition ratio also depended on $V_s$, and the ratio was higher values of $V_s$. The corresponding c-Si/PEDOT:PSS heterojunction solar cell exhibited a relatively efficient $\eta$ of 12.5%, with a $J_{sc}$ of 28.79 mA/cm$^2$, a $V_{oc}$ of 0.548 V, and an FF of 0.71. Furthermore, a $\eta$ increased to 12.5% with a $J_{sc}$ of 35.6 mA/cm$^2$, a $V_{oc}$ of 0.53 V, and a FF of 0.67 using AR coating layer of CMD $MoO_x$ on isotropic textured c-Si substrate.

**Acknowledgments** This study was partially supported by a grant from the Japan Science and Technology Agency (JST) and by a Grant-in-Aid for Scientific Research from the Ministry of Education, Culture, Sports, Science and Technology of Japan. The authors wish to express their appreciation to Mr. Masayuki Sakurai and Mrs. Yoko Wasai of Horiba-Jobin Yvon Ltd., for the SE measurements. The authors also thank Mrs. Takashi Miyamoto and Koji Funato of Tokyo Direc Co. Ltd., for the allowance of the use of differential mobility analyzer. The one of authors (HS) wish to thank my colleague, Drs. Ishwor Khatri, Jaker Hossain, and Mrs. Shuji Funada, Koki Ichikawa, Kyohei Watanabe, Taiga Hiate, Kyohei Ishikawa, and Yoshinori Imamura for their efforts to this study.

## References


1. P.J. Cousins, D.D. Smith, H.C. Luan, J. Manning, T.D. Dennis, A. Waldhauer, K.E. Wilson, G. Harlrey, G.P. Muligan Gen III in Proceedings of the 35th IEE. Photovoltaic specialist conference, Honolulu, USA, p. 275–278 (2010)
2. K. Kinoshita, D. Fujishima, A. Yano, A. Ogane, S. Tohoda, K. Matsuyama, Y. Nakamura, N. Tokuoka, H. Kanno, H. Sakata, M. Taguchi, E. Maruyama, Y. Nakamura In: Proceedings of the 26th European photovoltaic solar energy conference, Hamburg, Germany, p. 871 (2011)
3. Y. Tsunomura, Y. Yoshimine, M. Taguchi, T. Baba, T. Kinoshita, H. Kanno, H. Sakata, E. Maruyama, M. Tanaka, Sol. Energy Mater. Sol. Cells **96**, 032105 (2010)
4. Q. Wang, M.R. Page, E. Iwanicko, Y. Xu, L. Royabal, R. Bauer, B. To, H.–.C. Yuan, A. Duda, F. Hasioon, Y.F. Yan, D. Levi, D. Meier, H.M. Branz, T.H. Wang, Appl. Phys. Lett. **96**, 013507 (2010)
5. K. Masuko, IEEE J.Photovoltaics **4**, 1433 (2014)
6. N. Mingirulli, J. Haschke, R. Gogolin, R. Ferré, T. Schulze, J. Düsterholt, N.-P. Harder, L. Korte, R. Gogolin, R. Rech, Phys. Status Solidi RPL **5**, 159 (2011)
7. L. He, C. Jiang, H. Wang, D. Lai, Rusli, Appl. Phys. Lett. **100**, 073503 (2012)
8. I. Khatri, Z. Tang, Q. Liu, R. Ishikawa, K. Ueno, H. Shirai, Appl. Phys. Lett. **102**, 063508 (2013)
9. Z. Tang, Q. Liu, I. Khatri, R. Ishikawa, K. Ueno, H. Shirai, Phys. Status Solidi **C9**, 2075 (2012)
10. J.Y. Chen, M.-H. Yu, S.-F. Chang, K.W. Sun, Appl. Phys. Lett. **103**, 133901 (2013)
11. A. Elschner, S. Kirchmeyer, W. Lovenich, U. Merkerm, K. Reuter, *PEDOT -Principle and Applications of an Intrinsically Conductive Polymer* (CRC Press, Taylor and Francis Group, New York, 2011)
12. Y. Xia, J. Ouyang, J. Mater. Chem. **21**, 4297 (2011)
13. H. Okuzaki, M. Ishihara, S. Ashizawa, Synth. Met. **137**, 947 (2003)
14. D. Zielke, C. Niehaves, W. Lövenicj, A. Elschner, M. Hörteis, J. Schmidt, Ener Procedia **77**, 331 (2015)
15. S. Jäckle, M. Mattiza, M. Liebhaber, G. Bronstrup, M. Rommel, K. Lips, S. Christiansen, Sci. Rep. **13008** (2015)
16. D. Zielke, A. Pazidis, F. Wernet, J. Schmidt, Sol. Ener. Mater. & Sol. Cells **131**, 110 (2014)
17. Y. Xia, K. Sun, J. Ouyang, Adv. Mater. **24**, 2436 (2012)
18. N. Kim, S. Kee, S.H. Lee, B.H. Lee, Y.H. Kahung, Y.-R. Jo, B.-J. Kim, K. Lee, Adv. Mater. **26** (14), 2268–2272, 2109 (2014)
19. S. Mukher, R. Singh, S. Gopinathan, S. Murugan, S. Gawali, B. Saha, J. Biswas, S. Lodha, A. Kumar, ACS Appl. Mater. Interfaces **6**(20), 17792–17803 (2014)
20. L.A.A. Pettersson, F. Carlsson, O. Inagäs, H. Arwin, Thin Solid Films **313-314**, 356 (1998)
21. L.A.A. Pettersson, S. Ghosh, O. Inagäs, Org. Electron. **3**, 143 (2002)
22. Q. Liu, M. Ono, Z. Tang, R. Ishikawa, K. Ueno, H. Shirai, Appl. Phys. Lett. **100**, 183901 (2012)
23. X. Crispin, F.L. Kakobsson, A. Crispin, P.C. Grim, P. Andersson, A. Volodin, C. van Hasendonck, N. Van der Auweraer, W.R. Salaneck, M. Berggren, Chem. Mater. **18**, 4354 (2006)
24. H. Okuzaki, Y. Harashina, Eur. Polym. J. **45**, 256 (2009)
25. T.J. Wang, Y.Q. Qi, J.K. Xu, X.J. Hu, P. Chen, Appl. Surf. Sci. **250**, 188 (2005)
26. H. Fujiwara, M. Kondo, Phys. Rev. **B71**, 075109 (2005)
27. M. Yamashita, C. Otani, M. Shimizu, H. Okuzaki, Appl. Phys. Lett. **99**, 143307 (2011)
28. E. Shanthi, V. Dutta, A. Banerjee, K.L. Chopra, J. Appl. Phys. **51**, 6243 (1980)
29. J. Hossain, T. Ohki, K. Ichikawa, K. Fujiyama, K. Ueno, Y. Fujii, T. Hanajiri, H. Shirai, Jpn. J. Appl. Phys. **55**, 061602 (2016)
30. K. Fujita, T. Ishikawa, T. Tsutsumi, Jpn. J. Appl. Phys. **41**, L70 (2002)
31. X. Mo, T. Mizukoshi, A. Kobayashi, G. Chen, N. Tanigaki, H. Hiraga, Thin Solid Films **516**, 1663 (2008)

32. T. Fukuda, T. Suzuki, R. Kobayashi, Z. Honda, N. Kamata, Thin Solid Films **518**, 575 (2009)
33. T. Hiate, T. Ino, R. Ishikawa, K. Ueno, H. Shirai, Jpn. J. Appl. Phys. **51**, 10NE30 (2012)
34. Y. Kamada, T. Kahaharamura, H. Nishinaka, S. Fujita, Jpn. J. Appl. Phys. **45**, L857 (2006)
35. J.-Q. Lu, T. Kawaharamura, H. Nishinaka, Y. Kamada, T. Ohshima, S. Fujita, J. Cryst. Growth **299**(1), 1–10 (2007)
36. Y. Kamada, T. Kawaharamura, H. Nishinaka, S. Fujita, Mater. Res. Soc. Proc. **957**, K07–K27 (2007)
37. H. Nishinaka, T. Kawaharamura, S. Fujita, Jpn. J. Appl. Phys. **46**, 6811 (2007)
38. T. Ino, M. Ono, N. Miyauchi, Q. Liu, Z. Tang, R. Ishikawa, K. Ueno, H. Shirai, Jpn. J. Appl. Phys. **51**, 061602 (2012)
39. A. Uehara, The technical report in CHALLENGE Co. Ltd. (2010)
40. I. Khatri, T. Imamura, A. Uehara, R. Ishikawa, K. Ueno, H. Shirai, Phys. Status Solidi C **9**, 2134 (2012)
41. K. Shanmugasundaram, A study of the mist deposition and pattering of liquid precursor thin films, Doctor thesis in The Pennsylvania State University (2008)
42. T. Hiate, N. Miyauchi, Q. Liu, R. Ishikawa, K. Ueno, H. Shirai, J. Appl. Phys. **115**, 123514 (2014)
43. P.J. Flory, *Principles of Polymer Chemistry* (Cornell University Press, New York, 1953)
44. B. Vollmert, *Polymer Chemistry* (Springer, Berlin, 1973)

# Flexible, Stretchable, and Biodegradable Thin-Film Silicon Photovoltaics

**Xing Sheng, Shuodao Wang, and Lan Yin**

**Abstract** This chapter provides an overview of recent progress in the study of thin-film Si photovoltaic (PV) technologies, specifically devices in flexible, stretchable, and/or degradable formats for biomedical applications. First, various approaches for synthesis, fabrication, and assembly of different types of thin-film Si PV cells, ranging from amorphous Si, microcrystalline Si to single-crystalline Si cells, are reviewed. Performances of various cells are also discussed. Materials selection and mechanical considerations are presented, both analytically and numerically, for achieving design flexible and stretchable Si PV cells for lightweight, wearable, and/or implantable light detection and energy harvesting systems. Finally, strategies to form thin-film Si PV cells in a water-soluble and biodegradable "transient" format for bio-integration are discussed. Because of these novel characteristics, thin-film Si PV cells offer a promising solution for energy supply in emerging biomedical applications.

## 1 Introduction

Silicon (Si)-based solar cells have dominated most of the photovoltaic (PV) market, because of the abundance of Si and the mature technology associated with Si-based semiconductor devices. Si solar cells have achieved wide applications for terrestrial solar energy harvesting, from rooftop mounted solar panels to large utility-scale

X. Sheng (✉)
Department of Electronic Engineering, Tsinghua University, Beijing, China
e-mail: xingsheng@tsinghua.edu.cn

S. Wang
Department of Mechanical and Aerospace Engineering, Oklahoma State University, Stillwater, OK, USA
e-mail: shuodao.wang@okstate.edu

L. Yin
School of Materials Science and Engineering, Tsinghua University, Beijing, China
e-mail: lanyin@tsinghua.edu.cn



S. Ikhmayies (ed.), *Advances in Silicon Solar Cells*,
https://doi.org/10.1007/978-3-319-69703-1_6

solar farms. Compared to their bulk wafer counterparts, thin-film Si cells with thicknesses from tens of micrometers to a few hundred nanometers not only reduce the cell cost by using less material but also offer various attractive features associated with the thin-film device geometry. Advanced materials processes, mechanical design tools, and micro-/nanoelectronic fabrication methods enable us to build thin-film Si PV devices and systems in flexible, stretchable, and/or biodegradable formats. Because of their light weight and compatibility with soft, curvilinear, and/or even wet systems (especially, biological systems), such thin-film Si PV cells provide promising solutions to light detection and power generation for a variety of applications including wearable or implantable biomedical electronic and photonic systems.

## 2 Strategies of Forming Thin-Film Si PV Cells

In contrast to thick Si wafer-based solar cells that are produced by cutting and sawing process, thin-film Si cell fabrication requires advanced microelectronic processing methods. In addition, those methods should also be compatible with substrate materials that enable flexible and/or stretchable functionalities. In this section, synthesis and fabrication methods for making thin-film Si solar cells, including amorphous Si (a-Si), microcrystalline Si (μc-Si), as well as single-crystalline Si (c-Si), are summarized. The cell performance is also discussed and compared.

### *2.1 Amorphous and Microcrystalline Si PV Cells*

Vacuum-based deposition methods allow us to produce a-Si- and μc-Si-based thin-film single-junction and multi-junction solar cells on various substrates, as illustrated in Fig. 1a. These thin films are usually formed by plasma-enhanced chemical vapor deposition (PECVD) (Fig. 1b) [1]. Gases combining silane ($SiH_4$) and hydrogen ($H_2$) are introduced into a vacuum reactor chamber. Gas discharge plasma is initiated by using a voltage bias. Positively charged ions (e.g., $SiH_3^+$ and $H^+$) bombard the substrate and create hydrogenerated amorphous silicon (a-Si:H) that is deposited onto the substrate. In this process, the incorporation of hydrogen is crucial since hydrogen passivates dangling bond defects in a-Si by forming Si-H bonds, greatly reducing the defect density. Doping of a-Si:H can be accomplished by introducing gases like phosphine ($PH_3$) and diborane ($B_2H_6$) to form n-type and p-type a-Si:H layers, respectively. Therefore, by alternating the dopants, a vertical p-i-n junction can be created as the active device layer. One of the key advantages of using a-Si:H stems from its high optical absorption coefficients in the visible spectral range ($10^4$–$10^5$ $cm^{-1}$). Therefore, an a-Si:H p-i-n junction with a thickness of a few hundreds of nanometers is sufficient to absorb most of the solar energy

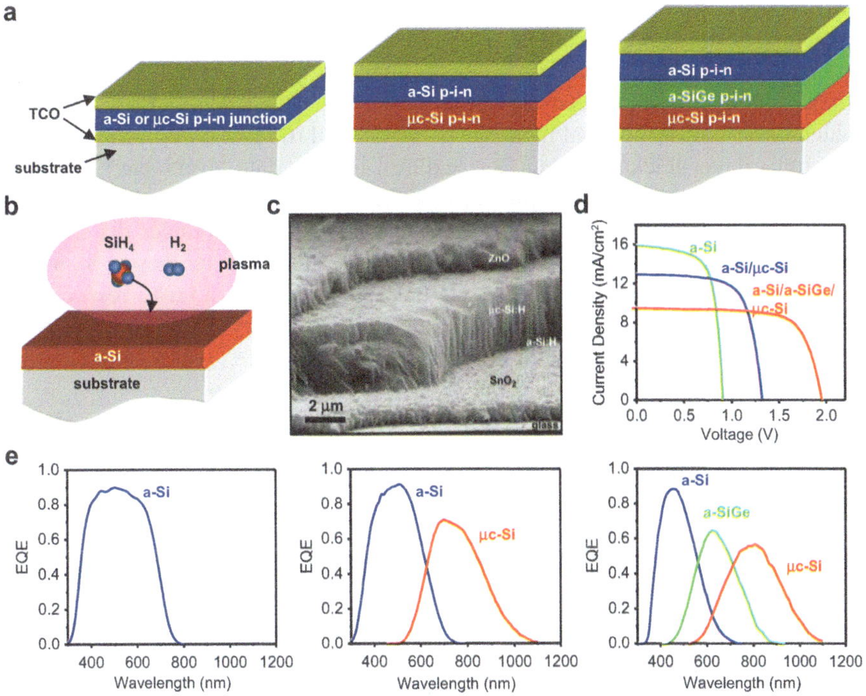


**Fig. 1** (**a**) Schematic illustration of thin-film a-Si, a-Si/μc-Si, and a-Si/a-SiGe/μc-Si solar cell structures. (**b**) PECVD process for forming thin-film a-Si. (**c**) SEM image of an a-Si/μc-Si cell [2]. (**d**) Current density–voltage curves for a-Si, a-Si/μc-Si, and a-Si/a-SiGe/μc-Si cells with world-record efficiencies under the AM1.5g spectrum [4–6]. (**e**) External quantum efficiency (EQE) spectra for a-Si, a-Si/μc-Si, and a-Si/a-SiGe/μc-Si cells with world-record efficiencies [4–6] (Reproduced with permission. Copyright 1999 Springer. Copyright 2013 Wiley-VCH. Copyright 2013 Elsevier)

above the optical bandgap of a-Si:H (~1.6 eV). Contact electrodes are fabricated by depositing transparent conductive oxides (TCOs) such as indium tin oxide (ITO), aluminum-doped zinc oxide (AZO), or fluorine-doped tin oxide (FTO). In the PECVD process, the induced plasma enables the film deposition at relatively low temperatures (200–400 °C), which makes the process compatible with various flexible substrates such as metal (stainless steel, aluminum, etc.) and plastic films.

Single-junction a-Si solar cells with a bandgap of about 1.6 eV covers the spectral range from about 300 nm to 800 nm (Fig. 1e). In order to utilize photons with lower energies and achieve higher cell efficiencies, multi-junction cell architectures can be realized by modifying the PECVD process and incorporate other semiconductor junctions. Microcrystalline (μc-Si) solar cells with a bandgap of ~1.1 eV can be formed by varying the PECVD process conditions (plasma frequency, chamber pressure, gas flow rate, etc.), which allow the fabrication of a-Si/μc-Si double-junction "micromorph" cells (Fig. 1a). Figure 1c shows a characteristic cross-sectional scanning electron microscopic (SEM) image of an a-Si/μc-Si double-junction cell [2]. In addition, germanium (Ge) can be introduced by adding germane ($GeH_4$) into the gas flow. The formed a-SiGe alloy has a tunable bandgap from 1.6 eV to 0.8 eV by varying the germanium concentration [3]. Therefore, triple-junction a-Si/a-SiGe/μc-Si (1.6 eV / 1.4 eV / 1.1 eV) cells can be fabricated, which cover a wider spectral range from 300 nm to 1100 nm. By adjusting the cell structures (layer thicknesses, surface textures, intermediate layers, etc.), cell performance can be optimized. Figure 1d plots the current–voltage characteristics for the world record a-Si, a-Si/μc-Si, and a-Si/a-SiGe/μc-Si cells under the standard AM1.5 g spectrum [4–6]. By far, maximum 1-sun efficiencies for these a-Si, a-Si/μ c-Si, and a-Si/a-SiGe/μc-Si cells are 10.2%, 12.7%, and 13.6%, respectively [7]. Typical spectral responses (external quantum efficiency, EQE) for these a-Si, a-Si/μc-Si, and a-Si/a-SiGe/μc-Si cells are plotted in Fig. 1e [4–6].

## 2.2 *Single-Crystalline Si PV Cells*

As discussed above, thin-film a-Si-based single- and multi-junction solar cell technologies have been explored for decades and gained great success in various commercial productions, from portable consumer devices to large scale on-grid and off-grid power plant systems [1, 8]. Despite these accomplishments, a-Si- and μc-Si-based thin-film cells still have efficiencies inferior to their single-crystalline (c-Si) counterparts, mostly because of the high defect densities associated with the dangling bonds and grain boundaries. In addition, a-Si-based cells exhibit degradation (Staebler-Wronski effect) due to light-induced metastable defects [1]. Therefore, it is desirable to explore thin-film Si cells in a c-Si format to obtain low-cost, mechanically flexible, and highly efficient modules at the same time. However, it is challenging to fabricate thin-film c-Si devices with a thickness less than 50 μm based on conventional wafer sawing and dicing techniques. In this part, an overview of some very recent progress on the fabrication of thin-film c-Si solar cells using unconventional "kerf-less" methods is provided. These methods enable the isolation of thin-film Si membranes or devices (from tens of nanometers to tens of micrometers thick) from thick Si wafers and all the Si membranes to be integrated with flexible supporting substrates for further uses.

One way to obtain free-standing thin-film Si devices is to start with silicon-on-isolator (SOI) wafer substrates. As illustrated in Fig. 2a, the $SiO_2$ insulating layer can be selectively removed by hydrofluoric acid (HF)-based wet etching process: $SiO_2 + 6HF \rightarrow 2H_2O + 2H^+ + SiF_6^{2-}$, while the Si top layer and the substrate remain intact. Specific photoresist structures serve as anchors, holding the released devices in suspended forms at their original locations after complete undercut [9]. For example, Fig. 2b shows the SEM image of a completely undercut SOI mesa structure, with photoresist anchor structures underneath the Si [9]. After etching, the unetched Si top layer can be released from the Si substrate and transferred onto various new substrates [10]. Based on this technique, a variety of

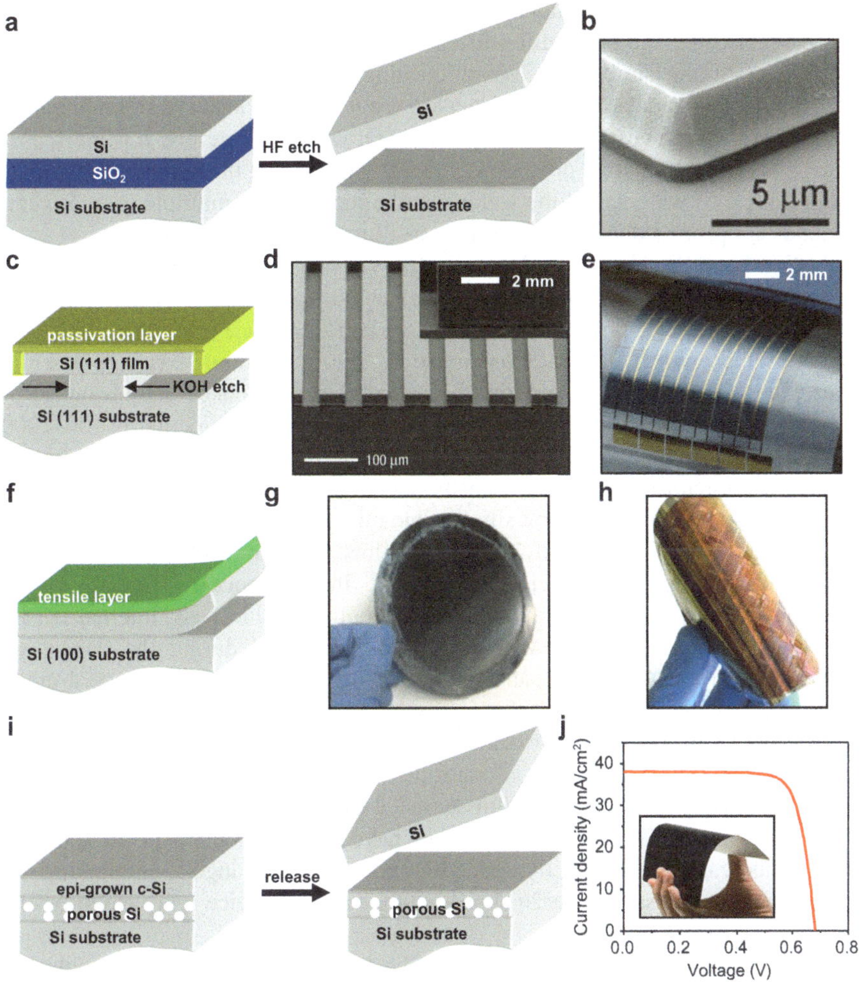


**Fig. 2** (**a**) Schematic illustration of releasing a c-Si thin film from a SOI substrate using HF undercut. (**b**) SEM image of a fully HF undercut c-Si film on a SOI substrate, with photoresist anchor structures underneath [9]. (**c**) Schematic illustration of releasing a c-Si thin film from a Si (111) substrate using KOH anisotropic etching. (**d**) SEM image of ribbon-shaped c-Si microcell array on a Si (111) wafer after KOH undercut. Inset: magnified cross-sectional SEM image of a typical c-Si cell, with a thickness of about 20 μm [12]. (**e**) Optical image of an interconnected c-Si microcell array printed on a flexible substrate after KOH undercut [12]. (**f**) Schematic illustration of forming a c-Si layer by controlled spalling process. (**g**) Optical image of exfoliated thin-film Si layers (11 μm thick, 4 inch diameter) [13]. (**h**) Optical image of a flexible c-Si-based circuit (~10 μm thick) by controlled spalling [14]. (**i**) Schematic illustration of releasing an epitaxially grown c-Si layer from an anodic porous Si surface. (**j**) Current density–voltage curves for a representative exfoliated c-Si cells (35 μm thick) under the AM1.5g spectrum [7]. Inset: image of a flexible c-Si cell [16] (Reproduced with permission. Copyright 2012 IOP Publishing. Copyright 2008 Nature Publishing Group. Copyright 2015 Wiley-VCH. Copyright 2013 American Chemical Society)

thin-film c-Si devices with thicknesses from tens of nanometer to tens of micrometers are fabricated and utilized for different applications [11].

In spite of the simple process of creating thin-film c-Si devices using HF undercut SOI substrates, it is impractical to use it for large area PV cell fabrication because of the high cost associated with the SOI wafers. Therefore, it is indispensable to develop alternative processes of directly producing thin-film c-Si materials using thick c-Si wafers, and the wafers can be recycled for multiple uses. Several methods are proposed and demonstrated, as illustrated in Fig. 2c, f, i.

In Fig. 2c, a Si (111) wafer is used as the starting material. Aligned along the [110] direction, ribbon-shaped solar cells are fabricated based on doping, lithographic, and etching processes, with top surfaces and sidewalls protected by a metallic layer. When the wafer is immersed into an alkaline-based (potassium hydroxide, KOH) aqueous solution, the undercut process takes the advantage of the anisotropic etching; the etching rate along the Si [110] direction is 100–1000 times higher than that along the Si [111] direction. Ribbon PV cells with different geometries (thickness, width, and density) can be obtained, allowing the formation of semitransparent and mechanically flexible microcell arrays (Fig. 2e) [12].

Another way to obtain exfoliated c-Si films from Si wafers is to utilize a controlled spalling process (Fig. 2f) [13, 14]. In this process, a metallic layer (e.g., Ni) is coated onto the Si wafer, with a controlled thickness. Internal tensile stress within the coated metal film can be introduced during the deposition process or via the thermal expansion mismatch during annealing. Once a crack is initiated at the wafer edge, the fracture can propagate parallel to the wafer surface direction, creating an exfoliated flexible c-Si layer (Fig. 2g). The thickness of the exfoliated Si layer is controlled by the tensile stress as well as the thickness of the deposited metal film. Assisted by this controlled spalling process, c-Si films with a diameter up to 5 inches have been demonstrated. In addition, this method can also be directly applied on semiconductor ingots, eliminating the wafer sawing process. Figure 2h shows a fabricated Si device array based on the spalling process [14].

In addition, thin-film c-Si device can be formed by epitaxially growth and release c-Si layers from a porous Si surface [15, 16] (Fig. 2i). In this process, a porous Si layer is firstly formed on the Si substrate, usually by Si anodization in HF solution [15]. Subsequently, the anodized Si is used as a growth substrate, on which a c-Si film with a controlled thickness is grown using high-temperature chemical vapor deposition (CVD). After the PV cell fabrication and bonding onto a foreign substrate, the thin-film cell can be separated from the growth wafer by applying a mechanical force at the porous Si layer. The Si wafer can be polished and reused for new cell fabrication. The high-temperature deposition process ensures that the formed c-Si PV cells achieve high crystal and device quality. Based on this porous Si release method, a thin-film flexible c-Si cell (35 μm thick) with a 1-sun efficiency of 21.2% has been demonstrated [7, 16].

# 3 Flexible and Stretchable Thin-Film Si Solar Cells

By replacing thick, rigid, and hard Si substrates with thin-film Si, one of the key advantages is to achieve mechanically flexible PV cells and modules, since the film flexural rigidity is proportional to the cube of the film thickness [11]. In addition, mechanically stretchable cells and modules can be realized by the combination of material and mechanical design methods. In this section, material requirements and mechanical designs for device integration to obtain flexible and stretchable thin-film Si PV cells and modules are presented. Potential applications for flexible and stretchable cells are discussed.

## *3.1 Substrate Requirements and Applications*

Unlike conventional thin-film PV cells that are formed onto thick glass or metal substrates, flexible and/or stretchable cells require thin metal or polymer substrates to obtain desired mechanical properties. For a-Si-based single- and multi-junction cells, the deposition process (PECVD) requires that cell substrates remain stable at elevated temperatures (200–400 °C), as discussed previously. Cell fabrication has been realized on substrates made by materials like stainless steel and polyimide, as shown in Fig. 3a, b, respectively. In addition, flexible substrates enable the roll-to-roll cell production, greatly enhancing the process speed and reducing the cost. By using the roll-to-roll process, PV modules with areas of few square meters can be formed, with efficiencies up to 12% [7]. Compared to cells on rigid substrates, large area flexible PV modules have advantages in transportation and installation and are more compatible with building integrated power systems (Fig. 3d). Small area PV cells can be mounted onto wearable and portable systems like back bags, clothes, and cellphones (Fig. 3e) and even onto the surface of human skin (Fig. 3f), working as power sources for both military and civilian uses.

The use of thin-film Si layer release and transfer processes that further expands candidate materials that can be explored as thin-film cell substrates, because the cell process can be performed at room temperature. Flexible thin-film c-Si cells laminated on transparent polymer substrates have shown efficiencies of more than 21% [7]. Furthermore, thin-film c-Si cells integrated onto elastomer-based substrates (e.g., silicones) can achieve both mechanical flexibility and stretchability, making it more compatible with biological systems such as soft human tissues. For example, Fig. 3c illustrates a c-Si photodiode array printed onto polydimethylsiloxane (PDMS)-based substrates [17]. The unusual materials and mechanical design ensure that the device can be elongated by more than 20% without mechanical fracture. With careful designs, such thin-film microscale solar cells can be integrated within the epidermal electronic circuit systems (Fig. 3f), serving as a wireless power source for various electronic and photonic skin-mounted biological sensors [18].

**Fig. 3** (**a**) Photograph of flexible a-Si solar cells deposited onto stainless steel substrates, prepared by roll-to-roll process [19]. (**b**) Optical image of a flexible Si solar cell printed on a glass window surface [20]. (**c**) Optical image of c-Si photodiode arrays printed onto PDMS substrates [17]. (**d**) Photograph of flexible a-Si solar modules for building integration [19]. (**e**) Flexible solar cells mounted on a portable device [20]. (**f**) Optical image of epidermal electronic circuits (Reproduced with permission. Copyright 2008 Wiley-VCH. Copyright 2012 Nature Publishing Group. Copyright 2008 PNAS)

## *3.2 Mechanical Considerations*

Silicon is an extremely brittle material which breaks when stretched by less than 1% of strain; therefore careful designs are required to overcome this intrinsic mechanical limit. This is especially challenging when the fabrication and processing involve the integration of mechanically dissimilar materials such as in the stretchable and flexible systems shown in Figs. 2e, h and 3. The key strategy in realizing these systems is to design these composite structures such that large deformation occurs in the elastomeric materials while the brittle components barely deform. Many advanced mechanics ideas and fabrication processes have been developed for a wide range of applications based on stretchable inorganic electronics, represented by the work of Ko et al. [21], Baca et al. [12], and Kim et al. [18]. Many of these mechanics analyses can be adapted for the designs of flexible and stretchable PV systems due to the similarities in materials and geometric layouts.

Here we limit our discussions on designs and mechanics specifically related to PV applications with the example shown in Fig. 2e. This flexible PV module involves a composite structure consisting of a planarizing/adhesive layer (NOA61; thickness ~ 30 μm), which also serves as the substrate, arrays of μ-cells and metal interconnects, and a polymer encapsulation layer (NOA61; thickness ~ 30 μm). The mechanically flexible μ-cell module is modeled as a composite beam as shown in Fig. 4a, where $W$, $W_{\mathrm{Si}}$, and $W_{\mathrm{NOA}}$ are the widths of the beam, silicon μ-cell, and the distance between adjacent μ-cells, respectively, and $t$, $t_{\mathrm{m}}$, $b$, and $(a\text{-}t)$ are the thicknesses of the μ-cell, metal interconnect layer, and NOA layers above and below the μ-cell. The Young's modulus of silicon, metal (Au), and NOA are denoted by $E_{\mathrm{Si}}$, $E_{\mathrm{Au}}$, and $E_{\mathrm{NOA}}$, respectively. The strain in the beam is given by $\varepsilon_{yy} = (z - z_0)/R$, where $R$ is the bending radius of the beam and $z_0$ is the position of the neutral mechanical plane measured from the bottom edge. Using the geometric and material parameters shown in Fig. 4a, $z_0$ is obtained analytically by

$$z_0 = \frac{a-t}{2}\left(1+\frac{b}{a-t}\right)^2 + 2\frac{b}{a-t}\frac{t+t_{\mathrm{m}}}{a-t} + \frac{W_{\mathrm{NOA}}}{W}\frac{t}{a-t}\left(2+\frac{t}{a-t}\right)$$
$$+\frac{E_{\mathrm{Si}}t}{E_{\mathrm{NOA}}(a-t)}\frac{W_{\mathrm{Si}}}{W}\left(2+\frac{t}{a-t}\right) + \frac{E_{\mathrm{Au}}t_{\mathrm{m}}}{E_{\mathrm{NOA}}(a-t)}\left(2+\frac{2t+t_{\mathrm{m}}}{a-t}\right)1+\frac{b}{a-t} \quad (1)$$
$$+\frac{W_{\mathrm{NOA}}}{W}\frac{t}{a-t} + \frac{E_{\mathrm{Si}}t}{E_{\mathrm{NOA}}(a-t)}\frac{W_{\mathrm{Si}}}{W} + \frac{E_{\mathrm{Au}}t_{\mathrm{m}}}{E_{\mathrm{NOA}}(a-t)}.$$

The above analytical modeling can be used to tune the geometric layout (especially the thicknesses of different materials) such that the neutral mechanical plane is placed near the center of the Si μ-cells, which then ensures the maximum strains in the silicon and metal interconnects remain small even when the entire structure undergoes extreme deformation. For the experimental data shown in Fig. 4b, the above equations predict that the strain in the silicon is less than 0.3% even for severe bend radii less than 5 mm, for bending in any direction. Finite element modeling, with representative results shown in Fig. 4c, d, confirms that the maximum strain in silicon for the inward and outward bending is around 0.03%, and the maximum strain in the metal layer is around 0.13%, located near the silicon corner for both inward and outward bending as shown in Fig. 4d. Module performance, evaluated in outward bending along and perpendicular to the cell length under AM1.5 g illumination, shows behaviors consistent with expectations on the basis of mechanics analysis and relative insensitivity of the degree of illumination across the modest area of the module, for the bend radii examined here. For example, at bending radii of 12.6, 8.9, 6.3, and 4.9 mm, the module efficiency (~ 6.0%) and fill factor (~ 0.60) remain unchanged as summarized in Fig. 4e, f. The small strain in active device components not only ensures consistent module performance but also prevents fracture when these systems are used in environments that impose repetitive loadings. Fatigue tests, with bending up to 200 cycles, also show little change in performance, as summarized in Fig. 4g.

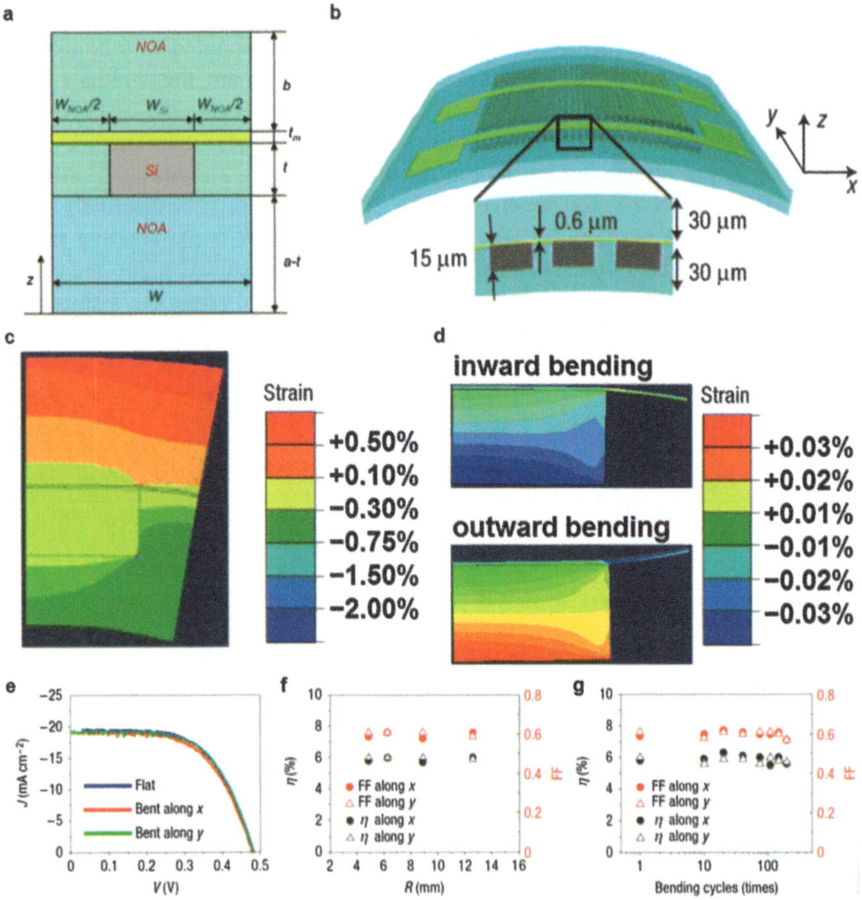


**Fig. 4** (**a**) Cross-sectional schematic illustration of a model composite structure composed of silicon μ-cell, and polymer encapsulation layer and a metal layer. (**b**) Schematic illustration of an optimized design in which the neutral mechanical plane is positioned near the center of the μ-cells (gray) through judicious choices of thickness for the polymer (blue) substrate and overcoat. (**c**) Color contour plot of calculated bending strains through the cross-section of a mechanically flexible μ-cell module, bent along the cell width direction at $R = 4.9$ mm. The calculations use symmetry boundary conditions for evaluation of a single unit cell of the system. The black lines delineate the boundaries of the μ-cell and metal interconnect line (top). (**d**) Color contour plot of calculated bending strains through the cross-section of a mechanically flexible μ-cell module, bent inward and outward along the cell width direction at $R = 4.9$ mm, without metal layer. (**e**) $J$–$V$ data from a module under AM1.5 g illumination in a flat configuration and bent along the cell width ($x$) and length ($y$) directions, both for $R = 4.9$ mm. (**f**) Plot of $\eta$ and fill factor ($FF$) under AM1.5 g illumination for $R = 12.6, 8.9, 6.3$, and 4.9 mm. (**g**) Plot of $\eta$ and $FF$ as a function of bending cycles up to 200 times at $R = 4.9$ mm [12] (Reproduced with permission. Copyright 2008 Nature Publishing Group)

# 4 Biodegradable Thin-Film Si Photovoltaic Cells

Recently, bio-integrated electronic and optical devices that can dissolve naturally in physiological conditions have attracted considerable attentions because of their potential applications in biomedical surgery, diagnosis, and therapy [22–27]. Comprising fully biodegradable materials (metals [25, 26], semiconductors [27, 28], dielectrics [29], and polymers [30, 31]), implantable devices dissolve in biological environments in a controlled manner after use. Such devices and systems are of particular interest since they eliminate the risk of secondary surgery after implantation. The operation of these systems also requires fully dissolvable energy devices as a power supply. In this section, we summarize some of the recent progress about the development of fully degradable thin-film Si photovoltaic cells, which provide a promising solution for remote and wireless powered energy source for these biodegradable and implantable electronic and photonic devices.

## *4.1 Materials Degradation*

The hydrolysis process of thin-film Si materials can be simply described as the reaction of Si with water to form silicic acid: $Si + 4H_2O \rightarrow Si(OH)_4 + 2H_2$, in which the Si dissolution rate varies with lots of factors like the Si crystallinity, doping, pH levels, temperature, and the solution chemistry [28]. Figure 5a plots the dissolution rates of monocrystalline Si (c-Si), polycrystalline Si (poly-Si), and amorphous Si (a-Si) in buffer solutions with different pH levels (pH 6, pH 7, pH 7.4, pH 8, and pH 10) at physiological temperature (37 °C) [23, 27]. In these experiments, c-Si thin-film membranes are prepared by transfer printing process, while poly-Si and a-Si layers are formed by chemical vapor deposition (CVD) methods. Dissolution rates range from about 1 nm/day to about 1 nm/sec, depending on pH levels in the solution. Additional details about dissolution for Si with different doping levels in different physiological environments (deionized water, phosphate-buffered saline, bovine serum, etc.) have also been extensively studied [28]. As an example, the dissolution process of a c-Si membrane in bovine serum (pH 7.4) at 37 °C is presented as the evolution of atomic force microscopic (AFM) images in Fig. 5b [28]. Dissolution behaviors of biodegradable metals (Mg, Fe, Zn, Mo, W, etc.), dielectric materials (ZnO, $SiO_2$, $Si_3N_4$, etc.), and polymers (silk, PLGA, PLA, etc.) are also investigated and discussed [25–31]. Combining these materials with Si-based junctions, thin-film Si PV cells can be formed in a fully degradable format. Figure 5c illustrates the functional transience of such a degradable solar cell during hydrolysis, made from an a-Si:H pn junction with ZnO- and Mg-based contacts [27]. Structural evolution for the a-Si:H solar cell is shown in Fig. 5d [27]. It is observed that the Mg and ZnO contacts dissolve within a few hours in water, followed by the destruction of the a-Si:H layer. Complete dissolution of the rest a-Si:H material occurs after several days.

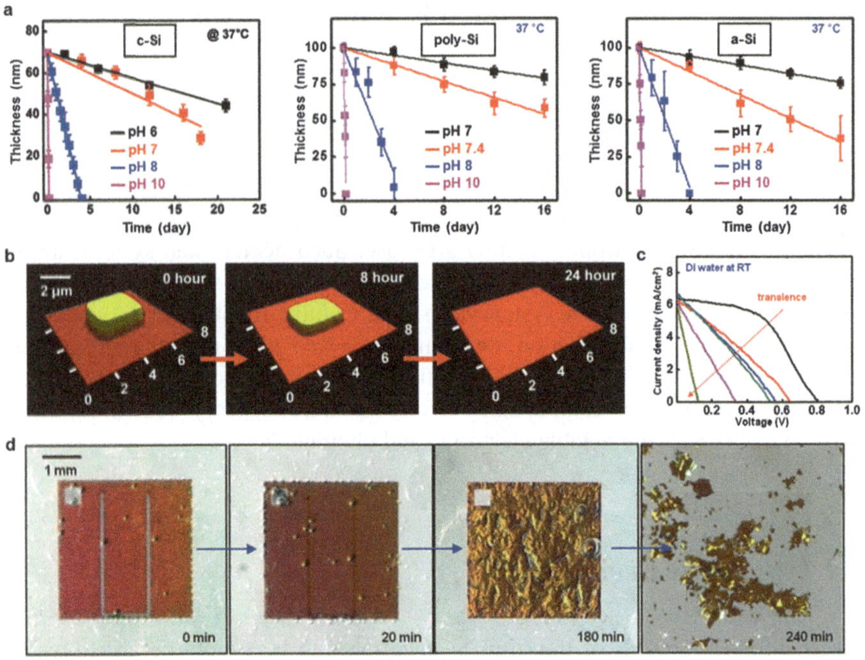


**Fig. 5** (**a**) Theoretical (lines) and experimental (symbols) dissolution rates for different types of thin-film Si (c-Si, poly-Si, and a-Si) in buffer solutions with different pH at physiological temperature (37 °C) [23, 27]. (**b**) AFM images of a c-Si thin-film layer at various dissolution stages at pH of 7.4 [28]. (**c**) IV characteristics and performance degradation of a thin-film a-Si:H solar cell, made by fully degradable materials [27]. (**d**) Optical images showing the dissolution of a thin-film a-Si:H solar cell in deionized water at room temperature [27] (Reproduced with permission. Copyright 2014 Wiley-VCH. Copyright 2015 American Chemical Society. Copyright 2014 American Chemical Society)

## *4.2 Biological Compatibility*

In order to evaluate the biocompatibility of these biodegradable electronic devices and ultimately use them for implantable biomedical applications, in vitro and in vivo cytotoxicity tests are necessary. Fluorescence images in Fig. 6a show the viability of L929 mouse fibroblast cell lines cultured on the surface of poly-Si thin films [27]. Results show that the cell viability reaches more than 95% after 7 days. Similar results are obtained for cells grown on other semiconductor surfaces including a-Si, SiGe, Ge, and mono-Si, summarized in Fig. 6b [27]. Cell culture studies on other materials like $SiO_2$ are also explored in previous works [29]. Furthermore, the constituent materials (e.g., Si, Mg, Zn, etc.) used to form the thin-film PV cells have been proved to be biocompatible, with the amounts far below the daily intake limits for clinical use [22, 25]. To demonstrate the biocompatibility for those thin-film Si-based devices, a representative device array of thin-film Si

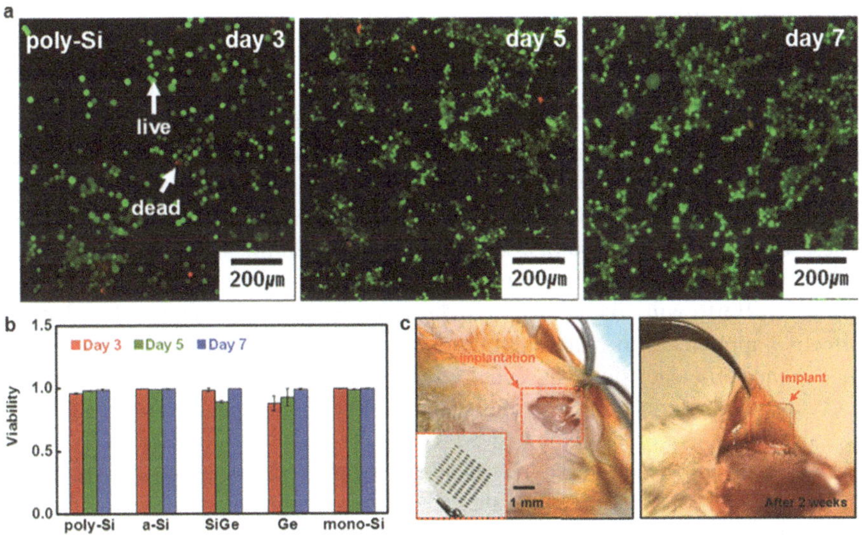


**Fig. 6** (**a**) Fluorescence images illustrating cell viability on a poly-Si thin film, where live and dead cell assays are labeled as green and red, respectively [27]. (**b**) Viability for cells cultured on surfaces of poly-Si, a-Si, SiGe, Ge, and mono-Si at day 3, day 5, and day 7 [27]. (**c**) In vivo demonstration of a thin-film c-Si electronic device array implanted in the dorsal region of a mouse. The devices disappear in 3 weeks after the surgery [24] (Reproduced with permission. Copyright 2015 American Chemical Society. Copyright 2013 Wiley-VCH)

CMOS transistors (with Mg contacts) on silk substrates is implanted underneath the skin of a living mouse (Fig. 6c). The implanted region is examined after 2 weeks, revealing that the device array has fully degraded and absorbed by biological tissues [24].

Additional considerations for the use of implantable Si PV cells include the optical effects of biological tissues (skin, skull, fat, etc.) and the solar cell device optimization. Unlike solar cells working directly under the sun, the implantable devices are only operational within the transparency window of the biological tissue, which is usually in the red and near-infrared spectral range. At these wavelengths, thin-film Si PV cells usually have low efficiencies due to low absorption coefficients. Therefore, the Si film thickness should be optimized to obtain a trade-off among power generation, dissolution time, and mechanical flexibility. In such thin-film Si PV cells, light-trapping schemes could be adapted to optimize the cell optical absorption in the red and near-IR range [32]. In addition, advanced encapsulation and packaging strategies can be implemented to realize a more controllable degradation process. These full degradable thin-film Si PV cells provide a viable and promising approach to realize the wireless energy transfer for implantable biomedical systems, with potential applications including drug delivery [31], optogenetic stimulation [33], temperature monitoring [34], etc.

## 5 Conclusions

In this chapter, it is shown that advanced thin-film Si PV cells not only provide a viable solution for large scale solar energy utilization but also foreshadow a promising future in healthcare applications. Some of the recent progress about advanced thin-film Si solar cells is overviewed, with specific foci on their flexible, stretchable, and biodegradable formats and the associated fabrication schemes. These advanced cell technologies envision new possibilities in broad biomedical applications for wearable and implantable systems. In the future, fundamental research efforts will likely focus on the developments of high-yield and cost-effective approaches to make high-efficiency thin-film c-Si cells that reach thermodynamic efficiency limits for Si cells (one sun efficiency ~ 25%) [32]. Contacting and substrate materials that provide mechanical flexibility and/or biocompatibility are also to be explored, combined with advanced mechanical design strategies. Furthermore, advanced integration schemes need to be investigated to interconnect the thin-film Si PV cells with other electronic components and circuits, like batteries, antennas, transistors, and various sensors, to form a fully functional biomedical device system for sensing, diagnosis, and therapy.

## References


1. R.E.I. Schropp, M. Zeman, *Amorphous and Microcrystalline Silicon Solar Cells: Modeling, Materials and Device Technology* (Kluwer Academic Publishers, Boston, 1998)
2. H. Keppner, J. Meier, P. Torres, D. Fischer, A. Shah, Appl. Phys. A Mater. Sci. Process **69**, 169 (1999)
3. C.C. Wang, C.Y. Liu, S.Y. Lien, K.W. Weng, J.J. Huang, C.F. Chen, D.S. Wuu, Curr. Appl. Phys. **11**, S50 (2011)
4. T. Matsui, H. Sai, T. Suezaki, M. Matsumoto, K. Saito, I. Yoshida, M. Kondo, in *Proceedings of the 28th European Photovoltaic Solar Energy Conference and Exhibition*, 2013, p. 2213
5. T. Matsui, H. Sai, K. Saito, M. Kondo, Prog. Photovolt. Res. Appl. **21**, 1363 (2013)
6. S. Kim, J.W. Chung, H. Lee, J. Park, Y. Heo, H.M. Lee, Sol. Energy Mater. Sol. Cells **119**, 26 (2013)
7. M.A. Green, K. Emery, Y. Hishikawa, W. Warta, E.D. Dunlop, Prog. Photovolt. Res. Appl. **23**, 805 (2015)
8. J. Poortmans, V. Arkhipov, *Thin Film Solar Cells Fabrication, Characterization and Applications* (Wiley, Chichester, 2006)
9. H. Keum, A. Carlson, H. Ning, A. Mihi, J.D. Eisenhaure, P.V. Braun, J.A. Rogers, S. Kim, J. Micromech. Microeng. **22**, 055018 (2012)
10. A. Carlson, A.M. Bowen, Y. Huang, R.G. Nuzzo, J.A. Rogers, Adv. Mater. **24**, 5284 (2012)
11. J.A. Rogers, M.G. Lagally, R.G. Nuzzo, Nature **477**, 45 (2011)
12. J. Yoon, A.J. Baca, S.I. Park, P. Elvikiss, J.B. Geddes, L. Li, R.H. Kim, J. Xiao, S. Wang, T.H. Kim, M.J. Motala, B.Y. Ahn, E.B. Duoss, J.A. Lewis, R.G. Nuzzo, P.M. Ferreira, Y. Huang, A. Rockett, J.A. Rogers, Nat. Mater. **7**, 907 (2008)
13. S. Wang, B.D. Weil, Y. Li, K.X. Wang, E. Garnett, S. Fan, Y. Cui, Nano Lett **13**, 4393 (2013)
14. D. Shahrjerdi, S.W. Bedell, Nano Lett. **13**, 315 (2013)

15. H.S. Radhakrishnan, R. Martini, V. Depauw, K. Van Nieuwenhuysen, M. Debucquoy, J. Govaerts, I. Gordon, R. Mertens, J. Poortmans, IEEE J. Photovoltaics **4**, 70 (2014)
16. M.M. Moslehi, Thin-silicon, low-cost solar photovoltaic modules using kerfless epitaxial silicon lift-off technology. Solexel Inc (2012)
17. D.H. Kim, J. Song, W.M. Choi, H.S. Kim, R.H. Kim, Z. Liu, Y. Huang, K.C. Hwang, Y. Zhang, J.A. Rogers, Proc. Natl. Acad. Sci. **105**, 18675 (2008)
18. D.H. Kim, N. Lu, R. Ma, Y.S. Kim, R.H. Kim, S. Wang, J. Wu, S.M. Won, H. Tao, A. Islam, K.J. Yu, T. Kim, R. Chowdhury, M. Ying, L. Xu, M. Li, H.J. Chung, H. Keum, M. McCormick, P. Liu, Y. Zhang, F.G. Omenetto, Y. Huang, T. Coleman, J.A. Rogers, Science **333**, 838 (2011)
19. M. Pagliaro, G. Palmisano, R. Ciriminna, *Flexible Solar Cells* (Wiley-VCH Verlag GmbH & Co. KGaA, Weinheim, 2008)
20. C.H. Lee, D.R. Kim, I.S. Cho, N. William, Q. Wang, X. Zheng, Sci. Rep. **2**, 1000 (2012)
21. H.C. Ko, M.P. Stoykovich, J. Song, V. Malyarchuk, W.M. Choi, C.-J. Yu, J.B. Geddes III, J. Xiao, S. Wang, Y. Huang, J.A. Rogers, Nature **454**, 748 (2008)
22. S.W. Hwang, H. Tao, D.H. Kim, H. Cheng, J.K. Song, E. Rill, M.A. Brenckle, B. Panilaitis, S.M. Won, Y.S. Kim, Y.M. Song, K.J. Yu, A. Ameen, R. Li, Y. Su, M. Yang, D.L. Kaplan, M.R. Zakin, M.J. Slepian, Y. Huang, F.G. Omenetto, J.A. Rogers, Science **337**, 1640 (2012)
23. S.W. Hwang, G. Park, H. Cheng, J.K. Song, S.K. Kang, L. Yin, J.H. Kim, F.G. Omenetto, Y. Huang, K.M. Lee, J.A. Rogers, Adv. Mater. **26**, 1992 (2014)
24. S.W. Hwang, D.H. Kim, H. Tao, T.I. Kim, S. Kim, K.J. Yu, B. Panilaitis, J.W. Jeong, J.K. Song, F.G. Omenetto, J.A. Rogers, Adv. Funct. Mater. **23**, 4087 (2013)
25. L. Yin, H. Cheng, S. Mao, R. Haasch, Y. Liu, X. Xie, S.W. Hwang, H. Jain, S.K. Kang, Y. Su, R. Li, Y. Huang, J.A. Rogers, Adv. Funct. Mater. **24**, 645 (2014)
26. S.K. Kang, S.W. Hwang, S. Yu, J.H. Seo, E.A. Corbin, J. Shin, D.S. Wie, R. Bashir, Z. Ma, J.A. Rogers, Adv. Funct. Mater. **25**, 1789 (2015)
27. S.K. Kang, G. Park, K. Kim, S.W. Hwang, H. Cheng, J. Shin, S. Chung, M. Kim, L. Yin, J.C. Lee, K.M. Lee, J.A. Rogers, ACS Appl. Mater. Interfaces **7**, 9297 (2015)
28. S.W. Hwang, G. Park, C. Edwards, E.A. Corbin, S.K. Kang, H. Cheng, J.K. Song, J.H. Kim, S. Yu, J. Ng, J.E. Lee, J. Kim, C. Yee, B. Bhaduri, Y. Su, F.G. Omennetto, Y. Huang, R. Bashir, L. Goddard, G. Popescu, K.M. Lee, J.A. Rogers, ACS Nano **8**, 5843 (2014)
29. S.K. Kang, S.W. Hwang, H. Cheng, S. Yu, B.H. Kim, J.H. Kim, Y. Huang, J.A. Rogers, Adv. Funct. Mater. **24**, 4427 (2014)
30. H.L. Hernandez, S.K. Kang, O.P. Lee, S.W. Hwang, J.A. Kaitz, B. Inci, C.W. Park, S. Chung, N.R. Sottos, J.S. Moore, J.A. Rogers, S.R. White, Adv. Mater. **26**, 7637 (2014)
31. H. Tao, S.W. Hwang, B. Marelli, B. An, J.E. Moreau, M. Yang, M.A. Brenckle, S. Kim, D.L. Kaplan, J.A. Rogers, F.G. Omenetto, Proc. Natl. Acad. Sci. **111**, 17385 (2014)
32. X. Sheng, L.Z. Broderick, L.C. Kimerling, Opt. Commun. **314**, 41 (2014)
33. T.I. Kim, J.G. McCall, Y.H. Jung, X. Huang, E.R. Siuda, Y. Li, J. Song, Y.M. Song, H.A. Pao, R.H. Kim, C. Lu, S.D. Lee, I.S. Song, G. Shin, R. Al-Hasani, S. Kim, M.P. Tan, Y. Huang, F.G. Omenetto, J.A. Rogers, M.R. Bruchas, Science **340**, 211 (2013)
34. R.C. Webb, A.P. Bonifas, A. Behnaz, Y. Zhang, K.J. Yu, H. Cheng, M. Shi, Z. Bian, Z. Liu, Y.S. Kim, W.H. Yeo, J.S. Park, J. Song, Y. Li, Y. Huang, A.M. Gorbach, J.A. Rogers, Nat. Mater. **12**, 938 (2013)

# Silicon Nanocrystal-Based Organic/Inorganic Hybrid Solar Cells

**Yi Ding and Tomohiro Nozaki**

**Abstract** Comparing to bulk silicon, silicon nanocrystals (Si NCs) possess particularly interesting properties and have further broadened applications in optics, microelectronics, photovoltaics, and other fields. In this chapter, novel fabrication process of Si NCs by using a plasma will be introduced firstly; next, some basic properties of resulted Si NCs, such as crystallinity, optical, and electrical properties, have been studied extensively; then, its application in organic/inorganic hybrid solar cells has been explored; structure design, device fabrication, and performance characterization of Si NC-based organic/inorganic hybrid solar cells have been described finally. In addition, effects of Si NCs on device performance are also discussed extensively.

## 1 Introduction

Fabrication and application of freestanding semiconductor nanocrystals (NCs) are hot topics in recent years because of the many benefits that do not exist in the corresponding bulk materials [1, 2], such as (1) quantum confinement effect, whereby the particle bandgap starts to increase with size decreasing when the size is comparable with its Bohr radius [3–5]. This effect that enables tuning of material optical properties has potential applications in solar cells [6, 7] and light-emitting devices (LEDs) [8, 9]. In addition, continuous states shrink to discrete states, which can suppress thermal relaxation process effectively and enable the realization of multiple exciton generation processes (MEG). The latter one is commonly recognized as a substantial way to dramatically enhance solar cell

Y. Ding (✉)
Nankai University, College of Electronic Information and Optical Engineering, Nankai, Tianjin, China
e-mail: yiding@nankai.edu.cn

T. Nozaki
Tokyo Institute of Technology, Department of Mechanical Sciences and Engineering, Tokyo, Japan



S. Ikhmayies (ed.), *Advances in Silicon Solar Cells*,
https://doi.org/10.1007/978-3-319-69703-1_7

photoconversion efficiency (PCE), as one photon with high energy can excite two or more electron-hole pairs [10, 11]; (2) freestanding semiconductor NCs are solution processible. Hence, inorganic semiconductor films can be fabricated easily at low cost by spin coating, drop casting, and other processes [12, 13]. Moreover, such films are usually mechanically flexible when one chooses an elastic substrate [14, 15]; (3) it has huge surface area. Therefore, the surface becomes extremely important: it dramatically influences particle properties [16, 17].

Among freestanding semiconductor NCs, silicon nanocrystals (Si NCs) are particularly interesting because silicon is an abundant, nontoxic, and environmentally benign material [18–20]. Si NCs can also have further broadened applications in optics [21–23], microelectronics [24, 25], photovoltaics [26, 27], and other fields [28]. Several groups have already reported pioneering studies of Si NC optical and electrical properties [29–32].

In this chapter, the detailed fabrication process of Si NCs will be descripted firstly, and then some basic interesting properties of resulted Si NCs will be shown. Finally, the discussion will be mainly focused on the design, fabrication, and performance characterization of Si NC-based organic/inorganic hybrid solar cells. Effects of Si NCs on device performance are also studied extensively.

## 2 Silicon Nanoparticle Synthesis

Silicon nanoparticles (Si NPs) can be fabricated by several methods, such as aqueous solution synthesis while this is relatively cumbersome [20, 33] and mechanical milling or laser ablation from bulk silicon, especially the latter one which is commonly used for NPs fabrication. However, all these processes are affected by a low yield, uneven particle size, or poor optical quality of the material [34, 35]. In the current work, Si NPs were synthesized from silicon tetrachloride ($SiCl_4$) by using a very high frequency (VHF) nonthermal plasma. This process is capable to control particle size and crystallinity, and also mass production can be easily realized [36, 37].

A VHF (70 MHz) nonthermal plasma is generated by two copper electrodes surrounding a quartz reactor tube (with inner diameter of 48 mm) with a variable matching network used to match the plasma impedance [32, 38]. The basic configuration of the plasma is illustrated in Fig. 1. The plasma dissociates the $SiCl_4$ through electron impaction allowing for nucleation and subsequent growth of NCs, while hydrogen scavenges the excess chlorine, and argon is used to dilute and sustain the plasma. NCs synthesized in the plasma were collected downstream of the plasma on a mesh. All synthesis and handling of Si NCs were conducted without exposure to air. The effects of the plasma operating conditions on the properties of resulted NCs were studied. An ideal system would allow for independent control of NC size, size distribution, crystallinity, and surface termination, while this reactor

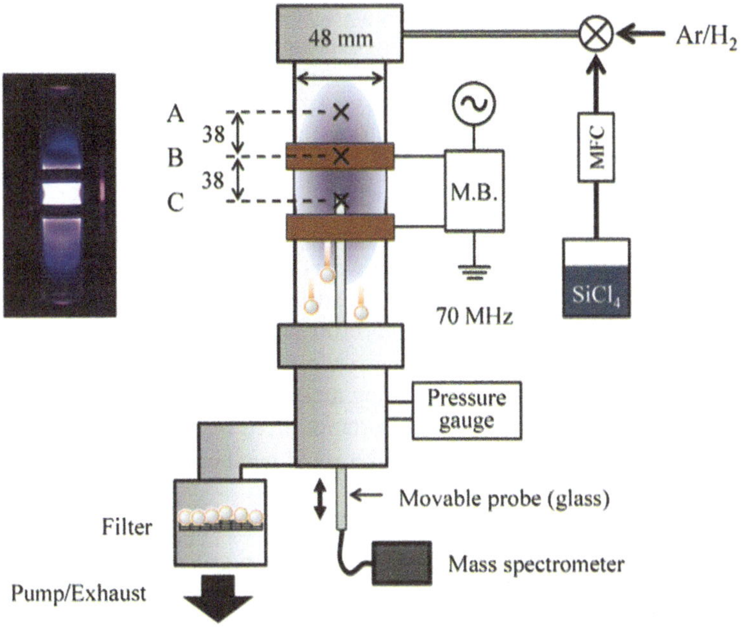


**Fig. 1** Schematic diagram of experimental setup (Reprinted with permission from Ref. [37]. Copyright 2014 IOP Publishing Ltd)

design allows for control of gas composition, flow rates, system pressure, and input power. Therefore, the influence on the resulting material was studied as a function of these reactor controls.

## 3 Si NC Surface Modification

As-produced particles are terminated by a large number of chlorine atoms because $SiCl_4$ is employed as the precursor and hereinafter known as the Cl:Si NCs [37]. Cl: Si NCs, with the high electronegativity of the halogen at the surface, polarize the electron density away from NC core. This termination can have advantages in processing of Cl:Si NCs, for example, Wheeler et al. reported that Si–Cl surface groups facilitate in hypervalent interaction with hard donor groups of certain solvents which provides colloidal stability as well as effectively doping of Si NCs [39]. However, the chlorine terminated surface is highly reactive and has been proven to be detrimental to device performance [37, 38]. Hence surface chlorine was removed afterward through surface modification in order to achieve devices with high performance.

Experimentally, as-produced, chlorine-terminated Si NCs were intentionally exposed to open air for 30 min. During air exposure, chlorine atoms will be removed completely, and particles were oxidized quickly, providing a convenient method to produce O:Si NCs. Then, by exposing O:Si NCs to hydrofluoric acid (HF) vapor, the oxide is etched, and the surface will be mainly terminated by hydrogen (H:Si NCs). Specifically, O:Si NCs were transferred on a Teflon supporting mesh and put it above 50% HF acid in a sealed container. HF vapor etching was conducted at room temperature and etching time was carefully controlled.

Figure 2a shows Fourier transform infrared spectroscopy (FTIR) spectra of Cl:Si NCs, O:Si NCs, and HF-etched Si NC powders for 0.5, 1, 2, and 5 h, respectively. For Cl:Si NCs, a peak around 560 $cm^{-1}$ is attributed to Si–Cl surface specie [40], and weak peaks around 630, 900 and 2100 $cm^{-1}$ are attributed to the Si–$H_x$ wagging, bending, and stretching modes, respectively [41]. The broad peak around 1080 $cm^{-1}$ is related to Si–O–Si stretching mode, which comes from unintentional oxidization during fabrication or transportation from reactor to a nitrogen-purged glove box [42, 43]. Therefore, except for a small number of hydrogen and oxygen

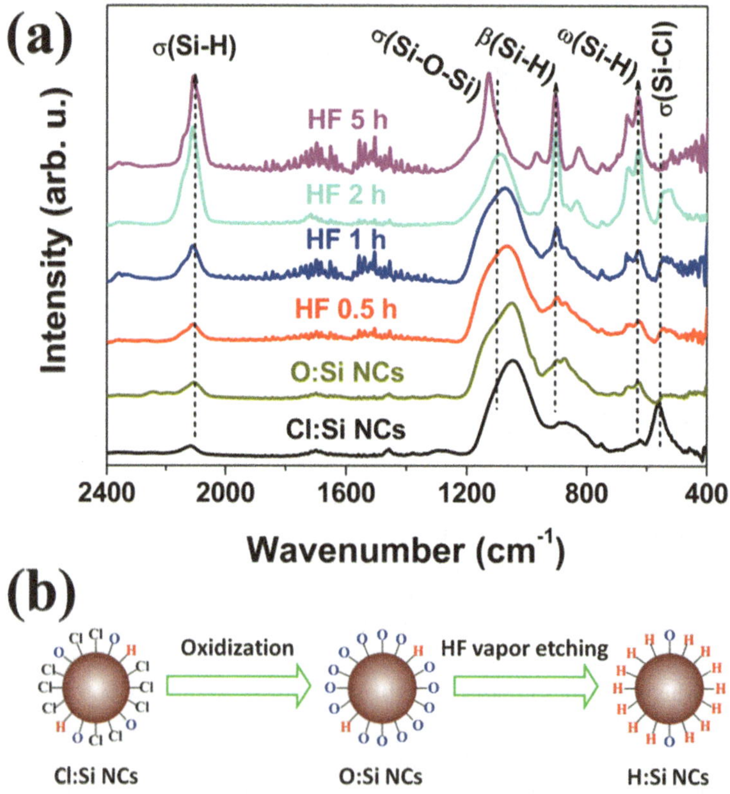


**Fig. 2** (**a**) FTIR spectra of as-produced Cl:Si NCs, O:Si NCs, and HF-etched NCs for 0.5, 1, 2, and 5 h, respectively. (**b**) Schematic depiction of Si NC surface evolution during treatment (Reprinted with permission from Ref. [37]. Copyright 2015 Elsevier Ltd)

atoms, as-produced Si NCs are mainly terminated by chlorine as illustrated in Fig. 2b. For O:Si NCs and HF-etched Si NCs, Si–H-related peaks increase gradually with etching time, while the relative intensity of Si–O–Si reduces, indicating the oxygen is gradually removed and replaced by hydrogen through HF etching as illustrated in Fig. 2b. However, oxygen still can be observed after etching for 5 h.

Therefore, HF vapor etching was conducted at room temperature for 24 h providing a thorough etching. Surface oxide will be removed completely during this process, and the particle surface will be mainly terminated with hydrogen, which is denoted as fresh Si NCs in the following. The FTIR spectrum of fresh Si NCs is shown in Fig. 3a. Peaks around 668 $cm^{-1}$ and 2360 $cm^{-1}$are assigned to $CO_2$ physisorbed during measurement, which do not influence particle properties and will be excluded from the following discussions [44]. The weak peak around 825 $cm^{-1}$ is related to Si–OH absorption [45]. In addition, strong Si–$H_x$ peaks attributed to wagging, bending, and stretching modes can be clearly distinguished. Meanwhile, no chlorine-related peak and very little occurrence related to Si–O stretching modes can be detected, indicating that fresh Si NCs are mainly terminated with hydrogen.

However, as shown in Fig. 3b, an obvious electron spin resonance (ESR) signal with a *g* factor of 2.006 appears in fresh Si NCs, which is due to the silicon dangling bonds [34]. Therefore, it is becoming clear that although the fresh Si NC surface is mainly terminated with hydrogen, still some surface sites are not passivated adequately, where electrically active dangling bonds will be formed and act as trap states, degrading particle electrical properties. Thus, proper passivation methods should be introduced in order to eliminate dangling bonds and improve electrical properties.

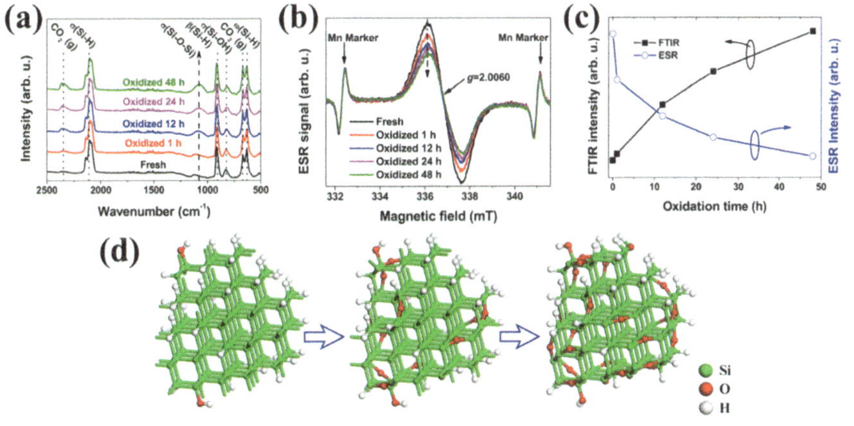


**Fig. 3** (**a**) FTIR and (**b**) ESR spectra of Si NCs oxidized for different times. (**c**) Si–O-related FTIR and dangling bond-related ESR intensities as a function of oxidation time. (**d**) Possible surface passivation process during controlled oxidation, actual particle size is larger than the schematic diagram (Reprinted with permission from Ref. [46]. Copyright 2015 Elsevier Ltd)

For bulk silicon, either thermal annealing in a hydrogen atmosphere or wet chemical processing in HF solution can provide a good hydrogen-passivated surface [47]. However, both methods cannot be adopted for Si NCs. Thermal annealing at temperatures greater than 800 C has a high risk of sintering the particles together. Wet chemical processes make it difficult to collect particles from aqueous solutions and control the reaction precisely. Therefore, other gentler methods have been explored, and controlled reoxidization under well-controlled conditions is proposed in the following to passivate the particles conveniently and efficiently with oxygen [46].

Fresh Si NCs were exposed in air at room temperature, and the humidity was controlled around 30%. This guarantees that oxidation evolves slowly over time. FTIR spectra of Si NCs oxidized for different times are shown in Fig. 3a. The Si–O stretching peak increases gradually with oxidation time, whereas Si–OH- and Si–$H_x$-related peaks stay nearly constant. In addition, the ESR signal related to dangling bonds decreases with oxidation time, as shown in Fig. 3b. The Si–O stretching peak and ESR signal intensities are derived and replotted in Fig. 3c as a function of oxidation time. It is explicit that Si–O stretching absorption increases, whereas the dangling bond signal decreases monotonically with oxidation time, demonstrating that, as illustrated in Fig. 3d, Si NCs are passivated effectively with oxygen through controlled reoxidation.

# 4 Si NP Characterization

## *4.1 Size of Si NPs*

The gas residence time in plasma has a strong influence on the size of resulted NCs, while the hydrogen concentration and reactor pressure have a minor influence on the size [32, 48]. Particle size can be adjusted from 3 to 15 nm through the gas residence time, where the standard deviation of the NC sizes is less than ~20%. Statistical size distribution of Si NCs was analyzed by counting more than 300 NCs in atomic force microscope (AFM) image. Si NCs were deposited on a polished silicon substrate by spin casting Si NCs solution dissolved in benzonitrile. Typical AFM image of Si NPs is shown in Fig. 4a; the corresponding size distributions are illustrated in Fig. 4b. It's clear that here Si NCs have an average size of ~7.2 nm [49].

## *4.2 Crystallinity of Si NPs*

Controlling over the crystallinity of NPs can be achieved through the plasma power adjustment, where a range of nearly entirely amorphous to entirely crystalline content is possible. This ability to control crystallinity is also desirable as it is

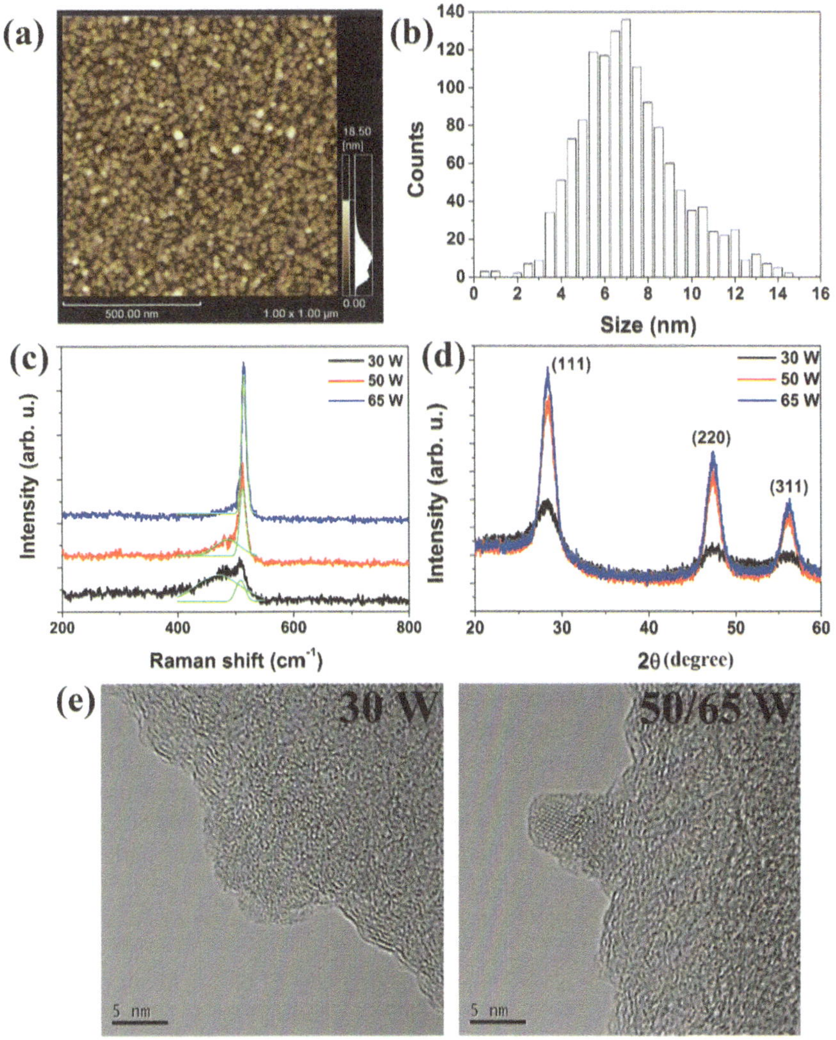


**Fig. 4** (**a**) Typical AFM image and (**b**) size distribution of Si NPs; (**c**) Raman spectra, (**d**) XRD patterns and (**e**) corresponding TEM images of particles fabricated under different VHF power (Reprinted with permission from Ref. [49]. Copyright 2015 AIP Publishing LLC)

another tunable dimension that influences the material properties. As an example, three kinds of particles were fabricated by changing VHF power to 30, 50, and 65 watts, respectively, and corresponding Raman spectra of particles are shown in Fig. 4c. The Raman peak can be deconvoluted carefully into two peaks centered around 480 $cm^{-1}$ and 520 $cm^{-1}$, which are related to amorphous and crystalline silicon phase, respectively [32]. Therefore, it's becoming obvious that the crystallinity of Si NPs increases with VHF power, as the peak intensity related to amorphous phase decreases with VHF power and almost disappears when the power increases to 65 W. In addition, X-ray diffraction (XRD) patterns of corresponding particles were also shown in Fig. 4d; three peaks can be observed, which are attributed to (111), (220), and (311) planes of silicon. It shows consistent results with Raman spectra that crystallinity of Si NPs increases with VHF power. Transmission electron microscopy (TEM) images of particles synthesized at different VHF power are shown in Fig. 4e. Consistent with the above analysis, no crystals can be observed in NPs synthesized at 30 W, while similar single-crystallized NCs can be observed in both particles synthesized at 50 and 65 watts.

## 4.3 Optical Properties of Si NPs

Light absorption spectra of Si NPs with different crystallinity are shown in Fig. 5a. Si NPs have a strong light absorption especially in the short-wavelength region. In addition, particles with lower crystallinity have stronger absorption. This implies that amorphous particles can absorb more light, and by integrating them in solar cells, higher solar cell efficiency might be expected. Figure 5b shows the light absorption spectra of pure PTB7 and the corresponding blends of Si NPs/PTB7. PTB7 is a polymer semiconductor, which is commonly employed in organic solar cells as electron donor material. Spectra were initialized with the value around 670 nm, where PTB7 has the strongest absorption, while Si NPs absorb a little as confirmed in Fig. 5a. It is undoubted that PTB7 has a weak absorption in short-wavelength region. Fortunately, combining with Si NPs can compensate the light absorption in this region. Moreover, as shown in the Fig. 5b, the blend containing Si NPs with lower crystallinity has better absorption in short-wavelength region; this is also consistent with the foregoing statements.

Figure 6 shows the photoluminescence (PL) spectra from Si NCs. Strong PL with a near Gaussian profile is observed for Si NCs after exposure to air, while weak PL with a similar profile is observed from as-produced NCs without oxidation. It was not expected that even weak PL would be observed for Si NCs which have a surface partially covered by chlorine, as halogens can effectively quench PL [50]. This would indicate that the non-radiative mechanism is not just dependent on the presence of chlorine. The exact PL mechanism of Si NCs is difficult to determine through steady-state PL measurements [51]. Similar steady-state PL has also been observed from both interface defect states and NC core recombination [52]. Additionally, red PL has been attributed to Si = O-related emission centers in porous Si and Si-rich oxides [45]. While here the PL peak wavelength can be adjusted through the residence time, which has been shown above to adjust the NC size. Here we observed peak wavelengths from ~650 to 900 nm with full-width half maxima of 100–120 nm. Additionally, the size of NCs observed in TEM agrees well with the predicted values from peak position due to the quantum confinement effects.

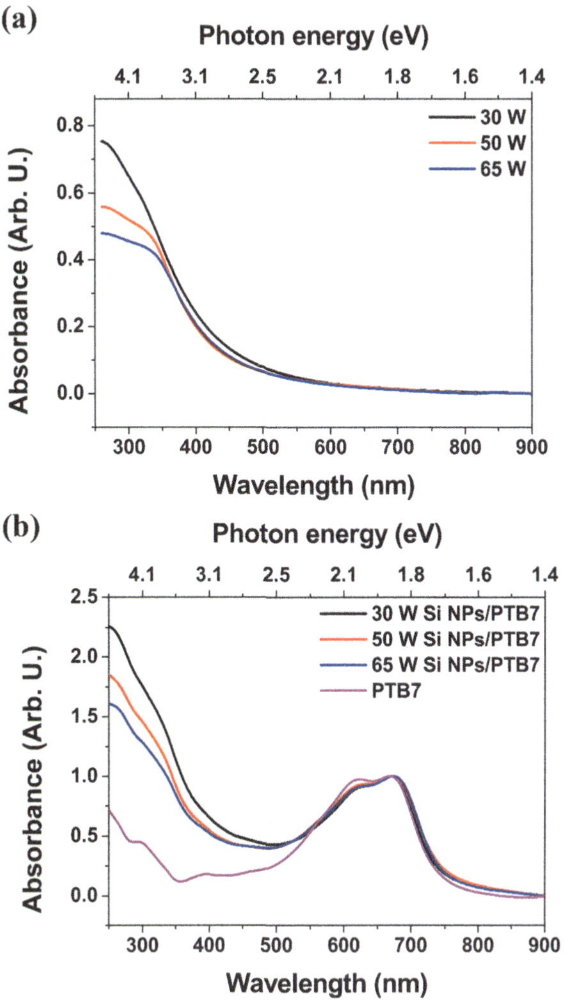


**Fig. 5** Light absorption spectra of (**a**) Si NPs-only, (**b**) PTB7-only, and Si NPs/PTB7 blends. Si NPs were synthesized under different VHF power of 30, 50, and 65 watts, respectively (Reprinted with permission from Ref. [49]. Copyright 2015 AIP Publishing LLC)

## *4.4 Electrical Properties of Si NPs*

Electrical properties of Si NPs were evaluated individually by fabricating Si NP-based thin film transistors (TFTs), from which, carrier mobility of particles can be derived. Si NPs films with the thickness of ~40 nm were deposited on p-type silicon substrate by spin-coating method, above which a 150 nm-thick thermal oxide layer had been prepared in advance and worked as the insulating layer. Two aluminum electrodes were evaporated finally on the top, and channel *W*/*L* ratio of 2000/20 μm was defined consequently as illustrated in Fig. 7. Figure 8 shows the typical output and transform curves of Si NP-based TFTs, which were measured with Keithley 4200 analyzer in glove box [46, 49]. All TFTs show n-channel depletion type.

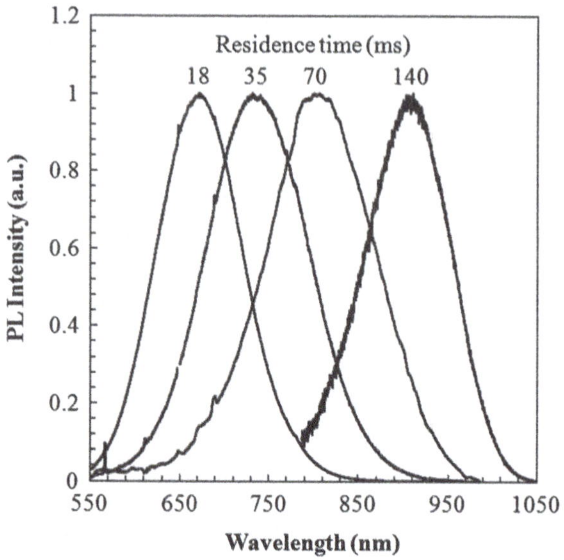


**Fig. 6** PL of oxidized Si NCs as a function of residence time. Estimated NC size from left to right is 3.4, 4, 5.4, and 8 nm (Reprinted with permission from Ref. [32]. Copyright 2011 IOP Publishing Ltd)

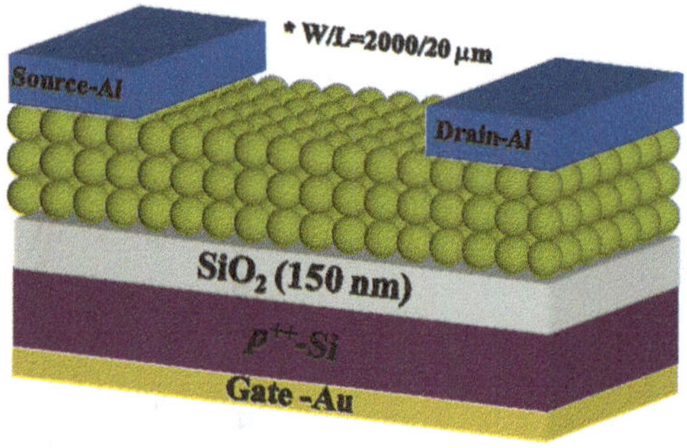


**Fig. 7** Schematic diagram of Si NP-based TFTs (Reprinted with permission from Ref. [46]. Copyright 2015 Elsevier Ltd)

Carrier mobility, $\mu$, and the threshold voltage, $V_{th}$, can be derived from the $V_g$–$I_d^{1/2}$ curve by fitting the linear region with a line as illustrated in Fig. 8. $V_{th}$ equals the x intercept of the linear function, and $\mu$ can be calculated through the following relations [53]:

$$\mu = \frac{2L}{WC_{inv}} \left( \frac{\partial \sqrt{I_d}}{\partial V_g} \right)^2 \tag{1}$$

and

$$C_{inv} = \frac{k\epsilon_0}{t}, \tag{2}$$

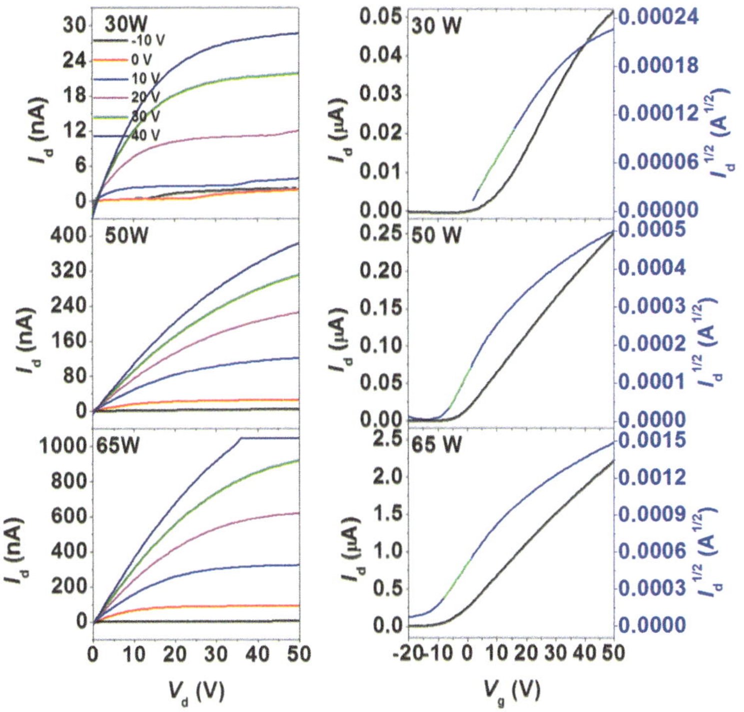


**Fig. 8** Output and transfer curves of Si NP-based TFTs (Reprinted with permission from Ref. [49]. Copyright 2015 AIP Publishing LLC)

where $L$ and $W$ are the channel length and width, respectively. $k$ is the relative dielectric constant of the material (3.9 for thermal oxide), $\varepsilon_0$ is the permittivity of free space (8.85 $\times$ $10^{-14}$ F/cm), and $t$ is the thickness of the capacitor oxide insulator (here, 150 nm).

Carrier motilities of Si NPs derived from TFTs are listed in Table 1, and it increases with particle crystallinity. This is reasonable as more dangling bonds and defects should be observed in particles with poor crystallinity; in addition, carrier scattering effect due to the disorder nature also affects the carrier transportation. The relationship between film conductivity and crystallinity is empirically expressed as follows [54]:

$$\sigma \propto \exp(\mathrm{X_C}), \tag{3}$$

where $\sigma$ and $X_C$ represent film conductivity and crystallinity, respectively.

**Table 1** TFT parameters derived from corresponding transfer curves [49]

| TFTs | Mobility [$cm^2\ V^{-1}\ s^{-1}$] | $V_{th}$ [V] | Off current [A] | On current [A] | On/off ratio |
|---|---|---|---|---|---|
| 30 W | $3.36 \times 10^{-5}$ | −0.48 | $1.06 \times 10^{-10}$ | $5.16 \times 10^{-8}$ | 487 |
| 50 W | $1.95 \times 10^{-4}$ | −8.36 | $1.78 \times 10^{-10}$ | $2.51 \times 10^{-7}$ | 1407 |
| 65 W | $9.42 \times 10^{-4}$ | −15.41 | $5.37 \times 10^{-9}$ | $2.22 \times 10^{-6}$ | 414 |

# 5 Hybrid Solar Cell Fabrication

As the attractive candidates of acceptor materials, inorganic semiconductor NCs are being increasingly introduced into organic-based bulk heterojunction solar cells [7, 55]. Compared to the organic counterparts, normally fullerene derivatives, NCs have promising potential in more efficient light harvesting while keeping the superiorities which organic materials possess such as light weight, flexibility, and solution processable. In addition, NCs have several unique advantages. First, the shape can be tailored through synthesis methods; this has been demonstrated useful for light trapping, exciton dissociation, as well as carrier transportation [56, 57]. Also, when the size is comparable or smaller than its Bohr radius, the quantum effect makes NC work totally different from its bulk material. Bandgap increases with decreasing size; this fact changes light absorption properties and provides the possibility of choosing the spectral window of the complementary absorption profile. Meanwhile, continuous energy bands start to shrink into discrete states. This suppresses the energy loss through thermal relaxation and facilitates MEG process, an effective strategy expected to promote solar cell PCE as mentioned above.

Recently, remarkable accomplishments have been achieved in NC/polymer hybrid solar cells (HSCs). Among them, CdS- and PbS-based HSCs with PCE of 4.1% and 8.5% have been achieved, respectively [58, 59]. Employing wide bandgap metal-oxide NCs such as $TiO_2$ and ZnO has also achieved a PCE as high as 2% [60]. In addition to those discussed above, Si NC has been extensively studied and shown feasibility as a proper acceptor material in HSCs. As discussed above, silicon is environmentally benign and has strong light absorption especially in the UV region. Furthermore, instead of the complicated solution process commonly employed for NC synthesis (e.g., ZnO, PbS, and CdS.), mass production can be realized conveniently by means of the nonthermal plasma. In this section, the fabrication, characterization, factors affecting device performance, and working principles of Si NC-based organic/inorganic hybrid solar cells are investigated extensively.

## *5.1 Si NC-Based Hybrid Solar Cell Fabrication*

Hybrid solar cell structure is illustrated in Fig. 9. Device fabrication and characterizations were mainly performed in a nitrogen-purged glove box (<1 ppm water and oxygen). Commercially available ITO glass with sheet resistance of 15 Ω/□

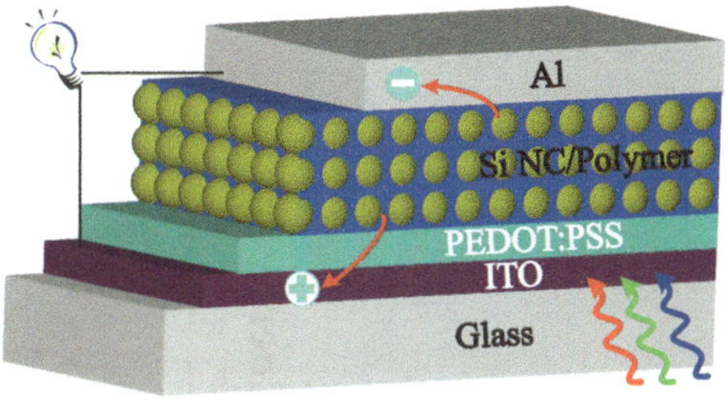


**Fig. 9** Schematic diagram of a hybrid solar cell with structure of ITO/PEDOT:PSS/ Si NC-polymer blend/ Al (Reprinted with permission from Ref. [46]. Copyright 2015 Elsevier Ltd)

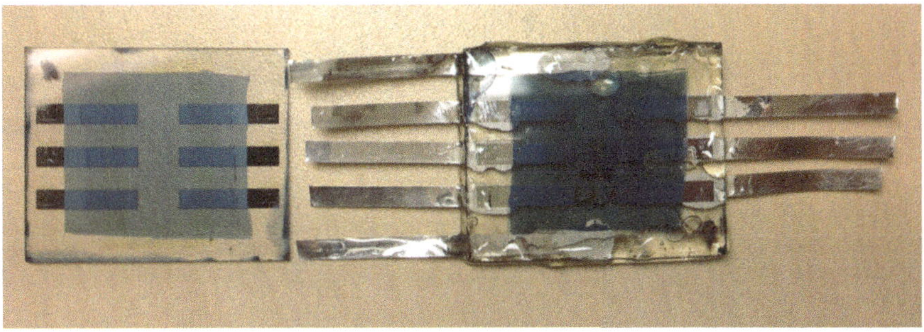

**Fig. 10** Photos of devices before and after encapsulation (Reprinted with permission from Ref. [39]. Copyright 2015 Elsevier Ltd)

and thickness of 250 nm was adopted as the substrate. The ITO was patterned with photolithography, and wet etching was used to develop the electrodes. Poly(3,4-ethylenedioxylenethiophene):poly(styrenesulfonic acid) (PEDOT:PSS, CLEVIOS PH 1000, Heraeus) mixed with 5 wt% of dimethyl sulfoxide (DMSO, Sigma Aldrich) and 0.1 wt% of Zonyl (FS-300, Sigma Aldrich) was spin casted on it at 4000 rpm for 1 min and annealed at 130 °C for 5 min. Si NC/polymer blend solutions were prepared and stirred for 24 h prior to use and then spin casted at 1200 rpm. Finally, Al electrodes with thickness of 100 nm were evaporated through a shadow mask (active area 4.6 $mm^2$). Six devices were prepared to obtain the standard deviations. Devices were annealed at 110 °C for 2 h before evaluation. Photocurrent density–voltage (*J*–*V*) curves were measured under 1 sun (AM 1.5G, 100 $mW/cm^2$) illumination with a Keithley 2400 digital source meter. Light source was calibrated by Asahi Spectra Co., Ltd. Incident light power is checked before every measurement by using a standard silicon photodiode. Devices were also encapsulated to confirm *J*–*V*, and external quantum efficiency (EQE) spectra were also obtained on a Hypermonolight system in open air (Bunkoukeiki, CEP-25BX). Encapsulation avoids detrimental influences from oxygen and moisture. Photos of devices before and after encapsulation can be found in Fig. 10.

Possible band alignment of the device is shown in Fig. 11, where, as an example, P3HT was used as the polymer semiconductor. Type II staggered band alignment without obvious potential barrier is established, enabling efficient charge separation, transportation, and collection. The conduction band of bulk silicon is ~4 eV

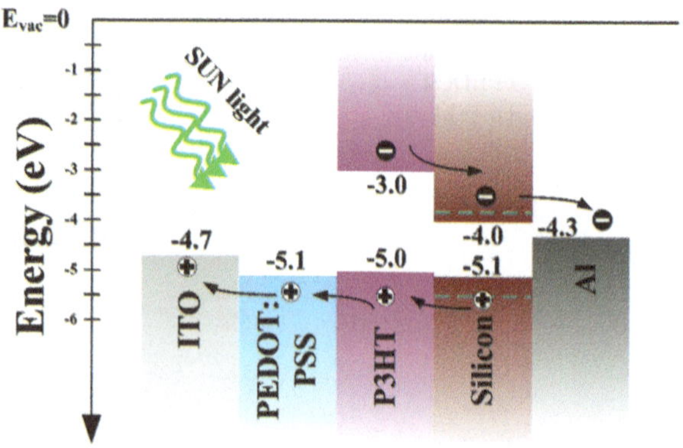


Fig. 11 Possible band alignment of Si NC/P3HT hybrid solar cells (Reprinted with permission from Ref. [38]. Copyright (2013) The Japan Society of Applied Physics)

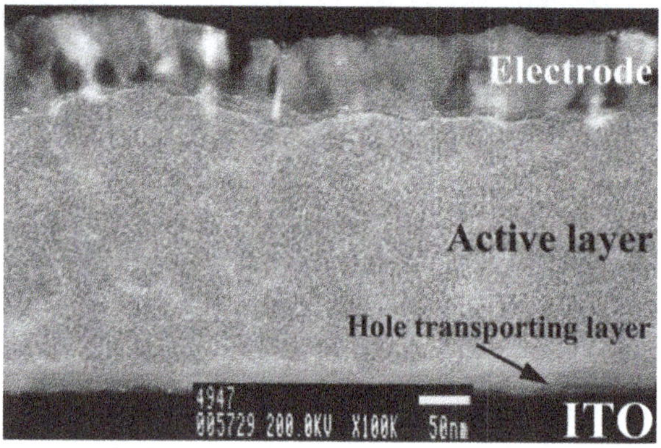


Fig. 12 Typical cross-sectional TEM image of a hybrid solar cell with structure of ITO/PEDOT: PSS/Si NC-polymer blend/ Al (Reprinted with permission from Ref. [37]. Copyright 2015 Elsevier Ltd)

with respect to vacuum and has a bandgap of ~1.1 eV. P3HT is expected to act as the electron donor material and silicon as the electron acceptor. When the size of Si NCs is less than the exciton Bohr radius (~4.1 nm), quantum confinement effects increase the bandgap of Si NCs with respect to the bulk. Previously it was found that for quantum confined Si NCs, the band shift of the valence band is as twice large as the conduction band as illustrated with dotted line [61].

Nanoscale morphology arrangement plays a crucial role in polymer-related devices because of its high exciton binding energy (~100 meV) as well as short diffusion length (~10 nm) [62]. Materials need to percolate with each other to form bulk heterojunction in order to achieve sufficient exciton dissociation. Figure 12 shows the typical cross-sectional TEM image of device; Si NCs are embedded uniformly in conjugated polymer without remarkable agglomerations providing sufficient opportunity for exciton dissociation around Si NC/polymer interfaces. This has been further verified by elemental mapping as shown in Fig. 13. It is worth to mention here that the uniform structure can be realized for all Si NCs with different terminations examined here. Although significant agglomeration is absent, a pathway of touching NCs is expected for this type of homogenous blend at proper Si NCs ratios.

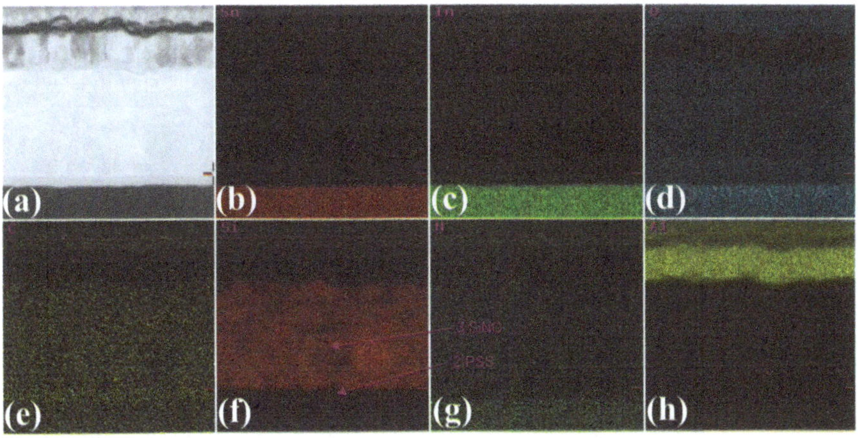


**Fig. 13** (**a**) Typical cross-sectional TEM image of device. (**b–h**) And corresponding elemental mappings of Sn, In, O, C, Si, N, and Al on the same area. ITO, PEDOT:PSS, active layer, and Al electrode can be clearly distinguished. Each layer has a uniform morphology and well-defined interface. Si NCs distribute uniformly in the whole active region as shown in (**f**) (Reprinted with permission from Ref. [39]. Copyright 2015 Elsevier Ltd)

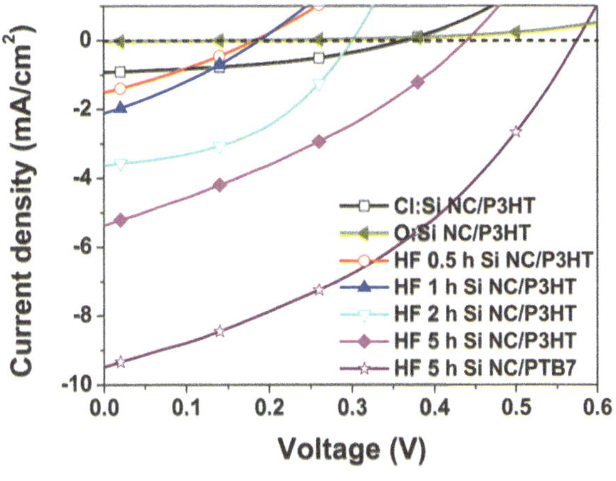


**Fig. 14** *J–V* curves of hybrid solar cells under 1 sun illumination; see legend for description of curves (Reprinted with permission from Ref. [37]. Copyright 2015 Elsevier Ltd)

## 5.2 The Influences of Particle Surface Elements

Figure 14 shows the *J–V* curves for devices fabricated with Cl:Si NCs, O:Si NCs, and HF-etched Si NCs measured under AM 1.5G solar-simulated light at room temperature. Device performance parameters with standard deviations of six devices are summarized in Table 2. Devices with Cl:Si NCs have a low short-circuit current ($J_{SC}$) of 0.7 mA/cm$^2$; this value drops to almost zero when Si NCs are totally oxidized and then increases dramatically after etching. It can be enhanced to

**Table 2** Hybrid solar cell performance with standard deviations [37]

| Device | PCE [%] | $J_{SC}$ [mA/cm$^2$] | $V_{OC}$ [mV] | FF [%] | $R_{SH}$ [Ω·cm$^2$] | $R_S$ [Ω·cm$^2$] |
|---|---|---|---|---|---|---|
| Cl:Si NC/P3HT | 0.10 ± 0.05 | 0.70 ± 0.36 | 347 ± 13 | 39.9 ± 1.7 | 1663 ± 113 | 154 ± 13 |
| O:Si NC/P3HT | N/A | N/A | N/A | N/A | N/A | N/A |
| HF 0.5 h Si NC/P3HT | 0.06 ± 0.01 | 1.21 ± 0.21 | 179 ± 14 | 29.0 ± 0.8 | 248 ± 42 | 112 ± 20 |
| HF 1 h Si NC/P3HT | 0.10 ± 0.01 | 1.79 ± 0.19 | 192 ± 5 | 29.9 ± 0.8 | 228 ± 23 | 78 ± 12 |
| HF 2 h Si NC/P3HT | 0.45 ± 0.06 | 3.63 ± 0.18 | 299 ± 4 | 41.9 ± 3.8 | 520 ± 144 | 29 ± 1 |
| HF 5 h Si NC/P3HT | 0.52 ± 0.15 | 3.73 ± 0.98 | 439 ± 7 | 31.4 ± 0.8 | 201 ± 33 | 68 ± 17 |
| HF 5 h Si NC/PTB7 | 2.25 ± 0.18 | 9.97 ± 0.41 | 579 ± 2 | 38.9 ± 1.8 | 148 ± 30 | 24 ± 1 |

as high as 1.2 mA/cm$^2$ by only 0.5-h HF etching. This enhancement is because of the removal of chlorine, which assists in effectively highly doping the NCs. However, after a 0.5-h etching, open-circuit voltage ($V_{OC}$) drops to 0.18 V. This is because that Si NC surface is still covered by plenty of oxide even after 0.5-h etching as verified in Fig. 2a. The surface oxygen is predicted to increase the electron affinity and lower the conduction band [63]. It is therefore reasonable that the $V_{OC}$ is lower due to the presence of a significant amount of oxide as the $V_{OC}$ is mainly defined by the energy-level mismatch between the highest occupied molecular orbital (HOMO) state of polymer and conduction band of Si NC [64]. For longer etching times, $J_{SC}$ and $V_{OC}$ increase, and a PCE of 0.5% has been achieved after a 5-h HF etching, which is at least three times higher than the Cl: Si NC-based device.

### 5.3 The Influences of Polymer Semiconductors

P3HT has been commonly employed as the donor material so far. It is also widely used in other types of HSCs, because of its efficient light absorption and high hole mobility [65]. However, it may not be the best choice in Si NC-based HSCs due to its wide bandgap of 2.0 eV. Absorbance spectra of Si NCs etched for 5 h, P3HT, and Si NC/P3HT blend films are shown in Fig. 15a semiquantitatively. P3HT shows a strong absorption peak centered around 550 nm. Si NC films start to absorb light gradually from the IR to the UV region. Absorption spectrum of Si NC/P3HT blend film shows complementary absorption of the individual components. Unfortunately, light with wavelengths greater than 650 nm is not effectively absorbed and hence cannot contribute to the generation of excitons. Therefore, in Si NC-

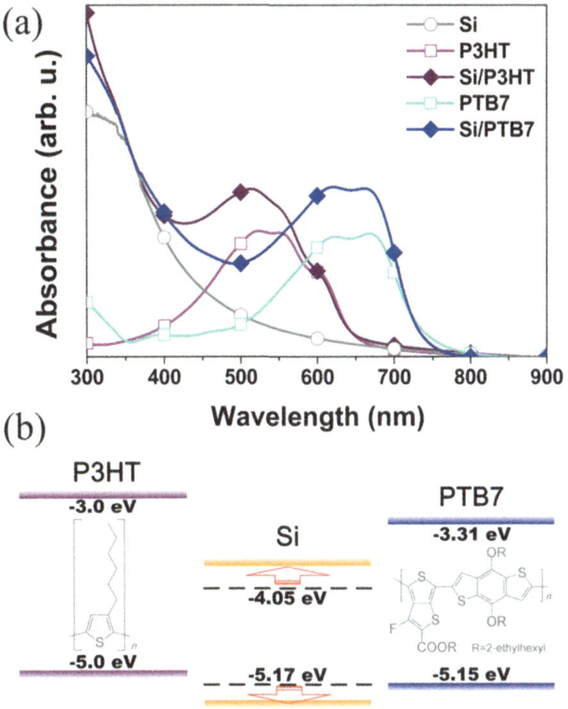


**Fig. 15** (**a**) Absorbance spectra of Si NCs, P3HT, PTB7, Si NC/P3HT blend, and Si NC/PTB7 blend films. (**b**) Chemical structures and energy-level diagrams of Si NCs, P3HT, and PTB7 (Reprinted with permission from Ref. [37]. Copyright 2015 Elsevier Ltd)

based HSCs, a donor material with a smaller bandgap is preferred, allowing light with longer wavelengths to be converted.

PTB7 should be a more suitable donor material than P3HT with an effective bandgap of ~1.8 eV [66]. Si NC/PTB7 blend films have sufficient complementary absorption over the range of 300 to 800 nm as shown in Fig. 15a, which is expected to produce higher photocurrents. Figure 15b shows the molecular structures and corresponding energy levels of P3HT, PTB7, and Si NCs. The alignment of the energy levels of P3HT and Si NCs is expected to form a type II heterojunction, which is favorable for exciton dissociation at the interface. A similar alignment is expected for PTB7 and Si NCs with a slightly large offset due to the HOMO level of PTB7 being about 0.15 eV lower than that of P3HT. This enlarges the difference between the HOMO level of PTB7 and the conduction band of Si NCs and leads to a large $V_{OC}$ theoretically.

A typical $J$–$V$ curve of Si NC/PTB7 device with Si NCs etched for 5 h is also shown in Fig. 14. Compared to Si NC/P3HT devices, $J_{SC}$ is enhanced dramatically, which is because of the improved light harvest as mentioned above. In addition, $V_{OC}$ has a remarkable promotion. Besides the contribution from large offset between HOMO level of PTB7 and the conduction band of Si NCs, enhanced $J_{SC}$ also leads to a larger $V_{OC}$ according to the relation:

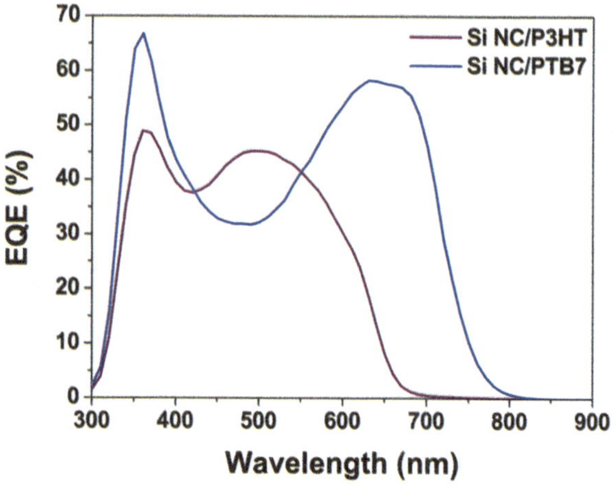


**Fig. 16** Typical EQE spectra for Si NC/P3HT and Si NC/PTB7 devices (Reprinted with permission from Ref. [37]. Copyright 2015 Elsevier Ltd)

$$V_{OC} = \frac{nkT}{q} \ln\left(\frac{J_{SC}}{J_0} + 1\right), \tag{4}$$

where $J_0$ is the dark saturation current, n is the ideality factor, $k$ is the Boltzmann constant, $q$ is the charge of an electron, and $T$ is the temperature. Finally, a PCE of 2.2% has been achieved, which is four times higher than Si NC/P3HT device with same Si NCs and at least ten times higher than Cl:Si NC-based device.

EQE spectra of typical Si NC/P3HT and Si NC/PTB7 devices are shown in Fig. 16. Si NC/P3HT devices show photocurrent response to ~650 nm with a maximum EQE of 50%, while Si NC/PTB7 devices extend the photocurrent response to ~800 nm with a maximum EQE as high as 70%. Photocurrent response ranges are consistent with the corresponding absorption spectra, and the theoretical $J_{SC}$ values obtained by integrating the product of the EQE are in good agreement with the measured value under 1 Sun.

## *5.4 The Influences of Si NC/PTB7 Mass Ratio*

Figure 17 illustrates the device parameters as a function of Si NCs and PTB7 weight ratio. A higher weight ratio indicates that more Si NCs have been introduced into the device. Shunt resistance ($R_{sh}$), typically related to device defects, remains almost stable, indicating that no more carrier traps have been introduced with more Si NCs, which suggests that particle defects have been passivated effectively after proper surface treatment, whereas the series resistance ($R_s$) decreases gradually with the Si NC/PTB7 weight ratio. $R_s$ is commonly recognized as the internal load of device and decided mainly by carrier mobility. It implies that the device

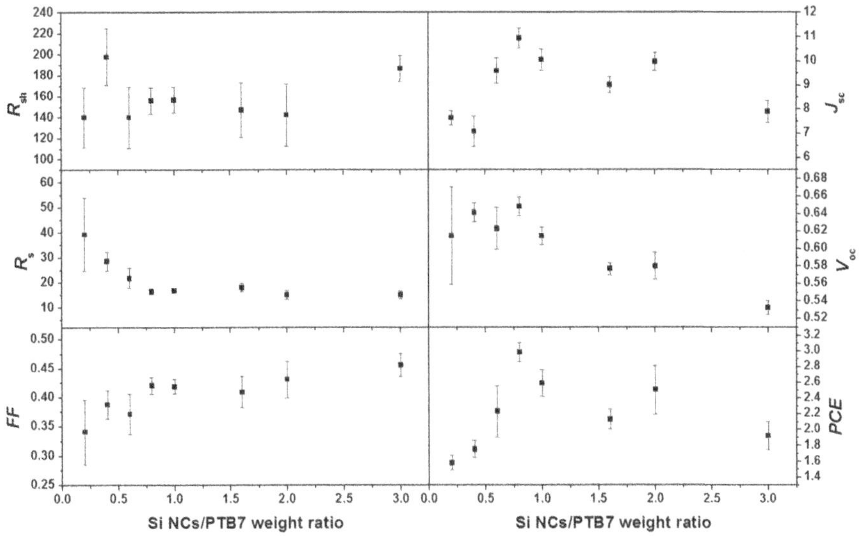


**Fig. 17** Dependencies of device parameters on Si NC/PTB7 weight ratio (Reprinted with permission from Ref. [67]. Copyright 2015 IOP Publishing Ltd)

carrier mobility has been improved with more Si NCs. A possible reason might be because that Si NCs come closer mutually with the increasing of Si NC/PTB7 mass ratio. Consequently, the carrier hopping efficiency through NCs will be enhanced dramatically, which in turn results in increased carrier mobility. In addition, the fill factor (FF) increases gradually with the mass ratio; this is convinced and mainly benefits from unaltered defect concentration along with enhanced carrier transportation, as discussed above.

Device $J_{SC}$ increases until the weight ratio reaches to 0.8. The conjugated polymer has relatively higher exciton binding energy hence an electric field; in other words, a junction is necessary for exciton dissociation. By introducing Si NCs, a new junction is established between PTB7 and Si NCs. Therefore, Si NCs promote exciton dissociation process. As more Si NCs are incorporated, a larger interface can be formed, and more excitons could be separated effectively. Therefore, $J_{SC}$ increases with the weight ratio. However, excess Si NCs will reduce the total light absorption of device because Si NCs have weaker light harvest compared to PTB7. This is also responsible for the following reduction of current when this ratio exceeds 0.8. $V_{OC}$ has a similar tendency with $J_{SC}$, which is reasonable because $V_{OC}$ is proportional to $J_{SC}$ according to the relation (4) shown above. Device PCE, as the consequence of all parameters discussed, increases first and then decreases when the ratio exceeds 0.8, yielding a maximum PCE as high as 3.0%.

Figure 18a shows the typical normalized EQE spectra of devices with different mass ratios. Each spectrum was multiplied by a certain factor to normalize the value

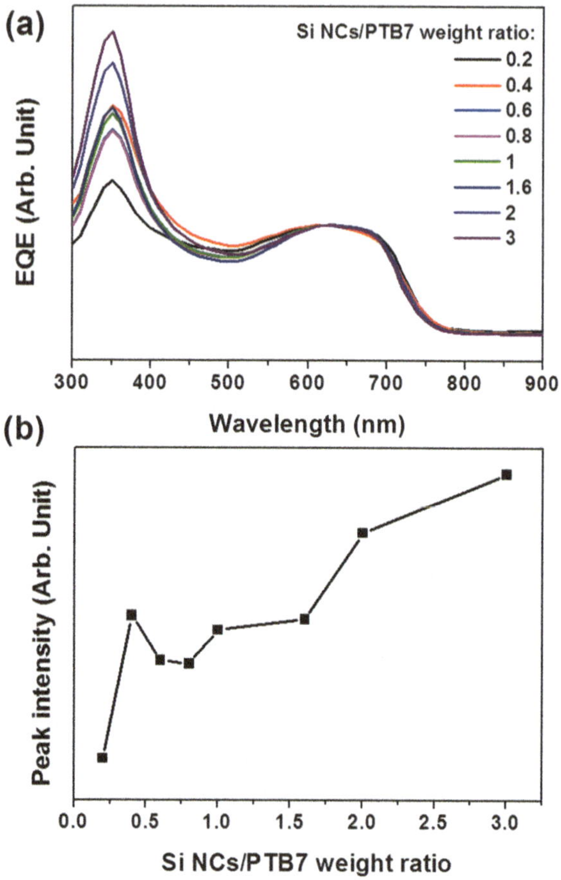


**Fig. 18** (**a**) Normalized EQE spectra of devices with different Si NC/PTB7 weight ratio. (**b**) Peak intensity at 350 nm as the function of Si NC/PTB7 weight ratio (Reprinted with permission from Ref. [67]. Copyright 2015 IOP Publishing Ltd)

based on the peak intensity at 650 nm. Each spectrum, respectively, shows two peaks centered around 350 nm and 650 nm. The peak at 650 nm is known as the contribution from PTB7. If one presumes that the contributions of PTB7 in all devices are the same, the peak intensity around 350 nm will increase almost linearly with the weight ratio as shown in Fig. 18b, suggesting that this peak is attributed to the light absorption of Si NCs.

## *5.5 The Influences of Si NCs Crystallinity*

Figure 19 shows the typical *J–V* curves of hybrid solar cells where Si NPs with different crystallinity were integrated. Device parameters derived from *J–V* curves

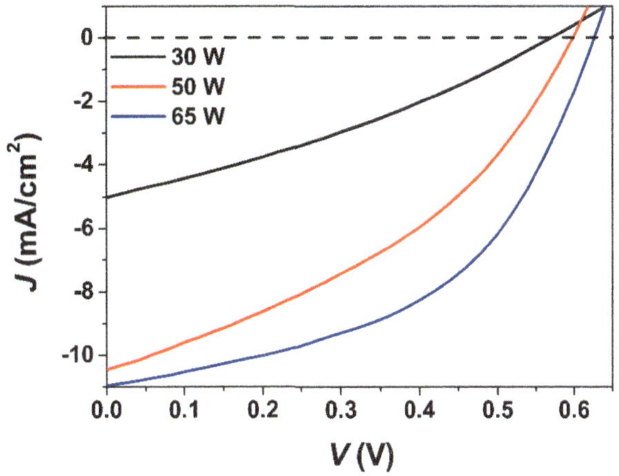


**Fig. 19** Typical *J–V* curves of Si NP-based hybrid solar cells devices, where Si NPs with different crystallinity were integrated (Reprinted with permission from Ref. [49]. Copyright 2015 AIP Publishing LLC)

**Table 3** Solar cell parameters derived from corresponding *J–V* curves [49]

| Solar cells | PCE [%] | $J_{SC}$ [mA/cm$^2$] | $V_{OC}$ [mV] | FF [%] | $J_0$ [mA/cm$^2$] |
|---|---|---|---|---|---|
| 30 W | 0.98 | 5.00 | 567 | 32.0 | −0.097 |
| 50 W | 2.38 | 10.43 | 597 | 38.2 | −0.419 |
| 65 W | 3.36 | 10.94 | 628 | 43.4 | −0.329 |

are summarized in Table 3. $J_{SC}$, $V_{OC}$, FF, and PCE increase with particle crystallinity. And device with amorphous Si NPs has the lowest PCE although they possess the highest light absorption as shown hereinabove (Sects. 4.3 and 4.4). This is mainly because that although amorphous particles demonstrate better light absorption especially in short-wavelength region, however, photogenerated carriers cannot be collected to the electrode efficiently, and most of them will be trapped and recombined again in particles. As a result, corresponding solar cells show the worse performance.

## 5.6 *The Influences of Si NCs Surface Passivation*

Typical *J–V* curves of devices with Si NCs oxidized for different time are shown in Fig. 20. Solar cell parameters derived from corresponding *J–V* curves are summarized in Table 4. In all devices, FF and $V_{OC}$ do not change drastically with oxidation time and stay around 50% and 0.6 V, respectively.

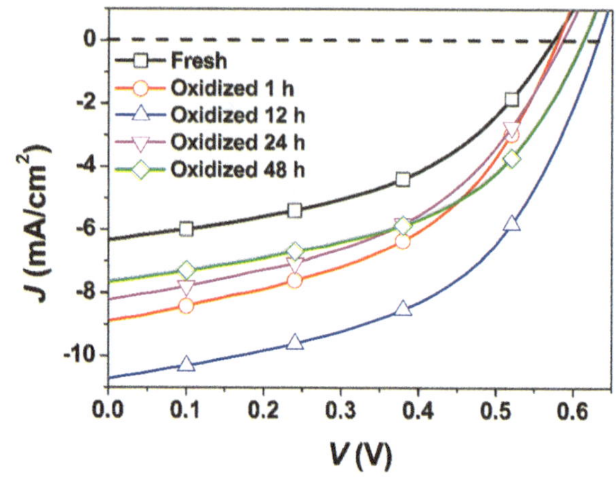


**Fig. 20** Typical photo *J*–*V* curves of solar cells fabricated with Si NCs oxidized for different time (Reprinted with permission from Ref. [46]. Copyright 2015 Elsevier Ltd)

On the other hand, as shown in Fig. 20, $J_{SC}$ and PCE increase until the oxidation time exceeds 12 h. The average PCE has been improved considerably from 1.7% in a device with fresh Si NCs to 3.3% after 12-h oxidization, which is enhanced almost twofold, and undoubtedly, due to effective oxygen passivation. In addition, the best PCE of 3.6% has been achieved eventually.

Both $J_{SC}$ and PCE then decrease with further oxidization. This tendency is similar to that of electron mobility of Si NCs as shown in Fig. 21 and Table 5, because the electron mobility has a strong relationship with device performance. Unbalanced electron and hole mobility in solar cells should be avoided as it leads to carrier accumulation, space-charge-limited photocurrent, and finally, results in deteriorated device performance [68]. Hole mobility of PTB7 can normally be as high as ~$5 \times 10^{-3}$ $cm^2$ $V^{-1}$ $s^{-1}$ [69], which is at least five times higher than the best electron mobility of Si NCs we have achieved (~$9.4 \times 10^{-4}$ $cm^2$ $V^{-1}$ $s^{-1}$). Therefore, in present Si NC/PTB7 solar cells, unbalanced charge transport occurs, and solar cell performance is still limited significantly by Si NC electron mobility.

## 6 Conclusions

In this chapter, advanced properties and fabrication methods of Si NCs were described. Detailed synthesis process of Si NCs by using nonthermal plasma was also discussed. Some basic interesting properties of resulted Si NCs, such as optical and electrical properties, have been listed afterward. Finally, investigations focusing on the design, fabrication, and performance characterization of Si NC-based organic/inorganic hybrid solar cells were carefully carried out. Moreover, effects of Si NCs on device performance were also studied extensively.

**Table 4** Si NC/PTB7 hybrid solar cell performance with standard deviations [46]

| Solar cell | Best PCE [%] | PCE [%] | $J_{SC}$ [mA/cm$^2$] | $V_{OC}$ [mV] | FF [%] | $R_{SH}$ [Ω·cm$^2$] | $R_S$ [Ω·cm$^2$] |
|---|---|---|---|---|---|---|---|
| Fresh | 1.74 | 1.68 ± 0.06 | 6.34 ± 0.04 | 582 ± 5 | 45.6 ± 1.0 | 323 ± 61 | 26 ± 2 |
| Oxidized 1 h | 2.79 | 2.43 ± 0.21 | 8.86 ± 0.68 | 583 ± 5 | 47.1 ± 1.1 | 240 ± 34 | 17 ± 2 |
| Oxidized 12 h | 3.57 | 3.34 ± 0.14 | 10.83 ± 0.39 | 628 ± 4 | 49.0 ± 0.7 | 244 ± 19 | 15 ± 1 |
| Oxidized 24 h | 2.59 | 2.28 ± 0.17 | 8.27 ± 0.57 | 591 ± 11 | 46.6 ± 0.9 | 248 ± 26 | 20 ± 2 |
| Oxidized 48 h | 2.59 | 2.30 ± 0.25 | 7.66 ± 0.83 | 618 ± 6 | 48.5 ± 0.7 | 285 ± 43 | 18 ± 2 |

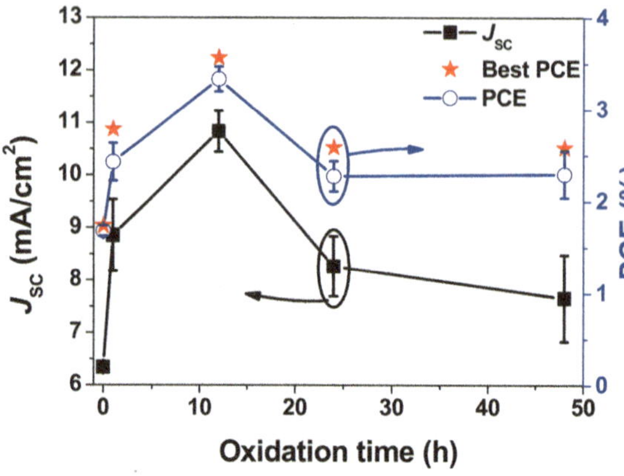


**Fig. 21** Dependences of device $J_{SC}$ and PCE on Si NC oxidation time. Best PCEs of devices for each condition are also marked with asterisk (Reprinted with permission from Ref. [46]. Copyright 2015 Elsevier Ltd)

**Table 5** Parameters of TFTs fabricated with Si NCs oxidized for different time [46]

| TFT | Mobility [cm$^2$ V$^{-1}$ s$^{-1}$] | $V_{th}$ [V] | Off current [A] | On current [A] | On/Off ratio |
|---|---|---|---|---|---|
| Fresh | 3.00E-4 | −11.3 | 1.34E-9 | 7.70E-7 | 575 |
| Oxidized 1 h | 5.57E-4 | −13.5 | 2.32E-9 | 1.30E-6 | 564 |
| Oxidized 12 h | 9.42E-4 | −15.4 | 5.37E-9 | 2.22E-6 | 413 |
| Oxidized 24 h | 5.97E-4 | −14.7 | 3.58E-9 | 1.45E-6 | 405 |
| Oxidized 48 h | 4.66E-4 | −14.1 | 2.72E-9 | 1.09E-6 | 400 |

**Acknowledgments** The authors would thank all contributors to the publication, who have been important for this work. This work was financially supported by the National Natural Science Foundation of China (No. 61504069), the 111 Project of China (B16027), and Grant-in-Aid for Scientific Research (b) of Japan (No. 26289045).

## References


1. W.K. Leutwyler, S.L. Bürgi, H. Burgl, Semiconductor clusters, nanocrystals, and quantum dots. Science **271**(5251), 933–937 (1996)
2. A.P. Alivisatos, Perspectives on the physical chemistry of semiconductor nanocrystals. J. Phys. Chem. **100**(31), 13226–13239 (1996)
3. F.W. Wise, Lead salt quantum dots: the limit of strong quantum confinement. Acc. Chem. Res. **33**(11), 773–780 (2000)
4. W.E. Buhro, V.L. Colvin, Semiconductor nanocrystals: shape matters. Nat. Mater. **2**(3), 138–139 (2003)
5. D.V. Melnikov, J.R. Chelikowsky, Quantum confinement in phosphorus-doped silicon nanocrystals. Phys. Rev. Lett. **92**(4), 046802 (2004)
6. I. Gur, N.A. Fromer, M.L. Geier, A.P. Alivisatos, Air-stable all-inorganic nanocrystal solar cells processed from solution. Science **310**(5747), 462–465 (2005)
7. W.U. Huynh, J.J. Dittmer, A.P. Alivisatos, Hybrid nanorod-polymer solar cells. Science **295** (5564), 2425–2427 (2002)

8. V. Colvin, M. Schlamp, A. Alivisatos, Light-emitting diodes made from cadmium selenide nanocrystals and a semiconducting polymer. Nature **370**(6488), 354–357 (1994)
9. M. Schlamp, X. Peng, A. Alivisatos, Improved efficiencies in light emitting diodes made with CdSe (CdS) core/shell type nanocrystals and a semiconducting polymer. J. Appl. Phys. **82**(11), 5837–5842 (1997)
10. O.E. Semonin, J.M. Luther, S. Choi, H.-Y. Chen, J. Gao, A.J. Nozik, M.C. Beard, Peak external photocurrent quantum efficiency exceeding 100% via MEG in a quantum dot solar cell. Science **334**(6062), 1530–1533 (2011)
11. J.M. Luther, M. Law, M.C. Beard, Q. Song, M.O. Reese, R.J. Ellingson, A.J. Nozik, Schottky solar cells based on colloidal nanocrystal films. Nano Lett. **8**(10), 3488–3492 (2008)
12. B.A. Ridley, B. Nivi, J.M. Jacobson, All-inorganic field effect transistors fabricated by printing. Science **286**(5440), 746–749 (1999)
13. J. Ziegler, S. Xu, E. Kucur, F. Meister, M. Batentschuk, F. Gindele, T. Nann, Silica-coated InP/ZnS nanocrystals as converter material in white LEDs. Adv. Mater. **20**(21), 4068–4073 (2008)
14. D. Huang, F. Liao, S. Molesa, D. Redinger, V. Subramanian, Plastic-compatible low resistance printable gold nanoparticle conductors for flexible electronics. J. Electrochem. Soc. **150**(7), G412–G417 (2003)
15. R. Malakooti, L. Cademartiri, Y. Akçakir, S. Petrov, A. Migliori, G.A. Ozin, Shape-controlled Bi2S3 nanocrystals and their plasma polymerization into flexible films. Adv. Mater. **18**(16), 2189–2194 (2006)
16. Y. Wang, N. Herron, Nanometer-sized semiconductor clusters: materials synthesis, quantum size effects, and photophysical properties. J. Phys. Chem. **95**(2), 525–532 (1991)
17. M. Grätzel, Photoelectrochemical cells. Nature **414**(6861), 338–344 (2001)
18. L. Mangolini, E. Thimsen, U. Kortshagen, High-yield plasma synthesis of luminescent silicon nanocrystals. Nano Lett. **5**(4), 655–659 (2005)
19. N. Daldosso, G. Das, S. Larcheri, G. Mariotto, G. Dalba, L. Pavesi, A. Irrera, F. Priolo, F. Iacona, F. Rocca, Silicon nanocrystal formation in annealed silicon-rich silicon oxide films prepared by plasma enhanced chemical vapor deposition. J. Appl. Phys. **101**(11), 113510 (2007)
20. R.K. Baldwin, K.A. Pettigrew, J.C. Garno, P.P. Power, G.-y. Liu, S.M. Kauzlarich, Room temperature solution synthesis of alkyl-capped tetrahedral shaped silicon nanocrystals. J. Am. Chem. Soc. **124**(7), 1150–1151 (2002)
21. K. Littau, P. Szajowski, A. Muller, A. Kortan, L. Brus, A luminescent silicon nanocrystal colloid via a high-temperature aerosol reaction. J. Phys. Chem. **97**(6), 1224–1230 (1993)
22. R.J. Walters, G.I. Bourianoff, H.A. Atwater, Field-effect electroluminescence in silicon nanocrystals. Nat. Mater. **4**(2), 143–146 (2005)
23. Z. Ding, B.M. Quinn, S.K. Haram, L.E. Pell, B.A. Korgel, A.J. Bard, Electrochemistry and electrogenerated chemiluminescence from silicon nanocrystal quantum dots. Science **296** (5571), 1293–1297 (2002)
24. M. Ostraat, J. De Blauwe, M. Green, L. Bell, M. Brongersma, J. Casperson, R. Flagan, H. Atwater, Synthesis and characterization of aerosol silicon nanocrystal nonvolatile floating-gate memory devices. Appl. Phys. Lett. **79**(3), 433–435 (2001)
25. T. Lu, M. Alexe, R. Scholz, V. Talelaev, M. Zacharias, Multilevel charge storage in silicon nanocrystal multilayers. Appl. Phys. Lett. **87**(20), 202110 (2005)
26. C.-Y. Liu, Z.C. Holman, U.R. Kortshagen, Hybrid solar cells from P3HT and silicon nanocrystals. Nano Lett. **9**(1), 449–452 (2008)
27. G. Conibeer, M. Green, E.-C. Cho, D. König, Y.-H. Cho, T. Fangsuwannarak, G. Scardera, E. Pink, Y. Huang, T. Puzzer, Silicon quantum dot nanostructures for tandem photovoltaic cells. Thin Solid Films **516**(20), 6748–6756 (2008)
28. M.C. Beard, K.P. Knutsen, P. Yu, J.M. Luther, Q. Song, W.K. Metzger, R.J. Ellingson, A.J. Nozik, Multiple exciton generation in colloidal silicon nanocrystals. Nano Lett. **7**(8), 2506–2512 (2007)

29. K.-Y. Cheng, R. Anthony, U.R. Kortshagen, R.J. Holmes, High-efficiency silicon nanocrystal light-emitting devices. Nano Lett. **11**(5), 1952–1956 (2011)
30. V. Svrcek, D. Mariotti, T. Nagai, Y. Shibata, I. Turkevych, M. Kondo, Photovoltaic applications of silicon nanocrystal based nanostructures induced by nanosecond laser fragmentation in liquid media. J. Phys. Chem. C **115**(12), 5084–5093 (2011)
31. X. Pi, Q. Li, D. Li, D. Yang, Spin-coating silicon-quantum-dot ink to improve solar cell efficiency. Sol. Energy Mater. Sol. Cells **95**(10), 2941–2945 (2011)
32. R. Gresback, T. Nozaki, K. Okazaki, Synthesis and oxidation of luminescent silicon nanocrystals from silicon tetrachloride by very high frequency nonthermal plasma. Nanotechnology **22**(30), 305605 (2011)
33. N. Shirahata, T. Hasegawa, Y. Sakka, T. Tsuruoka, Size-tunable UV-luminescent silicon nanocrystals. Small **6**(8), 915–921 (2010)
34. A. Stegner, R. Pereira, K. Klein, R. Lechner, R. Dietmueller, M. Brandt, M. Stutzmann, H. Wiggers, Electronic transport in phosphorus-doped silicon nanocrystal networks. Phys. Rev. Lett. **100**(2), 026803 (2008)
35. V. Švrček, T. Sasaki, Y. Shimizu, N. Koshizaki, Blue luminescent silicon nanocrystals prepared by ns pulsed laser ablation in water. Appl. Phys. Lett. **89**(21), 213113 (2006)
36. Y. Ding, R. Yamada, R. Gresback, S. Zhou, X. Pi, T. Nozaki, A parametric study of non-thermal plasma synthesis of silicon nanoparticles from a chlorinated precursor. J. Phys. D. Appl. Phys. **47**(48), 485202 (2014)
37. Y. Ding, R. Gresback, Q. Liu, S. Zhou, X. Pi, T. Nozaki, Silicon nanocrystal conjugated polymer hybrid solar cells with improved performance. Nano Energy **9**, 25–31 (2014)
38. Y. Ding, R. Gresback, R. Yamada, K. Okazaki, T. Nozaki, Hybrid silicon nanocrystal/poly (3-hexylthiophene-2, 5-diyl) solar cells from a chlorinated silicon precursor. Jpn. J. Appl. Phys. **52**(11S), 11NM04 (2013)
39. L.M. Wheeler, N.R. Neale, T. Chen, U.R. Kortshagen, Hypervalent surface interactions for colloidal stability and doping of silicon nanocrystals. Nat. Commun. **4**, 2197 (2013)
40. S. Rivillon, F. Amy, Y.J. Chabal, M.M. Frank, Gas phase chlorination of hydrogen-passivated silicon surfaces. Appl. Phys. Lett. **85**, 2583 (2004)
41. M. Brodsky, M. Cardona, J. Cuomo, Infrared and Raman spectra of the silicon-hydrogen bonds in amorphous silicon prepared by glow discharge and sputtering. Phys. Rev. B **16**(8), 3556 (1977)
42. Y. Ding, D. He, H. Shirai, Deposition of low dielectric constant SiOC films by using an atmospheric pressure microplasma jet. J. Phys. D. Appl. Phys. **42**(12), 125503 (2009)
43. A. Grill, D.A. Neumayer, Structure of low dielectric constant to extreme low dielectric constant SiCOH films: Fourier transform infrared spectroscopy characterization. J. Appl. Phys. **94**(10), 6697–6707 (2003)
44. N. Benissad, K. Aumaille, A. Granier, A. Goullet, Structure and properties of silicon oxide films deposited in a dual microwave-rf plasma reactor. Thin Solid Films **384**(2), 230–235 (2001)
45. M. Wolkin, J. Jorne, P. Fauchet, G. Allan, C. Delerue, Electronic states and luminescence in porous silicon quantum dots: the role of oxygen. Phys. Rev. Lett. **82**(1), 197 (1999)
46. Y. Ding, M. Sugaya, Q. Liu, S. Zhou, T. Nozaki, Oxygen passivation of silicon nanocrystals: influences on trap states, electron mobility, and hybrid solar cell performance. Nano Energy **10**, 322–328 (2014)
47. G. Higashi, Y. Chabal, G. Trucks, K. Raghavachari, Ideal hydrogen termination of the Si (111) surface. Appl. Phys. Lett. **56**(7), 656–658 (1990)
48. R. Gresback, Y. Murakami, Y. Ding, R. Yamada, K. Okazaki, T. Nozaki, Optical extinction spectra of silicon nanocrystals: size dependence upon the lowest direct transition. Langmuir **29** (6), 1802–1807 (2013)
49. Y. Ding, S. Zhou, F.B. Juangsa, M. Sugaya, Y. Asano, X. Zhang, Y. Zhao, T. Nozaki, Optical, electrical, and photovoltaic properties of silicon nanoparticles with different crystallinities. Appl. Phys. Lett. **107**(23), 233108 (2015)

50. D. Selmarten, M. Jones, G. Rumbles, P. Yu, J. Nedeljkovic, S. Shaheen, Quenching of semiconductor quantum dot photoluminescence by a π-conjugated polymer. J. Phys. Chem. B **109**(33), 15927–15932 (2005)
51. M. Sykora, L. Mangolini, R.D. Schaller, U. Kortshagen, D. Jurbergs, V.I. Klimov, Size-dependent intrinsic radiative decay rates of silicon nanocrystals at large confinement energies. Phys. Rev. Lett. **100**(6), 067401 (2008)
52. S. Godefroo, M. Hayne, M. Jivanescu, A. Stesmans, M. Zacharias, O. Lebedev, G. Van Tendeloo, V.V. Moshchalkov, Classification and control of the origin of photoluminescence from Si nanocrystals. Nat. Nanotechnol. **3**(3), 174–178 (2008)
53. S.W. Lin, D.H. Chen, Synthesis of water-soluble blue photoluminescent silicon nanocrystals with oxide surface passivation. Small **5**(1), 72–76 (2009)
54. Y. He, C. Yin, G. Cheng, L. Wang, X. Liu, G. Hu, The structure and properties of nanosize crystalline silicon films. J. Appl. Phys. **75**(2), 797–803 (1994)
55. M. Wright, A. Uddin, Organic—inorganic hybrid solar cells: a comparative review. Sol. Energy Mater. Sol. Cells **107**, 87–111 (2012)
56. J. Weickert, R.B. Dunbar, H.C. Hesse, W. Wiedemann, L. Schmidt-Mende, Nanostructured organic and hybrid solar cells. Adv. Mater. **23**(16), 1810–1828 (2011)
57. A. Salant, M. Shalom, Z. Tachan, S. Buhbut, A. Zaban, U. Banin, Quantum rod-sensitized solar cell: nanocrystal shape effect on the photovoltaic properties. Nano Lett. **12**(4), 2095–2100 (2012)
58. S. Ren, L.-Y. Chang, S.-K. Lim, J. Zhao, M. Smith, N. Zhao, V. Bulovic, M. Bawendi, S. Gradecak, Inorganic–organic hybrid solar cell: bridging quantum dots to conjugated polymer nanowires. Nano Lett. **11**(9), 3998–4002 (2011)
59. Z. Yang, A. Janmohamed, X. Lan, F.P. García de Arquer, O. Voznyy, E. Yassitepe, G.-H. Kim, Z. Ning, X. Gong, R. Comin, Colloidal quantum dot photovoltaics enhanced by perovskite shelling. Nano Lett. **15**(11), 7539–7543 (2015)
60. A. Abrusci, I.-K. Ding, M. Al-Hashimi, T. Segal-Peretz, M.D. McGehee, M. Heeney, G.L. Frey, H.J. Snaith, Facile infiltration of semiconducting polymer into mesoporous electrodes for hybrid solar cells. Energy Environ. Sci. **4**(8), 3051–3058 (2011)
61. J. Von Behren, T. Van Buuren, M. Zacharias, E. Chimowitz, P. Fauchet, Quantum confinement in nanoscale silicon: the correlation of size with bandgap and luminescence. Solid State Commun. **105**(5), 317–322 (1998)
62. H. Hoppe, N.S. Sariciftci, Morphology of polymer/fullerene bulk heterojunction solar cells. J. Mater. Chem. **16**(1), 45–61 (2006)
63. Z. Zhou, L. Brus, R. Friesner, Electronic structure and luminescence of 1.1-and 1.4-nm silicon nanocrystals: oxide shell versus hydrogen passivation. Nano Lett. **3**(2), 163–167 (2003)
64. M.C. Scharber, D. Muhlbacher, M. Koppe, P. Denk, C. Waldauf, A.J. Heeger, C.J. Brabec, Design rules for donors in bulk-heterojunction solar cells-towards 10% energy-conversion efficiency. Adv. Mat.-Deerfield Beach Then Weinheim **18**(6), 789 (2006)
65. P. Vanlaeke, A. Swinnen, I. Haeldermans, G. Vanhoyland, T. Aernouts, D. Cheyns, C. Deibel, J. D'Haen, P. Heremans, J. Poortmans, P3HT/PCBM bulk heterojunction solar cells: relation between morphology and electro-optical characteristics. Sol. Energy Mater. Sol. Cells **90**(14), 2150–2158 (2006)
66. H.-L. Yip, A.K.-Y. Jen, Recent advances in solution-processed interfacial materials for efficient and stable polymer solar cells. Energy Environ. Sci. **5**(3), 5994–6011 (2012)
67. Y. Ding, R. Gresback, S. Zhou, X. Pi, T. Nozaki, Silicon nanocrystals synthesized using very high frequency non-thermal plasma and their application in photovoltaics. J. Phys. D. Appl. Phys. **48**, 314011 (2015)
68. P.W. Blom, V.D. Mihailetchi, L.J.A. Koster, D.E. Markov, Device physics of polymer: fullerene bulk heterojunction solar cells. Adv. Mater. **19**(12), 1551–1566 (2007)
69. P. Cheng, Y. Li, X. Zhan, Efficient ternary blend polymer solar cells with indene-C 60 bisadduct as an electron-cascade acceptor. Energy Environ. Sci. **7**(6), 2005–2011 (2014)

# Organic–Inorganic Hybrid Silicon Solar

**Yingfeng Li, Younan Luo, and Meicheng Li**

**Abstract** The real challenge of solar cells can be summarized to reduce the cost while increasing the conversion efficiencies. Organic–inorganic hybrid silicon solar cell is a newly developed type of solar cells which is expected to realize above requirement. It can inherit the high efficiency and reliability of the silicon material and the flexibility, light weight, and affordability of the organic materials. In this new type of devices, an organic layer is generally spread on the silicon to form the heterojunction, where the silicon wafer is adopted as the optical absorber. To reduce cost, ultrathin silicon wafer should be the best choice if only it can absorb enough sunlight to support the high efficiency. Therefore, light-trapping nanostructures, like silicon pyramid, nanowire, nanocone, and silver nanoparticle, have been widely utilized to improve the light absorption. As a consequence, the preparation of ultrathin silicon wafer and various silicon nanostructures is an important fundamental for the organic–inorganic hybrid silicon solar cells. In addition to form heterojunction with silicon, the other function of the polymer layer is to transport holes. So, the hole-transporting ability is one of the main criteria for choosing the polymer. In these days, PEDOT:PSS is most widely used and thought to be the best choice. Because silicon is inorganic, it is a challenge to form a perfect interface contact with the PEDOT:PSS; thus, there always exists abundance of defect structures and defect states on the organic–inorganic interface. These interface defects are another limitation for the efficiency of the organic–inorganic hybrid silicon solar cells. So, fabrication technology is another very important aspect to ensure the high efficiency of such type of hybrid solar cells. It includes modifying the surface wettability of silicon to improve the junction quality and importing

Y. Li · Y. Luo
State Key Laboratory of Alternate Electrical Power System with Renewable Energy Sources, North China Electric Power University, Beijing, China

M. Li (✉)
State Key Laboratory of Alternate Electrical Power System with Renewable Energy Sources, North China Electric Power University, Beijing, China

Chongqing Materials Research Institute, Chongqing, China
e-mail: mcli@ncepu.edu.cn



S. Ikhmayies (ed.), *Advances in Silicon Solar Cells*,
https://doi.org/10.1007/978-3-319-69703-1_8

effective passivation for the organic–inorganic interface to reduce the recombination centers. Considering these factors, in this chapter, firstly, some recent works on the preparation of crystal silicon membrane and several silicon nanostructures have been summarized by us; then, the basic process, especially the passivation technologies, in fabricating the heterojunction is given. We hope this summary is helpful for the future development of organic–inorganic hybrid silicon solar cells.

## 1 Introduction

The real challenge of solar cells can be summarized to reduce the cost while increasing the conversion efficiencies. Within current types of solar cells, silicon-based devices dominate the market with over 80% share due to high efficiency, reliability, and nontoxic Earth-abundant resources. However, the purification and fabrication processes of solar-grade silicon are costly, which make for approximate 75% of the total cost. Additionally, the purification process of silicon is very energy consumed and therefore caused deleterious environmental consequences. Despite the thin-wafer or thin-film technologies used recently in the photovoltaic industry, actually, they can't solve the problem significantly.

The organic solar cell is another potential kind of photon-electric conversion devices. They have several advantages, including mechanical flexibility, lightweight, and simple production process under low-temperature; can be produced in arbitrary colors; etc. The most crucial problem for this kind of photovoltaic device is the low carrier mobility and conductance of organic materials. These cause abundant carrier recombination thus low device efficiencies even for laboratory cells. Besides, stability is another problem that can't be neglected for the organic solar cells.

Organic–inorganic hybrid silicon solar cell offers a practical means to reduce fabrication cost while getting considerable power conversion efficiencies. The device structure of organic-inorganic hybrid silicon solar cells is rather straightforward. Conjugated polymers are spread on a silicon surface with low-temperature and soluble processes; thus, a heterojunction can be formed. Therefore, the power conversion efficiency of such type of photovoltaic devices is fundamentally determined by three main factors: (1) the optical absorption in the silicon absorber [1], (2) the hole transport in the organic layer [1], and (3) the carrier recombination on the silicon–organic interface [2].

Crystalline silicon thin membranes are generally adopted as the optical absorber in organic–inorganic hybrid silicon solar cells, in order to reduce cost; and for photon trapping, silicon nanostructures including silicon pyramids, silicon nanowires, and silicon nanocones have to be fabricated on the thin membrane [3, 4]. It has been demonstrated that the hole mobility in PEDOT:PSS is much higher than that of most other polymers [4]; so, PEDOT:PSS is the mostly used organic material in the organic–inorganic hybrid silicon solar cells, and actually, the thickness of the PEDOT:PSS layer is generally limited to about 100 nm to reduce the transport distance of hole in it as far as possible. The carrier recombination on the silicon–organic interface is a key scientific problem [2]; therefore, the passivation for this interface attracts increasing attentions [2, 5, 6]. Some passivation technologies, like natural oxidation, methylation, and amorphous silicon layer, have been proposed.

In this chapter, the preparation of crystalline silicon membrane and several silicon nanostructures, which are of great importance for the fabrication of flexible, low-cost, and high-efficiency organic–inorganic hybrid silicon solar cells, will be first introduced. Then, the basic fabrication process of organic–inorganic hybrid silicon solar cells will be given, especially the passivation technologies for the silicon–organic interface. For organic–inorganic hybrid silicon solar cell, the passivation technologies are the bottleneck to boost the efficiency.

# 2 Fabrication of Silicon Membrane and Silicon Nanostructures

## 2.1 Ultrathin Silicon Wafer

The ultrathin silicon wafer is the basis for the fabrication of low-cost organic–inorganic hybrid silicon solar cell. Till now on, the available fabrication technologies for ultrathin silicon wafer can be generally divided into two classes: mechanical cutting + grinding method and chemical etching method. However, using the former technology, it is difficult to obtain silicon wafers with a thickness less than 100 μm; and it will import damage layer 0.2–2.0 μm, as shown in Fig. 1 [7].

The chemical etching method is the most advanced method to fabricate ultrathin silicon wafer till now. It is a simple, low-cost, and low-damage method. Using this method, the silicon wafer can be reduced to be several micrometers, and the thickness can be easily controlled by just adjusting the etching conditions and/or

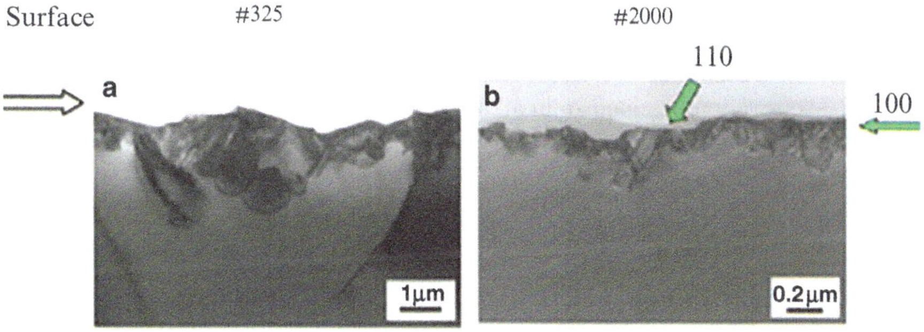


**Fig. 1** TEM image of monocrystalline silicon surface defects ground by traditional grinding method [7]. (**a**) Ground by 325 mesh powder; (**b**) ground by 2000 mesh powder

etching time. According to the composition of the etchant, the chemical etching method can be classified into acid etching and alkaline etching methods.

For acid etching, the etching solution contains HF, $AgNO_3$, $H_2O_2$, and $H_2O$. The actual etching is a metal-assisted chemical etching (MACE) process [8]. When the silicon wafer is immersed into the etching solution, the $Ag^+$ ions can seize electrons from the valence band of silicon and thus form Ag nucleation which gradually grows into Ag nanoparticles. After that, the formed Ag particles will act as the catalyst: the Ag particles will seize electrons from the regions of silicon wafer contacting with or around them; the seized electrons will reduce the $H^+$ at the surface of Ag, and the holes will be consumed by the oxidation of silicon. The reaction equations can be summarized as follows [8]:

Reaction at metal (cathode)

$$H_2O_2 + 2H^+ \rightarrow 2H_2O + 2h^+ \text{and } 2H^+ + 2e^- \rightarrow H_2 \uparrow \tag{1}$$

Reaction of Si etching (anode)

$$Si + 4h^+ + 4HF \rightarrow SiF_4 + 4H^+ \text{and } SiF_4 + 2HF \rightarrow H_2SiF_6 \tag{2}$$

From these equations, it can be seen that $H_2O_2$ can increase hole injection rate thus leading to high etching rate. Another important feature in the MACE of silicon is that the reaction rate shows great crystal orientation selectivity [9, 10]. This is because the bond density along the different directions of silicon is quite different. It has been demonstrated that the anisotropy in MACE of silicon can be effectively regulated by the etchant concentration [11, 12].

Through adjusting the ratio of the HF, $AgNO_3$, and $H_2O_2$, Bai et al. [13] have realized the fabrication of ultrathin silicon wafer using the MACE method. They have obtained ultrathin silicon wafer with a thickness about 30 μm, as shown in Fig. 2. However, since the intrinsic anisotropic etching feature, the surface quality of the obtained ultrathin silicon wafer is not very well. The average roughness is about 13 nm. Besides, some Ag nanoparticles will be left on the surface of the ultrathin silicon wafer after etching, which will affect the performance of solar cells fabricated.

Alkaline etching is a more simple method compared with the acid etching method [14–17]. In this method, the standard silicon wafer simply needs to be immersed in an alkaline solution for sometimes to fabricate the ultrathin silicon wafers. The reaction mechanism in alkaline etching process can be briefly summarized as [18]

$$KOH + H_2O \longrightarrow K^+ + 2OH^- + H^+ \tag{3}$$

$$Si + 2OH^- + 4H_2O \longrightarrow Si(OH)_6^{2-} + 2H_2 \uparrow \tag{4}$$

When we put the silicon wafer into the solution, the silicon atoms on the surface of the wafer will react with the $OH^-$ forming the complex $Si(OH)_6^{2-}$. Such complex is easy to be dissolved into water. What are given in Fig. 3 are the ultrathin

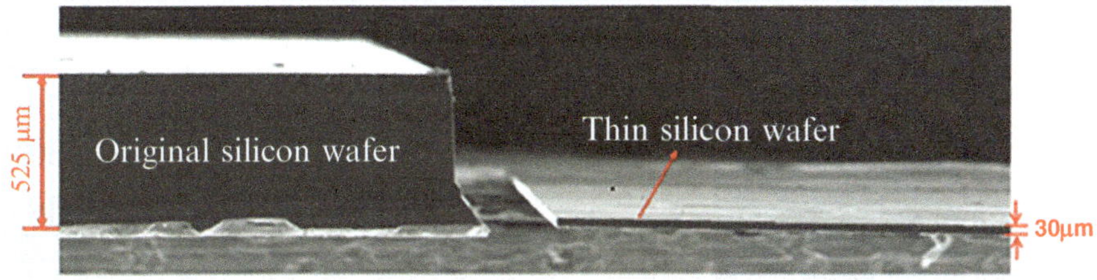


**Fig. 2** Cross-sectional SEM image of the contrast on the thicknesses of the original silicon wafer and the acid thinned ultrathin silicon wafer

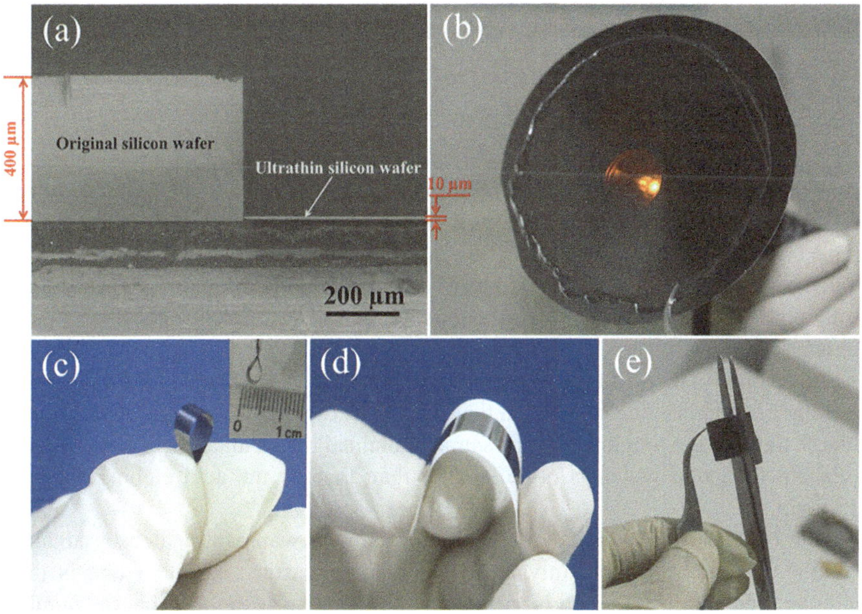


**Fig. 3** Contrast on the thicknesses of the silicon wafers before and after the alkaline solution thinning process and the physical properties of the ultrathin silicon wafer. (**a**) Cross-sectional SEM image of the silicon wafers before and after thinning; (**b**) 4-inch ultrathin silicon wafer; (**c**) flexibility of the ultrathin silicon wafer without substrate; (**d**) flexibility of the ultrathin silicon wafer with the substrate; (**e**) ultrathin silicon wafer's cutting

silicon wafers fabricated using the alkaline etching method with a thickness about 10 μm. It can be seen that the silicon wafer of this thickness can transmit orange light, and it shows very good mechanical flexibility. Its radius of curvature reaches about 2 mm. This indicates the organic–inorganic hybrid silicon solar cells based on such ultrathin silicon wafer can also have good mechanical flexibility.

Besides, using alkaline etching method, the ultrathin silicon wafer can be reduced to about 2 μm, and the thickness can simply be controlled by adjusting the concentration of the solution, the reaction temperature, and the time. Figure 4

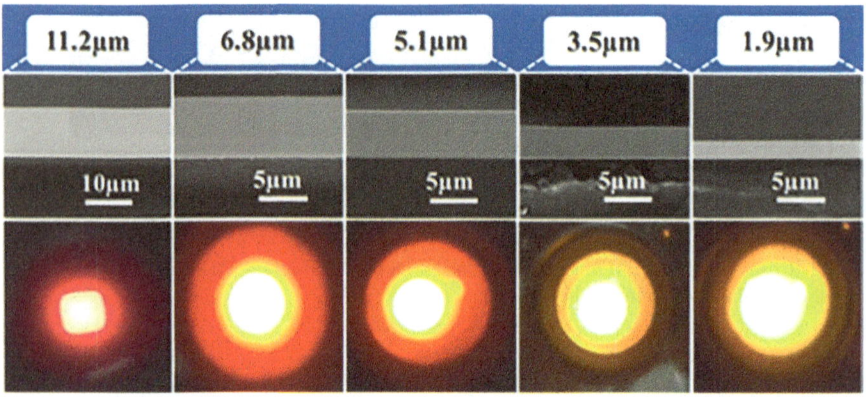


**Fig. 4** Corresponding relationship between the colors of the lights getting through the ultrathin silicon wafers and their thicknesses

shows silicon wafers with several thicknesses from 1.9 μm to 11.2 μm and their corresponding light transmission properties. The light transmission properties reflect the light absorption ability of the silicon wafers. This means that how much solar light can be effectively absorbed in the organic–inorganic hybrid silicon solar cells based on the ultrathin silicon wafer can be estimated according to these figures.

Additionally, the surface quality of ultrathin silicon wafer fabricated by alkaline etching method is much better than that fabricated by the acid method. As shown in Fig. 5, the average roughness of the sample obtained by the alkaline etching method is about 0.15 nm, which is quite smaller than that by the acid etching method, 13.8 nm.

The better surface quality of the ultrathin silicon wafer fabricated by alkaline etching method can be mainly attributed to two reasons. The first one is that, although the alkaline etching is anisotropic determined by the fact that the number of Si–Si bonds along different crystalline directions are different, the reaction is a homogeneous process. This is because there are no metal particles acting as the reaction centers during the etching process due to that the alkaline solution is only composed of potassium hydroxide and DI water. The second one is that the etching rate of alkaline etching is much lower than that of acid etching. For the acid etching, at a low temperature about 40 °C, the etching rate can also reach about 48 μm/min. However, for the alkaline etching, even at a high temperature about 90 °C, the etching rate is only about 1.3 μm/min.

It has demonstrated that adding isopropyl alcohol or n-butyl alcohol into the alkaline etching solution as protectant is also helpful for obtaining samples of high surface quality [14–17]. However, we have not found the similar results. The samples fabricated with no protectant, with isopropyl alcohol added and n-butyl alcohol added, are given in Fig. 6. The results show that the ultrathin silicon wafer with no protectant owns the best surface quality.

**Fig. 5** Optical micrograph and AFM image of ultrathin silicon wafer fabricated by the (**a**) acid etching and (**b**) alkaline etching method, respectively. The amplification factor of the optical micrographs is 200 times. Sq means the surface root mean square roughness

## *2.2 Silicon Nanostructures*

Fabrication of silicon nanostructures is another basic process for organic–inorganic hybrid silicon solar cells, which undertake the responsibility to improve light absorption. Silicon pyramid, silicon nanowire, and silicon nanocone are the three most popular silicon nanostructures for light absorption enhancement. Therefore, the recent progress on the fabrication of them will be summarized next.

### 2.2.1 Silicon Pyramid

Silicon pyramid is the anti-reflectance structure used currently in industrial crystal silicon solar cells. After the silicon pyramid being fabricated, the reflectance of the crystal silicon surface can be reduced to about 11% from 40% (polished silicon) [19]. Such an anti-reflectance effect is not very good, but there are many advantages for the silicon pyramid structure.

The main merit of silicon pyramid is that the fabrication process is very simple and cheap. It can be fabricated by immersing the crystalline wafer into alkaline

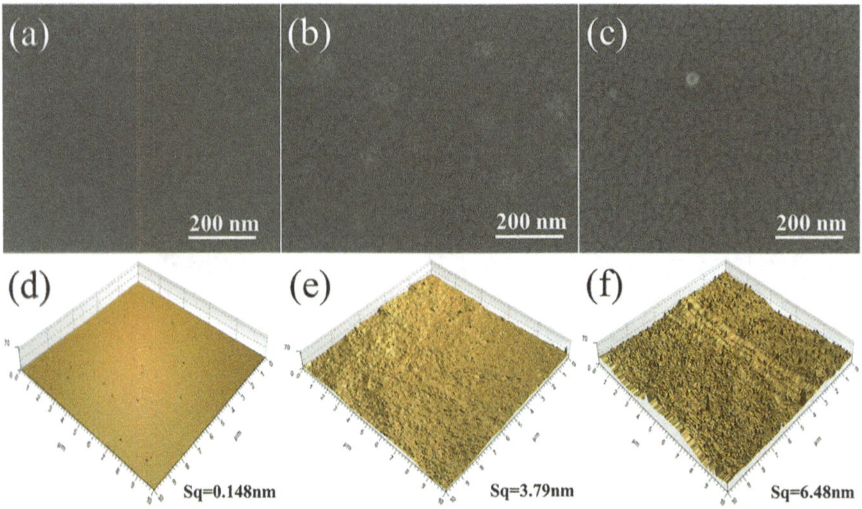


**Fig. 6** SEM (upper) and AFM images for ultrathin silicon wafers thinned without a protective agent and with different protective agents in alkaline solution. (**a**) and (**d**) without a protective agent; (**b**) and (**e**) with isopropyl alcohol; (**c**) and (**f**) with butyl alcohol

solutions with solutes NaOH, KOH, $NaHCO_3$, $NaCO_3$, and $NaSiO_3$. The details can be found everywhere. The formation of a pyramid structure occurs because the bond density on (100) crystal surface is much lower than that on (111) crystalline surface. Therefore, during the etching process, the silicon atoms on (100) surface easily tend to be removed but the ones on (111) surface tend to remain. Figure 7a shows the SEM image of silicon pyramid structure fabricated using the KOH solution. We also found that hierarchical silicon pyramid structure can be fabricated through controlling the concentration of KOH and the etching time, as given in Fig. 7b [20].

Another advantage of silicon pyramid utilized in organic–inorganic hybrid silicon solar cells is its better surface quality compared as other silicon nanostructures will be introduced below. This is attributed to the low height (the general height of the silicon pyramids is about 5 μm) and large tilt angle (the tilt angle is fixed to 30.3°). These features ensure that all the small facets of the pyramid are exposed and there is no sharp valley on the silicon surface. High surface quality is very helpful to obtain high conversion efficiency in photovoltaic devices.

Besides, the low height and large tilt angle can also contribute to fabricate high-quality organic–inorganic contact interface directly on the silicon pyramid surface. That is, an organic layer can be directly spun-coated on the silicon pyramid surface to fabricate the heterojunction. This is of great significance since the good filling of organic materials into silicon nanostructures is one of the biggest challenges for fabricating organic–inorganic hybrid silicon solar cells with high efficiency. In this point of view, the hierarchical silicon pyramid we fabricated should be of the promising application as the hierarchical structure can catch the organic materials more tightly.

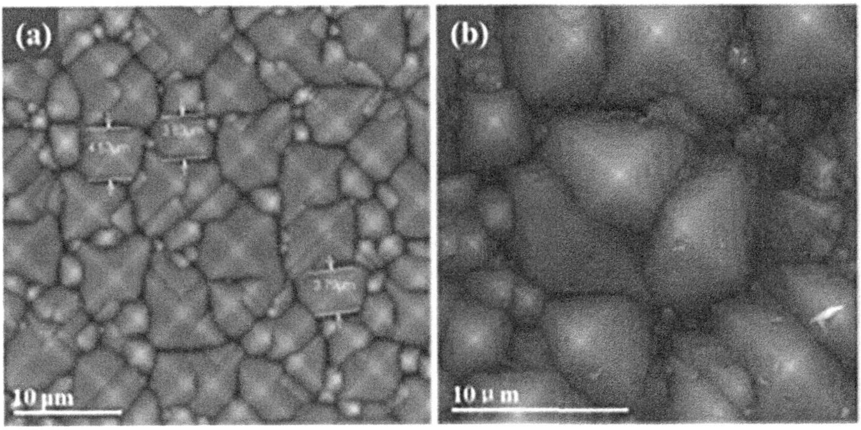


**Fig. 7** SEM images of (**a**) silicon pyramid and (**b**) hierarchical silicon pyramid structures etched in KOH solution

### 2.2.2 Silicon Nanowire

Silicon nanowire exhibits much better light-trapping efficiency than silicon pyramid structures. It can reduce the reflectance of the silicon surface to 2% even lower [20–23]. Such an excellent anti-reflectance function of silicon nanowire is obviously very important for solar cells, thus attracting great attentions.

For the fabrication of silicon nanowire, generally, there are two types of methods, the bottom-up and top-down ones. The "vapor–liquid–solid" (VLS) growth is a representative method of the bottom-up scheme [24–29]. Using this method, relatively high-quality silicon nanowires with controllable length and diameter can be fabricated. The top-down scheme can then be divided into dry and wet etching approaches. The inductively coupled plasma (ICP) etching is the most commonly used dry etching approach [30, 31], and using any of them, the density and size of silicon nanowires can be strictly controlled. However, no matter the VLS or the ICP methods require large facility and high energy consumption. Therefore, they are both not suitable for large-scale applications in preparing materials for the organic–inorganic hybrid silicon solar cells. The MACE method is a typical wet etching approach with special advantages for industrial applications [32–40]. It only requires very simple equipment, the reaction temperature is lower than 100 °C, and meanwhile the shape, size, and surface quality of the samples are controlled. As a consequence, the MACE technique is thriving day-by-day advancing to a level where silicon nanowire of the desired shape, size, and surface quality can be readily fabricated.

When using silicon nanowire for light trapping in solar cells, there are several factors that need to be considered. The surface density is the first important factor as silicon nanowire has excellent light concentration ability. According to our numerical simulations [41], a single silicon nanowire can collect light in the area more

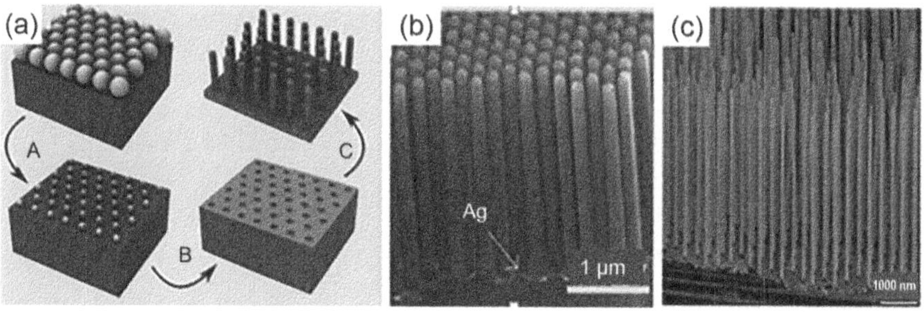


**Fig. 8** (**a**) An illustration of the template method in MACE using PS spheres as the mask [42] and silicon nanowire arrays of (**b**) high [43] and (**c**) low surface densities [44]

than 300 times of its geometrical cross-sectional area due to leaky mode excitation. This means the silicon nanowire array with fill fraction <5% can achieve near-unity broadband absorption. Therefore, it is unnecessary to fabricate compact silicon nanowire array to trap light for the organic–inorganic hybrid silicon solar cells, which will conversely hinder the filling of PEDOT:PSS in it, and bring in a large number of defects.

The surface density of silicon nanowires can be effectively controlled by the so-called template method in MACE, where PS sphere is the most popular used as the mask [42]. An overall process is illustrated in Fig. 8a, where (I) the PS spheres are firstly assembled on the silicon surface, (II) then an RIE process is carried out to form the spacing, (III) after that an Ag thin film is deposited and the PS spheres are removed, and (IV) finally the silicon nanowire array is obtained by MACE. In such template method, the surface density of the silicon nanowires can be controlled by the diameter of the PS spheres and the RIE time. Silicon nanowire arrays of high and low surface densities obtained by this method are shown in Fig. 8b, c [43, 44].

Diameter is another important parameter for the light-trapping effect of silicon nanowire array. Our simulation results demonstrated that silicon nanowire with diameter ~80 nm owns the best light-trapping ability [45]. As reflected in Fig. 9, the resonant wavelength (peak position) for silicon nanowire of diameter 80 nm corresponds to the maximum irradiance in the AM1.5G spectrum; and silicon nanowire with 80 nm diameter puts up the best light-concentration ability at the resonant wavelength (peak intensity).

Diameter can be also perfectly controlled by the above template method by adjusting the RIE time. However, it is unnecessary to use a so complex method to control the diameter of silicon nanowire. In MACE of silicon using no template, the diameter of the obtained nanowire can be generally controlled by adjusting the deposition process in fabricating the catalyst (silver film, usually fabricated by the magnetron sputtering). In our experiments and Leonard et al. report [46], the diameter of the silicon nanowires can be controlled to a range 50–100 nm [47]. Recently, through the more specific process, Um et al. [48] have controlled the diameter of the silicon nanowires to a range 60–90 nm.

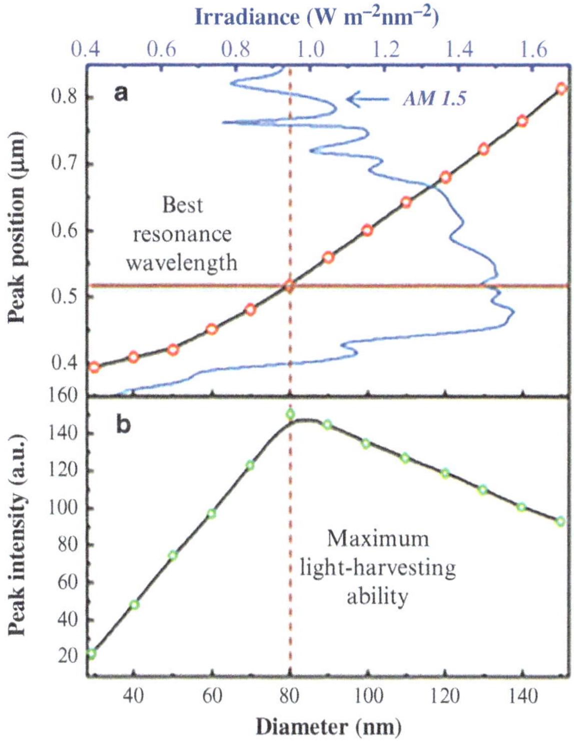


**Fig. 9** The (**a**) resonance wavelengths and (**b**) corresponding light-concentration abilities of silicon nanowires with different diameters

Length is also a very important parameter since it dramatically determines the amount of light being absorbed in the silicon nanowires themselves. For photovoltaic devices with a p–n junction in the bulk part of the silicon substrate, a balance between the light-trapping ability of the silicon nanowires and the amount of light being absorbed themselves must be considered. For solar cells with the radial junction, there will be a specific length of the silicon nanowire to achieve the optimized light absorption efficiency. The simulation results indicate that the silicon nanowire with length 3 μm owns the optimal light absorption efficiency [49]. By comparing the photovoltaic output characteristics of radial p–n junction solar cells fabricated on silicon nanowires with different lengths, Yang et al. demonstrated that the enhanced light absorption in silicon nanowires with length ranging from 2 to 5 μm can dominate over surface carrier recombination [50]. In our MACE, we found that the length of silicon nanowire is nearly linear with the reaction time even under different temperatures, as shown in Fig. 10. Therefore, the length of silicon nanowire can be simply controlled by the etching time.

Due to the excellent light-trapping ability, a silicon nanowire is of great promise to be used for conquering the inadequate light absorption in organic–inorganic hybrid silicon solar cells. Besides, the numerical simulation results have demonstrated that the collected light by silicon nanowire shows up very significant skin effect [21]. That is to say, most of the collected light will be localized near the

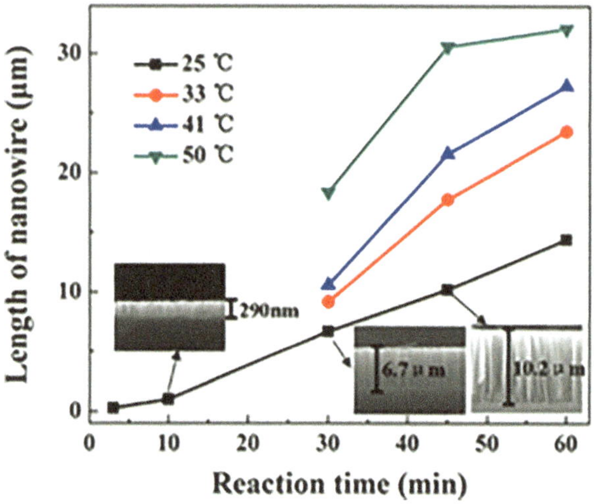


**Fig. 10** Relation of silicon nanowire length and reaction time at different temperatures

surface of the nanowire, which means silicon nanowire structure is very suitable to be used in radial junction photovoltaic devices. The PCE of organic–inorganic hybrid silicon solar cells with such radial junction architecture has reached to 13.6% [51]. However, since the inevitable metal residual and defects in silicon nanowire fabricated by MACE and the difficulty in filling organics in the nanowire structure, the fabrication of radial organic–inorganic hybrid silicon solar cells with higher efficiency is still a great challenge.

### 2.2.3 Silicon Nanocone

Silicon nanocone is another nanostructure of excellent light-trapping ability, which can be fabricated with the similar process as silicon nanowire. Besides, its shape is similar as silicon nanowire with great aspect ratio; thus, it is also called as tapered silicon nanowire elsewhere. The light-trapping effect of silicon nanocone is even better than nanowire [52, 53]; therefore, its fabrication and application attract great attentions.

Lin et al. [54] have developed a process to fabricate silicon nanocone with uniform shape and regular arrangement, as shown in Fig. 11a, b. However, this process is quite complex which needs a Ti/Au mesh film as the mask and several repeated processes of depositing catalysts and etching. As an improvement, Bai et al. [55] have developed a one-step template-free method to fabricate silicon nanocone. The only thing to do is to immerse the silicon wafer with deposited silver film in the $HF/H_2O_2$ solution. In their report, silicon nanocone array with length ~400 nm and continuously various diameters from 5–10 nm to 80–107 nm (tapering degree about 12.7) are obtained, as shown in Fig. 11c, d. The main mechanism of the tapering etching is the in situ oxidative dissolution of the silver network, which means the silver network sinks and dissolves gradually during the etching process.

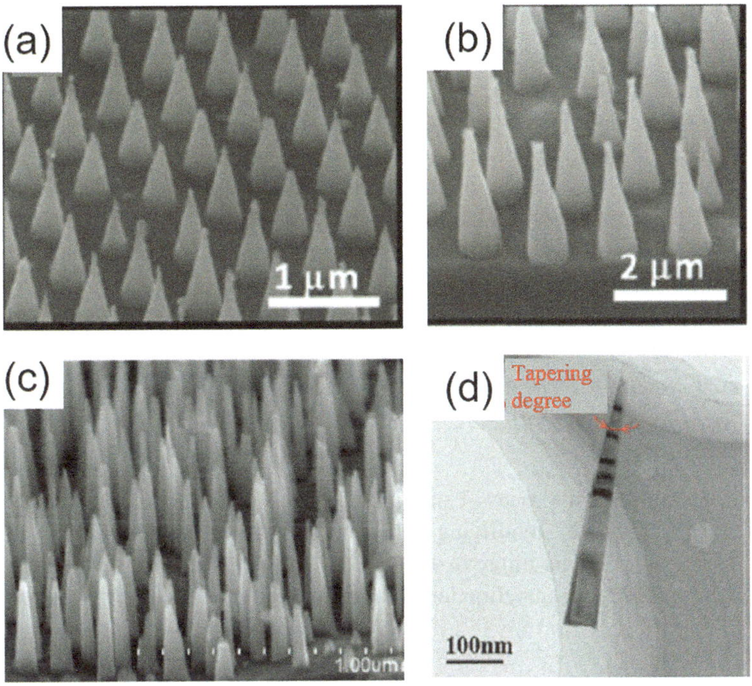


**Fig. 11** SEM images of the silicon nanocone fabricated by (**a** and **b** are reproduced from Ref. [54] with permission from The Royal Society of Chemistry) Lin et al. and (**c**) Bai et al. (**d**) TEM image of one individual silicon nanocone in (**c**) (Image **c** and **d** are reproduced from Ref. [55] with permission from The Royal Society of Chemistry)

Besides, Jung et al. [74–76] have demonstrated that silicon nanocone can be also fabricated by etching the silicon nanowire in KOH solution for about 60 s. This should be also a promising method as the processes for fabricating silicon nanowire are mature now.

According to our simulations, due to the continuous diameter variation and resonant wavelengths excited in nanocone, silicon nanocone shows better light-trapping ability than silicon nanowire [53]. Meanwhile, we have found that the average reflectance of a silicon nanocone array with length 1.47 μm in the waveband 300–1000 nm can reach about 1.7%. Taking into account that the tapered feature of silicon nanocone is beneficial for the filling of organics in the interval, silicon nanocone should be also of great application prospect in organic–inorganic hybrid silicon solar cells.

# 3 Device Fabrication and Passivation

As mentioned in the introduction, fabrication technology is very important to ensure the efficiency of such type of organic–inorganic hybrid solar cells. Following, the basic fabrication process of such hybrid device and point out, the key factors improving the quality of the organic–inorganic hetero junction will be given; and the passivation technologies of the organic–inorganic interface from the recent reports in the device also will be summarized, which is necessary to reduce the carrier recombination.

## *3.1 Fabrication Process of the Organic–Inorganic Hybrid Silicon Solar Cells*

The whole fabrication process can be generally divided into four main steps: cleaning the substrate, modifying the wettability of the silicon surface and PEDOT:PSS solution to improve the quality of heterojunction, fabricating the organic–inorganic heterojunction, and preparing the front and back electrodes.

### 3.1.1 Substrate Cleaning

Substrate cleaning is the first and very important process in fabricating the organic–inorganic hybrid silicon solar cells. In this step, the physical and chemical adsorbed impurities on the wafer will be removed. The impurities can not only affect the film quality of the organic layer but also bring in abundant carrier recombination. This will dramatically reduce the conversion efficiency and even affect the lifetime or stability of the fabricated organic–inorganic hybrid silicon solar cells. Therefore, a rigid cleaning scheme for the silicon substrate is quite necessary. What's given in Fig. 12 is a mature cleaning procedure technology we used to ensure the formation of the good organic–inorganic interface.

The silicon wafers (N (100), resistivity of 2–4 Ω.cm) are firstly cleaned by acetone (5 min), anhydrous ethanol (5 min), and DI water (10 min) under ultrasonic conditions step-by-step at room temperature. This step aims to remove the organic impurities.

After that, a standard RCA cleaning process is executed [56]. It contains four steps: (a) APM solution ($H_2SO_4$:30% $H_2O_2$ = 1:3) is used to clean the surface for 10–15 min, which can remove the organic particles; (b) SPM solution with ($NH_4OH$:$H_2O_2$:$H_2O$) ranging from (1:1:5) to (1:2:7) is used to clean the surface for 10–15 min, which can reduce metal atom pollution; (c) soak the silicon wafer with DHF(HF:$H_2O$) = 1:(2~10) solution for 5 min to remove the Al, Fe, Zn, Ni, etc. metal particles; and (d) HPM solution with (HCl:$H_2O_2$:$H_2O$) ranging from (1:1:6) to (1:2:8) is used to clean the surface for 10–15 min, to remove the $Al^{3+}$, $Fe^{3+}$, $Mg^{2+}$, $Zn^{2+}$, etc. Besides, in the middle of each cleaning step, the silicon wafer must be

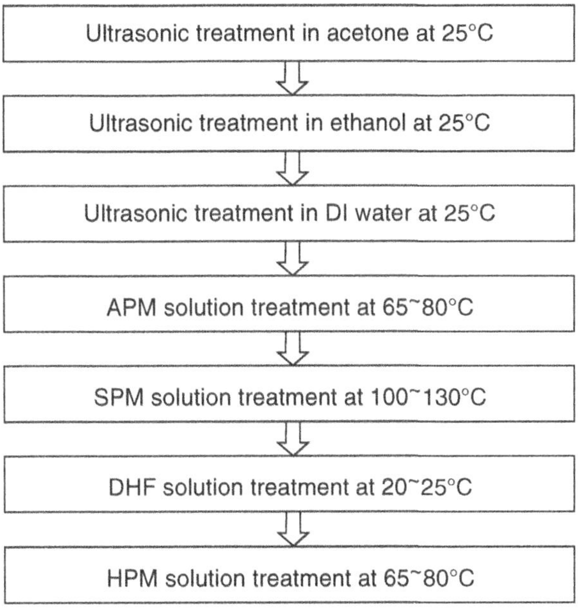


**Fig. 12** The cleaning procedure for the silicon wafer

rinsed clearly with super pure water (DI water). This rigid cleaning process can ensure the cleanliness of the silicon wafer reaching a very high degree, which lays a solid foundation for the next experiments. It should be noted that, if such cleaned silicon wafers have been stored for a long time, they are better to be treated by hydrofluoric acid again to remove the oxide layer.

### 3.1.2 Improve Quality of the Heterojunction

After being cleaned, the surface of the silicon wafer is hydrophobic. So, to form a good heterojunction contact with the hydrophilic PEDOT:PSS, it is required to modify the wettability of the silicon surface. Besides, as the PEDOT:PSS should be dissolved in DI water and then be spin-coated on the surface of the silicon wafer to form heterojunction with silicon, it is also possible to improve the quality of heterojunction by modifying the surface tension of the water solution of PEDOT: PSS.

The hydrophobicity of cleaned silicon surface can be directly reflected by the fact that water dip on the silicon surface exhibits almost spherical: the measured contact angle reaches about 70° [57]. This indicates that it is hard to form flat and homogeneous PEDOT:PSS film through spin coating PEDOT:PSS aqueous solution on untreated silicon. After being oxidized under natural circumstances for 12–24 h, the surface of the silicon wafer changes to be well hydrophilic: a water dip can be spread almost evenly on the silicon surface. This implies the PEDOT:PSS

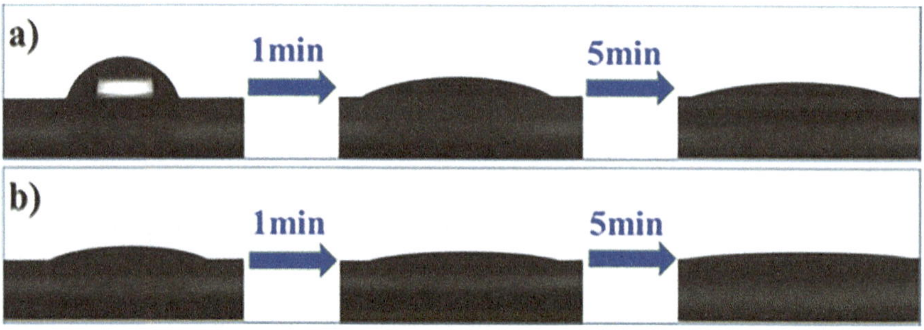


**Fig. 13** Influence of Zonyl FSH on the surface tension of PEDOT:PSS solution. (**a**) Before Zonyl FSH addition; (**b**) after Zonyl FSH addition

aqueous solution can be also spun-coated homogeneously on the silicon wafer, to form good heterojunction contact. Besides natural oxidation, the silicon wafer can be modified by the ultraviolet/ozone treatment technology. The ozone atmosphere can increase the oxidized rate dramatically and thus can obtain more uniform oxide layer. Such uniform oxide layer can not only improve the hydrophily of the silicon wafer but also is more effective in reducing the density of surface defects.

Modification on the surface tension of PEDOT:PSS aqueous solution is the other effective method to improve the quality of the heterojunction. According to the reports of Liu et al. [58], added Zonyl FSH (a kind of fluorocarbon surface-active agents) to the PEDOT:PSS solution with 0.1% mass fraction can effectively reduce the surface tension of the PEDOT:PSS solution. As shown in Fig. 13, the spreadability of the PEDOT:PSS aqueous solution can be obviously enhanced after the Zonyl FSH is added. As a result, the surface wetting, leveling, and adhesion of the PEDOT:PSS on the silicon surface are improved. This indicates that the quality of formed heterojunction will be improved after adding appropriate surface-active agents to the PEDOT:PSS water solution.

### 3.1.3 Fabrication of the Organic–Inorganic Heterojunction

The organic–inorganic heterojunction is generally fabricated by spin coating the PEDOT:PSS aqueous solution on the silicon surface. In this fabrication process, it is very important to control the thickness of the PEDOT:PSS layer. Since the only function of the PEDOT:PSS film is to form a heterojunction with the silicon substrate, the PEDOT:PSS layer should be deposited as thin as possible to reduce the carrier transport distance and to reduce the light absorption loss, so long as no leakage current. Depending on recent researches, the optimal thickness of the PEDOT:PSS layer is about 100 nm.

The thickness of the PEDOT:PSS layer fabricated by the spin-coat process can be controlled by the concentration of the solution and the spin rate, according to the following equation [59]:

$$h_{\mathrm{f}} = KC_0^2/\sqrt{\omega}, \quad (5)$$

where $h_f$ refers to the thickness of the obtained PEDOT:PSS layer, $C_0$ is the mass concentration of the solution, and $\omega$ is the spin rate. For the PEDOT:PSS aqueous solution with concentration of 1.4% (bought from Heraeus, PH1000), a PEDOT: PSS layer of thickness about 100 nm can be obtained under spin rate 3000 rev/min for 1 min according to our experience. A tip is that fabricating PEDOT:PSS layer in the $N_2$ atmosphere can enhance the performance of the organic–inorganic hybrid silicon solar cells (by about 71.3%) [60]. A reasonable explanation for this enhancement is that the $N_2$ atmosphere can protect the PEDOT:PSS layer from being polluted.

After the PEDOT:PSS being spin-coated on the silicon wafer, a subsequent annealing process is needed to remove the residual water vapor in the film. Notably, the water vapor has great influence on the performance of the devices. So, in this annealing process, we must ensure the water vapor in the PEDOT:PSS film is completely removed, but avoid the architecture of the organic layer being damaged. Therefore, a balance between eliminating the water vapor and maintaining the structure of PEDOT:PSS must be struck. The usually adopted process is annealing the samples under an $N_2$ atmosphere at 140–150 °C for about 15 min [61].

Under the same consideration, the annealed organic–inorganic heterojunction (and the final device) should be stored in conditions isolated from water vapor, as PEDOT:PSS has good hydrophilicity. From Fig. 14 it can be seen that the conversion efficiency of the organic–inorganic hybrid silicon solar cells reduces significantly with time when the devices are placed in the natural environment (wet circumstance). After 3 h, the efficiency of device preserved in wet circumstance decreases about 70.8% (from 4.32% to 1.26%). In contrary, the efficiency of device preserved in dry circumstance only decreases about 35.8% (from 4.32% to 2.77%). This comparison proves the great affection of storage environment on the performance of the solar cells.

Doping some functional materials in the PEDOT:PSS should be another effective way to improve the performance of the organic–inorganic hybrid silicon solar

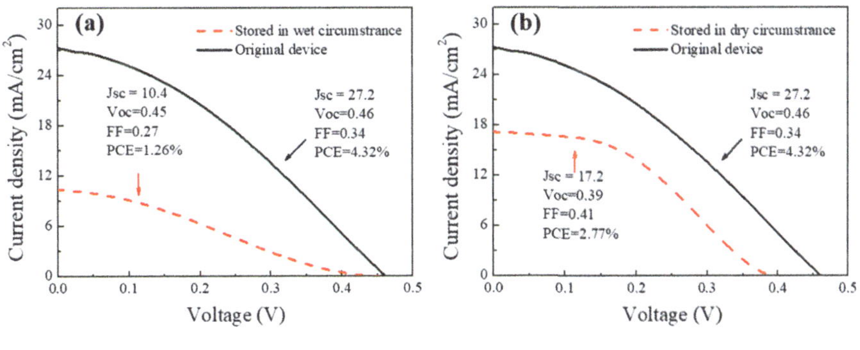


**Fig. 14** Comparison between the organic–inorganic hybrid silicon solar cells preserved in (**a**) wet and (**b**) dry circumstance for 3 h

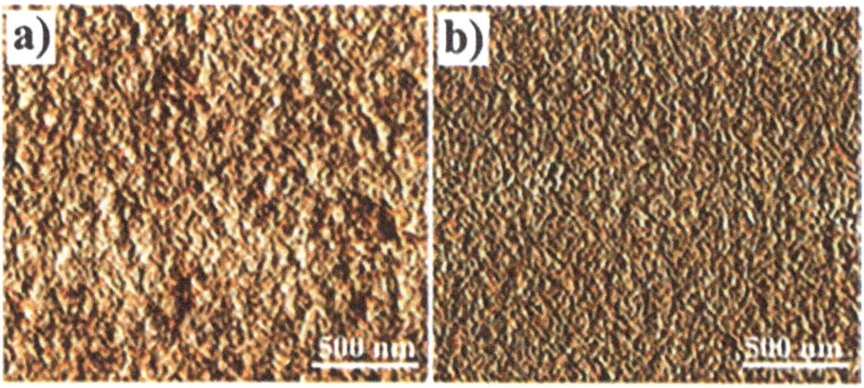


**Fig. 15** AFM images of the PEDOT:PSS film. (**a**) Without and (**b**) with DMSO doped

cell. For instance, dimethyl sulfoxide (DMSO) is widely used to dope in the PEDOT:PSS solution. It can make the PEDOT:PSS solution particles more clustered, thus improving the compactness of the organic film. This can be reflected by the average roughness of the film surface as shown in Fig. 15, where the surface quality of the PEDOT:PSS film doped with DMSO has been greatly improved. A compact PEDOT:PSS film is a benefit to forming tight contact with the silicon surface, thus improving the quality of the heterojunction.

The ability to conduct hole is a key property of PEDOT:PSS, which limits the efficiency of the organic–inorganic hybrid solar cells. It has been found that to dope some functional materials can dramatically improve the conductivity of PEDOT: PSS. As reported by Chou et al. [62], the conductivity of PEDOT:PSS can be increased by two orders of magnitude by doping the volume fraction of 5% of DMSO in the PEDOT:PSS solution. Meng et al. [63] have found that adding ZnP in PEDOT:PSS can also improve its ability in guiding hole. They reported that the $V_{oc}$ of the organic–inorganic hybrid silicon solar cell can be remarkably increased from 0.37 to 0.48 V after the PEDOT:PSS is doped with ZnP; meanwhile, the $J_{sc}$ can be increased from 1.94 to 3.81 mA/cm$^2$. Consequently, the conversion efficiency can be increased by more than two times. Sorbitol is also a widely used dopant in PEDOT:PSS, which can enhance the conductivity of PEDOT:PSS thin films by three orders of magnitude [64].

Some organic compounds can also protect PEDOT:PSS from being polluted by water vapor; thus, it is helpful to improve the stability of the solar cells. For instance, A.M. Nardes et al. [64] found that, if we add sorbitol in PEDOT:PSS aqueous solution, the sorbitol can be dissolved out from the PEDOT:PSS material and form an individual film during the annealing process. Such sorbitol film can significantly reduce the tendency to take up water from the air, which increases the environmental stability and improves the efficiency of solar cell. The conductivity of such fabricated PEDOT:PSS film can almost be maintained constant for 2 h up to 45% relative humidity [64].

### 3.1.4 Preparation of Electrodes

The last step to fabricate the organic–inorganic hybrid silicon solar cells is to prepare the front and back electrodes. The function of the front electrode is to collect holes; meanwhile, ensure the area of the plate is facing the sun as much as possible (not shelter by the front electrode). In the organic–inorganic hybrid silicon solar cell, since the hole-transporting capacity of the PEDOT:PSS is not as good as expected, an ITO layer should be deposited on the PEDOT:PSS layer firstly to avoid the lateral transport of holes. Besides it can improve the hole collection efficiency, and the ITO layer can also act to protect the PEDOT:PSS from the water vapor. After that, the silver finger electrode (usually with 50 μm in width, 20 μm in height) will be screen printed on the ITO layer.

The function of the back electrode is to collect electrons. The most simple and common method is to plate an Ag or Al film with a thickness of about 100 nm on the back of the solar cell directly. However, it is hard to fabricate solar cells of high conversion efficiency using this method. This is because the thickness of the silicon substrate in this type of organic–inorganic hybrid silicon solar cells is too thin; thus, a large number of light-generated holes will diffuse to the region near the back electrode. Therefore, there will be a large carrier recombination on the interface between the silicon substrate and the back electrode.

An effective method to reduce the carrier recombination at this interface is to introduce a back electric field to prevent the hole from spreading to the back electrode. Through fabricating a back surface heavy doping layer ($n^+$) before depositing the silver rear electrode, He et al. [51] realized a 13.6% efficiency on 20-μm-thick organic–inorganic hybrid silicon solar cell. Zhang et al. [6] have made many efforts on building the back electric field using simple solution process under low temperature (~150 °C). They found that inserting a Liq (8-hydroxyquinolinolato-lithium) layer ~2 nm between the rear side of Si and the Al electrode can effectively enhance the Schottky barrier height and built-in voltage, thus reducing charge recombination. This Liq insert can improve the PCE of solar cells 29.5% (from 9.4% to 12.2%). A cesium carbonate ($Cs_2CO_3$) layer has also been tried to insert between Si and the rear electrode Al. The charge carrier recombination rates are effectively suppressed at the rear surface due to the deflection of holes, and a high PCE of 13.7% is achieved.

It should be noted that the edges of the device must be cut out after the electrodes have been fabricated to avoid short circuit.

## *3.2 Passivation Technologies on the Organic–Inorganic Interface*

Passivation on the organic–inorganic interface is a very important step in fabricating organic–inorganic hybrid silicon solar cells. From transient and steady output characteristics measurement on the current density and voltage of the fabricated silicon/PEDOT:PSS devices, Sun et al. [2] have confirmed that the surface passivation dominated the charge recombination process, thus playing a critical role on the device performance. Since it is inevitable to introduce abundant defects when we directly spin coat the organics on the silicon, the carrier recombination can be effectively reduced by passivation technique. Recent researches demonstrated that effective passivation on the organic–inorganic interface should be the bottleneck for achieving solar cells with high efficiency, and many passivation methods have been attempted.

Native oxidizing passivation method is the most commonly adopted method in the past organic–inorganic hybrid silicon solar cells. The cleaned silicon wafer only should be put under clean air condition for 12–24 h, and a uniform silica layer about 1.5 nm can be formed on the silicon surface. Such a silica layer can effectively passivate the surface defects, which results in that the minority lifetime on the silicon surface can reach about 26 μs [5]. Ye et al. [65] have developed the surface quality of the silica passivation layer using a simple and repeatable wet oxidation method. They found that the oxidizing aqueous solution of $HNO_3$ can introduce a uniform and dense silica thin layer on the silicon surface, which can provide better surface passivation and stronger wettability than that formed after native oxide. The $HNO_3$-oxidized device they fabricated displays better performance than in the native oxide case, with the PCE is enhanced by 28.96% (from 8.53% to 11%).

Besides, Sun et al. [2] have found that organic compounds can also be used as the passivation layer for the silicon/PEDOT:PSS interface. They have fabricated a methyl/allyl organic monolayer on the silicon nanowires by immersing chlorine terminated silicon nanowire substrate in an organic solution to graft the methyl/allyl. The charge carrier lifetime of the hybrid devices can be extended to 29.3 μs, and correspondingly the PCE of the device reaches 10.2%. Most recently, Li et al. [5] have found that the intrinsic amorphous silicon (i a-Si) passivation technique widely used in HIT solar cells can also perform well for the organic–inorganic hybrid silicon solar cells. An i a-Si layer of about 2 nm has been inserted (PECVD deposition) between the silicon surface and the PEDOT:PSS and obtained a minority lifetime about 45 μs. The short current density of the fabricated solar cell reached 83.0% of the theoretical limit. This is the best passivation effect that has been reported and thus should be a practical and effective passivation technique for high-efficiency organic–inorganic hybrid silicon solar cells.

Using atomic layer deposition (ALD) to deposit an $Al_2O_3$ thin film as the passivation layer is widely used to fabricate the organic–inorganic hybrid silicon solar cells with the radial heterojunction. This is because a conformal thin film can be achieved on high aspect ratio surface using the ALD method. Notably, the thickness of the $Al_2O_3$ layer should be thinner than 2.5 nm to avoid hindering the carrier transport. By depositing an $Al_2O_3$ passivation layer with thickness ~0.7 nm between the silicon nanocone array and the PEDOT:PSS layer, Pudasaini et al. [66] found that the interfacial defect density decreased 31.3%, and the PCE is enhanced from 9.65% to 10.56%.

### *3.3 Key Points in the Device Fabrication Process*

As a summary, the fabrication of organic–inorganic hybrid silicon solar cells includes six key points.

1. The cleaning step is the basic process and the prerequisite for obtaining high-efficiency devices.
2. The modification of the silicon wafers and the PEDOT:PSS ensures the fabrication of good organic and inorganic heterojunction.
3. In the case that the above two steps have been perfectly completed, the preparation of the PEDOT:PSS film on the silicon substrate is one of the most important steps because the thickness of the film must balance the quality of heterojunction and electrical conductivity.
4. Storage of the silicon wafer with PEDOT:PSS layer isolated from water vapor is necessary since the PEDOT:PSS absorbs water which will seriously affect its performance.
5. The preparation of electrodes has great influence on the PCE of the devices obtained.
6. The passivation technique for the organic and inorganic interface should be paid special attentions on, which has been demonstrated to be the bottleneck limiting the PCE of the organic–inorganic hybrid silicon solar cells.

## 4 Conclusions

In this chapter, recent progress on the fabrication of ultrathin flexible crystal silicon membrane and several silicon nanostructures for light trapping and the detailed fabrication process including the key factors of organic–inorganic hybrid silicon solar cells are summarized. Preparation of crystalline silicon membrane is the premise as it is always adopted as the optical absorber in organic–inorganic hybrid silicon solar cells, while the silicon nanostructures are necessary to make up the insufficient light absorption in the ultrathin silicon. Till now, the fabrication processes of the ultrathin flexible crystal silicon membrane and the silicon nanostructures, using the wetting chemical etching methods, both have advanced to a level where desired thickness, size, and surface quality can be readily fabricated. The fabrication technique of the organic-inorganic hybrid silicon solar cells is very important to ensure the high efficiency. The fabrication steps are described in detail where six key points need to be paid attention are emphasized. For such organic–inorganic hybrid silicon solar cells, passivation techniques on the organic/silicon interface should be the bottleneck limiting the PCE. There are many efforts that have been made on this issue, but it still far away from being solved perfectly. We predict passivation must be the research highlight in the field of organic–inorganic hybrid silicon solar cells. Organic–inorganic hybrid silicon solar cells can reduce the cost, maintain high conversion efficiencies and good stability, and obtain good mechanism flexibility; therefore, they should be a promising type of solar cells. We hope that the above contents can make a little bit of contribution to the photovoltaic field.

**Acknowledgments** This work is supported partially by National High-Tech R&D Program of China (863 Program, No. 2015AA034601), National Natural Science Foundation of China (Grant nos. 91333122, 51372082, 51402106, and 11504107), PhD Programs Foundation of Ministry of Education of China (Grant no. 20130036110012), Par-Eu Scholars Program, Beijing Municipal Commission of Science and Technology Project (Z161100002616039), and the Fundamental Research Funds for the Central Universities (2016JQ01, 2015ZZD03, 2015ZD07, 2017ZZD02).

## References

1. P.C. Yu, C.Y. Tsai, J.K. Chang, C.C. Lai, P.H. Chen, Y.C. Lai, P.T. Tsai, M.C. Li, H.T. Pan, Y.Y. Huang, C.I. Wu, Y.L. Chueh, S.W. Chen, C.H. Du, S.F. Horng, H.F. Meng, ACS Nano **7**, 10780 (2013)
2. F. Zhang, D. Liu, Y. Zhang, H. Wei, T. Song, B. Sun, ACS Appl. Mater. Interfaces **5**, 4678 (2013)
3. M. Sharma, P.R. Pudasaini, F. Ruiz-Zepeda, D. Elam, A.A. Ayon, ACS Appl. Mater. Interfaces **6**, 4356 (2014)
4. J. He, P.Q. Gao, M.D. Liao, X. Yang, Z.Q. Ying, S.Q. Zhou, J.C. Ye, Y. Cui, ACS Nano **9**, 6522 (2015)
5. Y. Li, P. Fu, R. Li, M. Li, Y. Luo, D. Song, Appl. Surf. Sci. **366**, 494 (2016)
6. Y. Zhang, F. Zu, S.T. Lee, L. Liao, N. Zhao, B. Sun, Adv. Energy Mater. **4**, 2195 (2014)
7. W. Qiu, Dissertation, SiChuan Normal University, (2014)
8. F. Bai, M. Li, D. Song, H. Yu, B. Jiang, Y. Li, J. Solid State Chem. **196**, 596 (2012)
9. T. Huang, T. Yen, MRS Proceedings, 1302, mrsf10–1302-w1309–1304 , (2011)
10. C.Y. Chen, C.S. Wu, C.J. Chou, T.J. Yen, Adv. Mater. **20**, 3811 (2008)
11. L. Li, Y. Liu, X. Zhao, Z. Lin, C.-P. Wong, ACS Appl. Mater. Interfaces **6**, 575 (2013)
12. Z. Huang, T. Shimizu, S. Senz, Z. Zhang, N. Geyer, U. Gosele, J. Phys. Chem. C **114**, 10683 (2010)
13. F. Bai, M. Li, D. Song, H. Yu, B. Jiang, Y. Li, Appl. Surf. Sci. **273**, 107 (2013)
14. S. An, S.G. Lee, S.G. Park, E.H. Lee, O. Beom-Hoan, Sensor. Actuat. a-Phys. **209**, 124 (2014)
15. K.P. Rola, I. Zubel, J. Micromech. Microeng. **21**, 115026 (2011)
16. I. Zubel, M. Kramkowska, Sensor. Actuat. a-Phys. **101**, 255 (2002)
17. I. Zubel, K. Rola, M. Kramkowska, Sensor. Actuat. a-Phys. **171**, 436 (2011)
18. A. Splinter, J. Stürmann, W. Benecke, Mater. Sci. Eng. C **15**, 109 (2001)
19. K. Sato, M. Shikida, T. Yamashiro, M. Tsunekawa, S. Ito, Sensor. Actuat. a-Phys. **73**, 122 (1999)
20. B. Fan, F. Peng-Fei, C. Peng, H. Rui, L. Rui-Ke, Z. Zhi-Rong, Y. Hang, Z. Yan, S. Dan-Dan, L. Ying-Feng, J. Infrared Millim. W. **34**, 471 (2015)
21. K. Peng, Y. Xu, Y. Wu, Y. Yan, S.T. Lee, J. Zhu, Small **1**, 1062 (2005)
22. Y. Li, L. Yue, Y. Luo, W. Liu, M. Li, Opt. Express **24**, A1075 (2016)
23. Y. Li, M. Li, R. Li, P. Fu, T. Wang, Y. Luo, J.M. Mbengue, M. Trevor, Sci. Rep-Uk **6**, 24847 (2016)
24. I. Zardo, S. Conesa-Boj, S. Estradé, L. Yu, F. Peiro, P.R.i. Cabarrocas, J. Morante, J. Arbiol, A.F.i. Morral, Appl. Phys. A Mater **100**, 287 (2010)
25. S. Hofmann, C. Ducati, R. Neill, S. Piscanec, A. Ferrari, J. Geng, R. Dunin-Borkowski, J. Robertson, J. Appl. Phys. **94**, 6005 (2003)
26. D. Ma, C. Lee, F. Au, S. Tong, S. Lee, Science **299**, 1874 (2003)

27. J.R. Maiolo, B.M. Kayes, M.A. Filler, M.C. Putnam, M.D. Kelzenberg, H.A. Atwater, N.S. Lewis, J. Am. Chem. Soc. **129**, 12346 (2007)
28. Y. Wu, P. Yang, J. Am. Chem. Soc. **123**, 3165 (2001)
29. D. Kwak, H. Cho, W.-C. Yang, Phys. E. **37**, 153 (2007)
30. J. Bhardwaj, H. Ashraf, A. McQuarrie, Proc. Symp. Microstructures and microfabricated systems, ECS, 1–13 (1997)
31. R.M. Ng, T. Wang, F. Liu, X. Zuo, J. He, M. Chan, Elect. Device Lett. Ie **30**, 520 (2009)
32. K.-Q. Peng, Y.-J. Yan, S.-P. Gao, J. Zhu, Adv. Mater. **14**, 1164 (2002)
33. K.Q. Peng, Z.P. Huang, J. Zhu, Adv. Mater. **16**, 73 (2004)
34. K. Peng, J. Hu, Y. Yan, Y. Wu, H. Fang, Y. Xu, S. Lee, J. Zhu, Adv. Funct. Mater. **16**, 387 (2006)
35. H. Han, Z. Huang, W. Lee, Nano Today **9**, 271 (2014)
36. Z. Huang, N. Geyer, L. Liu, M. Li, P. Zhong, Nanotechnology **21**, 465301 (2010)
37. W. Chern, K. Hsu, I.S. Chun, B.P.D. Azeredo, N. Ahmed, K.-H. Kim, J.M. Zuo, N. Fang, P. Ferreira, X. Li, Nano Lett. **10**, 1582 (2010)
38. J. Tang, J. Shi, L. Zhou, Z. Ma, Nano-Micro. Lett. **3**, 129 (2011)
39. J. Huang, S.Y. Chiam, H.H. Tan, S. Wang, W.K. Chim, Chem. Mater. **22**, 4111 (2010)
40. X. Geng, M. Li, L. Zhao, P.W. Bohn, J. Electron. Mater. **40**, 2480 (2011)
41. Y.F. Li, M.C. Li, D.D. Song, H. Liu, B. Jiang, F. Bai, L.H. Chu, Nano Energy **11**, 756 (2015)
42. J. Yeom, D. Ratchford, C.R. Field, T.H. Brintlinger, P.E. Pehrsson, Adv. Funct. Mater. **24**, 106 (2014)
43. S. Su, L. Lin, Z. Li, J. Feng, Z. Zhang, Nanoscale Res. Lett. **8**(1), 405 (2013)
44. Z. Huang, H. Fang, J. Zhu, Adv. Mater. **19**, 744 (2007)
45. Y.F. Li, M.C. Li, R.K. Li, P.F. Fu, B. Jiang, D.D. Song, C. Shen, Y. Zhao, R. Huang, Opt. Commun. **355**, 6 (2015)
46. F. Leonard, A.A. Talin, Phys. Rev. Lett. **97**, 163 (2006)
47. F. Bai, Dissertation, Harbin institute of Technology, (2014)
48. H.D. Um, J.Y. Jung, H.S. Seo, K.T. Park, S.W. Jee, S.A. Moiz, J.H. Lee, Jpn. J. Appl. Phys. **49**, 04DN02 (2010)
49. Y.F. Li, M.C. Li, R.K. Li, P.F. Fu, L.H. Chu, D.D. Song, Appl. Phys. Lett. **106**, 091908 (2015)
50. E. Garnett, P.D. Yang, Nano Lett. **10**, 1082 (2010)
51. J. He, P. Gao, M. Liao, X. Yang, Z. Ying, S. Zhou, J. Ye, Y. Cui, ACS Nano **9**, 6522 (2015)
52. Y. Cho, M. Gwon, H.-H. Park, J. Kim, D.-W. Kim, Nanoscale **6**, 9568 (2014)
53. Y. Li, M. Li, P. Fu, R. Li, D. Song, C. Shen, Y. Zhao, Sci. Rep-Uk **5**, 11532 (2015)
54. H. Lin, H.-Y. Cheung, F. Xiu, F. Wang, S. Yip, N. Han, T. Hung, J. Zhou, J.C. Ho, C.-Y. Wong, J. Mater. Chem. A **1**, 9942 (2013)
55. F. Bai, M. Li, R. Huang, Y. Li, M. Trevor, K.P. Musselman, RSC Adv. **4**, 1794 (2014)
56. T. Ohmi, J. Electrochem. Soc. **143**, 2957 (1996)
57. H. Jin, C. Xingbi, Y. Chuanren, X. Wang, Semicond. Sci. Technol. **5**, 006 (1999)
58. Q. Liu, M. Ono, Z. Tang, R. Ishikawa, K. Ueno, H. Shirai, Appl. Phys. Lett. **100**, 183901 (2012)
59. Z. Wang, H. Yang, X. Xi, Q. Qiao, J. JI, G. Li, J. Funct. Mater. **38**, 393 (2007)
60. X. Shen, B. Sun, D. Liu, S.-T. Lee, J. Am. Chem. Soc. **133**, 19408 (2011)
61. Y. Kim, A.M. Ballantyne, J. Nelson, D.D. Bradley, Org. Electron. **10**, 205 (2009)
62. T.-R. Chou, S.-H. Chen, Y.-T. Chiang, Y.-T. Lin, C.-Y. Chao, J. Mater. Chem. C **3**, 3760 (2015)
63. Q.-L. Meng, F. Li, X. Xi, W. Qian, L. Que, J. JI, Y. Ding, G. Li, J. Artificial Crystal **39**, 670 (2010)
64. A.M. Nardes, M. Kemerink, M. De Kok, E. Vinken, K. Maturova, R. Janssen, Org. Electron. **9**, 727 (2008)
65. J. Sheng, K. Fan, D. Wang, C. Han, J. Fang, P. Gao, J. Ye, ACS Appl. Mater. Interfaces **6**, 16027 (2014)
66. P. Pudasaini, A. Ayon, J. Phys. Conf. Ser. **476**, 012140 (2013)

# Recent Advances in the Use of Silicon-Based Photocathodes for Solar Fuel Production

**Ahmad M. Mohamed, Basamat S. Shaheen, Aya M. Mohamed, Ahmad W. Amer, and Nageh K. Allam**

**Abstract** With the substantial decrease in the world's reservoirs of fossil fuels, the need to develop industrial-scale energy-harvesting systems that rely on more sustainable sources is dire. With solar energy being the cheapest, and most giving, research progress utilizing it to replace fossil fuels, as well as to counterbalance the effects of using such fuels, is divided into three tracks: finding cheap and efficient photoabsorbers, devising industrially compatible fabrication procedures, and developing the proposed systems for higher efficiency. Being an abundant element with well-known chemical and electrical properties and well-established fabrication procedures, silicon may be the quickest solution for developing efficient solar energy conversion systems. Indeed, for $H_2$ production and $CO_2$ reduction in particular, Si-based materials with different morphologies, structural forms, and combinations were studied for decades. In this chapter, the recent studies for Si photocathodes are demonstrated in a way to classify the different systems and compare their performance. The use of plain and decorated nanostructured Si, as well as SiC nanostructured crystalline photocathodes for solar $H_2$ production, is briefly presented. Brief insight about amorphous Si and its use for the same purpose is also discussed. Finally, light is shed on the use of Si photocathodes in $CO_2$ reduction.

## 1 Introduction

With the dependency of the world's energy economy on oil products faltering, challenges facing the oil economy today, such as sustainability and eco-stability, are no longer tolerable. It is thus a global ultimatum either to redesign the existing systems to clean up the carbon footprint from the Earth's atmosphere or to design

---

A.M. Mohamed · B.S. Shaheen · A.M. Mohamed · A.W. Amer · N.K. Allam (✉)
Energy Materials Laboratory (EML), School of Sciences and Engineering,
The American University in Cairo, New Cairo, Egypt
e-mail: nageh.allam@aucegypt.edu

 
S. Ikhmayies (ed.), *Advances in Silicon Solar Cells*,
https://doi.org/10.1007/978-3-319-69703-1_9

new systems capable of emission-less production of energy. Indeed, both endeavors have been under way for more than 40 years.

On one hand, hydrogen, being one of the main sources that is expected to maintain supplying energy to the world for the next decades [1], has captivated global attention since the Second World War. The major advantage of hydrogen is that abundant and clean water is both the source and the product. However, the extraction of hydrogen gas from water is an energy-hungry process [2]. Thus, ever since the 1972 Fujishima-Honda experiment [3], researchers have been optimizing devices that can split water using off-the-grid methods, such as solar energy. In such photoelectrochemical cells (PECs), photo-active electrodes (mainly photoanodes) are used to drive forward the redox reactions of water splitting using merely the energy of the sun. Current problems facing the industrialization of PECs can be summed up as the ability of designing photoelectrodes that can absorb most of the visible region of sunlight (~40%) with a high quantum efficiency. In addition, the need of applying external bias to push the splitting reactions forward persists in order to raise the work function of the metallic cathode and initiate the hydrogen evolution reaction [4]. Indeed, several trials were reported in the literature in order to get rid of this external bias through forming a p–n junction inside the PEC, i.e., incorporating a photo-active cathode instead of a metallic one [5–12]. However, general problems facing the use of photocathodes are their instability in aqueous media, which leads to photo-corrosion, or their wide bandgaps, which render their use as photo-active electrodes inefficient [13]. Researchers thus attempt the fabrication of multi-component, bio-mimicking systems, in which a non-stable, yet highly absorbing photocathode is covered with a protection layer that assists in pushing the electrons into the electrolyte [14–22]. On the other side, researchers are developing systems that can absorb the greenhouse gas $CO_2$ from the atmosphere to decrease its persisting side effects.

Silicon is considered a reference material in the solar industries due to its near-optimum bandgap [23], which offers Si a strong absorption in the visible region, as well as its fast charge dynamics [24] and very high diffusion length [25]. In addition, the commercially available fabrication technologies render Si a very promising material for photoelectrodes.

This chapter focuses on the use of crystalline Si (c-Si) and amorphous Si (a-Si) as photocathodes in PECs, as well as sheds light on the recent studies using Si for $CO_2$ reduction. Herein, some of the most important routes for the modification of crystalline Si to be a successful candidate in water splitting will be reviewed. Likewise, for amorphous Si, this chapter discusses its points of strength and weakness and the various modification pathways researched in the literature.

# 2 Plain Crystalline Silicon: Nanostructuring for Higher Efficiencies

## 2.1 Nanowires

Making advantage of nanotechnology, forming p-Si photocathodes in nanostructures was found to be a way for enhancing the absorption and charge separation. Nanowires (NWs) were shown to be a promising architecture for their ability to decouple the directions of absorption and charge transfer, the increased surface area, and effective light trapping. This can in turn allow the use of low-purity Si, which reduces the cost of the system, while keeping the efficiency of the cell intact [26]. Sim et al. [27] showed that nanostructuring single crystalline Si by metal-catalyzed electroless etching for different durations produces a change in the photocurrent density as shown in Fig. 1. Besides the photocurrent, overpotential is another important quantity that determines the photoelectrochemical behavior. Precious metals such as platinum were usually deposited on c-Si structures to lower the overpotential, which affects the cost of the device. In this study [27], porous nanostructures were formed at etching times shorter than 120 min, followed by the formation of NWs at longer times. At 120-min etching, both of the nanoporous layers and NWs existed.

Table 1 summarizes the results of the photocurrent, the onset potential (which is related to the overpotential), and the solar-to-hydrogen (STH) conversion efficiency obtained for the nanoporous and NW structures as compared to planar Si and structures decorated with Pt. STH conversion efficiency is calculated from an equation combining the onset potential ($V_{os}$), the photocurrent at short circuit ($J_{sc}$), and the fill factor (FF), i.e.,

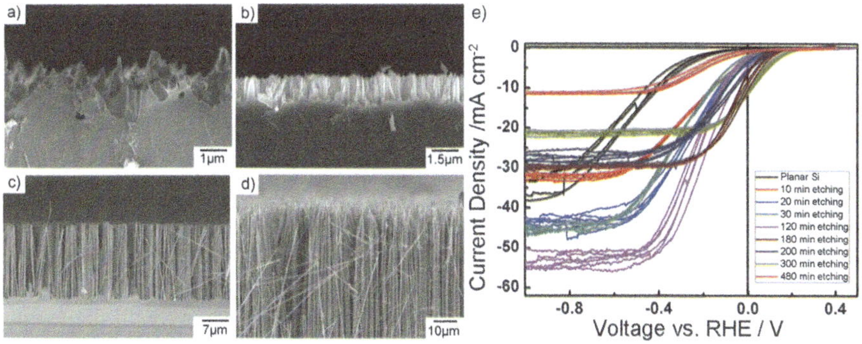


**Fig. 1** SEM micrographs of cross-sectional views of p-type (100) Si wafers etched in 5 MHF and 0.02-M $AgNO_3$ solution at different etching times: (**a**) 30 min, (**b**) 120 min, (**c**) 300 min, and (**d**) 480 min. (**e**) The photocurrent density of the Si electrodes formed at different etching times was measured at light intensity of 100 mW/cm$^2$ using an air mass 1.5 filter with 4 cycles for statistical reliability [27]

**Table 1** Photocurrent density, onset potential, and solar-to-hydrogen (STH) conversion efficiency obtained for Si nanoporous and nanowire structures as compared to planar Si and structures decorated with Pt [27]

| Structure | Limiting current density [mAcm¯2] | Current density at 0 V vs RHE [$mAcm^{-2}$] | Onset potential [V] | STH conversion efficiency [%] |
|---|---|---|---|---|
| Planar Si | −34.1 | −0.25 | 0.13 | 0.03 |
| Planar Si with Pt | −29.0 | −12.3 | 0.28 | 0.69 |
| Porous Si (10-min etching) | −32.1 | −3.21 | 0.19 | 0.10 |
| Porous Si with Pt (10-min etching) | −14.9 | −12.13 | 0.23 | 0.81 |
| Porous and 1.5-μm nanowire (120 min etching) | −54.1 | −3.02 | 0.23 | 0.12 |
| Porous and 1.5-μm nanowire with Pt (120-min etching) | −18.4 | −16.4 | 0.26 | 1.19 |
| 29.3-μm Si nanowire (300-min etching) | −21.4 | −10.65 | 0.38 | 0.48 |
| 29.3-μm Si nanowire with Pt (300-min etching) | −14.3 | −12.81 | 0.48 | 1.20 |
| 38.0-μm Si nanowire (480-min etching) | −11.1 | −0.86 | 0.25 | 0.11 |
| 38.0-μm Si nanowire with Pt (480-min etching) | −7.2 | −6.54 | 0.32 | 0.70 |

$$\text{STH conversion efficiency } (\%) = \frac{V_{\text{OS}} \times J_{\text{SC}} \times \text{FF}}{P_{\text{in}}} \times 100 \tag{1}$$

$$= \frac{V_{\max} \times J_{\max}}{P_{\text{in}}} \times 100. \tag{2}$$

The highest STH conversion efficiency (1.19%) was obtained for the nanostructures obtained at 120-min etching and deposited with Pt [27]. The deposition of Pt had the effect of increasing the STH conversion efficiency for all the structures. Without the usage of Pt catalyst, the efficiency of the 29.3 μm SiNWs increased to 70% of that of planar Si decorated with Pt, accompanied by an increase in the onset potential (decrease in the overpotential). It is important to note that the dimensions of the structure also have important effects on the PEC behavior. As it can be observed from Table 1, the limiting current density increases till the length of the NWs reaches 150 μm, after which a decrease in photocurrent density is observed.

This decrease was attributed to increased surface defects, wide depletion region, and the high rates of recombination, all factors that decrease the current density. These opposite effects of nanostructuring Si show the magnified effect that the

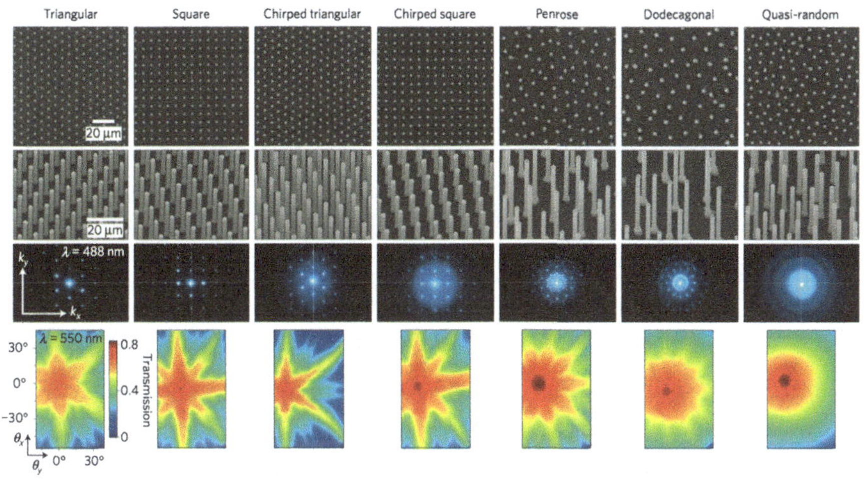


**Fig. 2** Representative composition and optical properties of each wire-array tiling pattern. The scale bars in the left column apply to all images across each row. Top row: SEM images of as-grown wire arrays viewed from a top-down perspective. Second row: SEM images viewed at a 20° angle. Third row: transmitted diffraction patterns of polymer-embedded wire arrays on a quartz slide, observed at $\lambda = 488$ nm. The axes indicate 4000 $cm^{-1}$ in the direction of $k_x$ and $k_y$. Bottom row: integrated transmission of each wire array observed at $\lambda = 550$ nm as a function of the beam incidence angle (${}_x$, ${}_y$) [28]

nanostructures and their dimensions have on the efficiency in general and on its components (photocurrent and onset potential) in specific.

In this sense, Kelzenberg et al. [28] studied the optical absorption of Si wire arrays grown by using the vapor-liquid-solid (VLS) process with seven different tilting patterns as shown in Fig. 2 (first two rows). The transmitted optical diffraction patterns and the angularly resolved measurements are shown in Fig. 2 (third and fourth rows), respectively. All the arrays had lower light absorption at normal angle of incidence than at other angles. Arrays having periodic arrangement of wires showed higher absorption than the arrays with quasiperiodic or random arrangements. The authors suggested the reduction of Si surfaces as well as randomization of the path of light between the wires as causes of maximizing the absorption.

## 2.2 Micro- and Nanoholes

Trying out different structures, Ji et al. studied Si microholes (MHs) as photocathodes for hydrogen production [29]. As shown in Fig. 3 panels a and c, the highly ordered MHs were fabricated with different diameters on a Si substrate by combining nanosphere lithography and metal-assisted etching processes. The length of the MHs was also controlled by the etching parameters as shown in Fig. 3 panels b

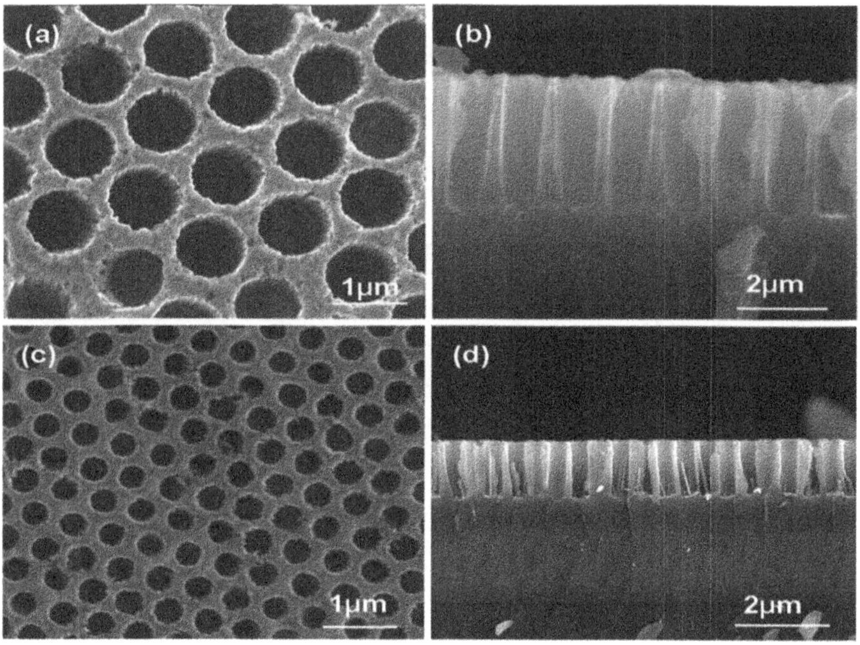


**Fig. 3** Typical top-view (**a**) and cross-sectional (**b**) SEM images of SiMH arrays fabricated using 1-μm nominal diameter of polystyrene nanospheres. Also shown is the typical top-view (**c**) and cross-sectional (**d**) SEM images of SiMH arrays fabricated using polystyrene nanospheres with 500-nm nominal diameter [29]

and d. Figure 4 shows the *J–V* curves for planar Si, Si MHs, and Pt–Si MHs composite electrodes. As shown in Fig. 4a, the MH structure provided antireflection property within the wavelength range of 300–1000 nm compared to planar Si. As a result, the measured photocurrent and efficiency was enhanced, and the overpotential was decreased compared to the planar structure as shown in Fig. 4 panels b and c.

Building on the same structure, Jung et al. [30] developed tapered Si nanoholes (NHs) that become narrow from top to bottom gradually, as shown in Fig. 5, by controlling the silver deposition and etching time in metal-assisted etching process. The tapered feature led to the enhancement in light absorption through gradation in the refractive indices between Si and air. Theoretical calculations and experimental measurements showed that the specular reflectance spectrum is dependent on the depth of the tapered NH. At depths of 150–200 nm, the tapered NHs showed high antireflection behavior compared to untapered NHs with lower depths. This effect was attributed to the formation of multiple dielectric layers as the depth increases. Although nanostructures promote the suppression of light reflection and better light management, their high surface area causes high recombination loss. Thus, Jung et al. studied the surface area and the filling ratios of the tapered NHs of depth 200 nm at different hole diameters to reach a balance between the surface area and

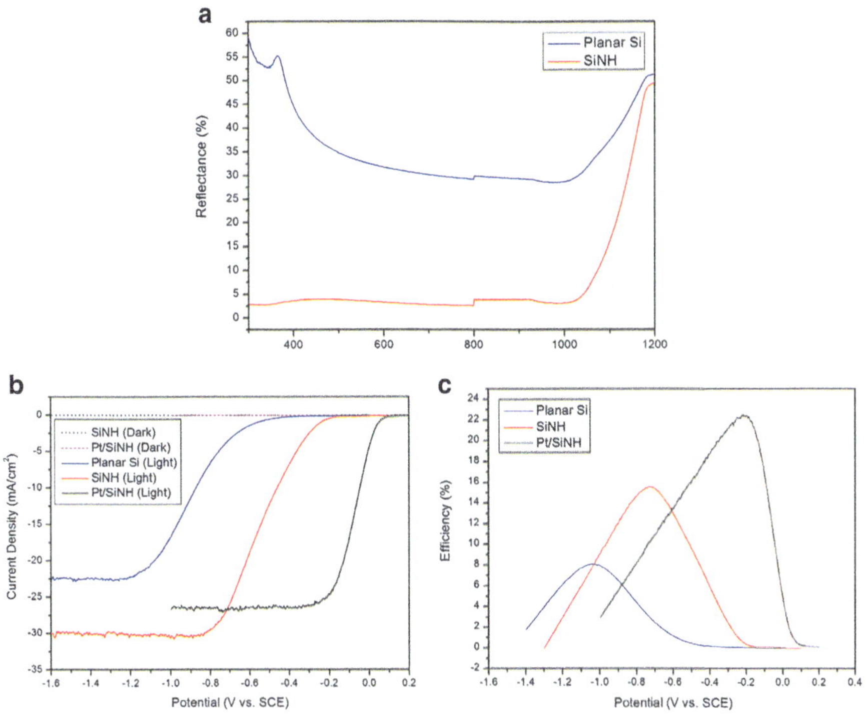


**Fig. 4** (**a**) UV-Vis-NIR reflection spectra of Planar Si and SiMHs showing the enhanced antireflection of SiMHs. (**b**) Photoelectrochemical *J*–*E* curves of planar Si, SiMH, and Pt/SiMH photocathodes during the water splitting reaction to produce hydrogen (both illuminated and dark) in aqueous solution of $H_2SO_4$ and 0.5-M $K_2SO_4$ (pH 1) and (**c**) photoconversion efficiencies vs bias potential of planar Si, SiMH, and Pt/SiMH at 100 mW/cm$^2$ of light power [29]

the reflectance aiming for the highest photocurrent. Figure 6a shows the change of the photocurrent measured for tapered Si NHs with different depths compared to planar Si. As shown, the highest photocurrent was obtained at a depth of 200 nm. For this particular depth, an Ag deposition time of 1 s showed the best photocurrent density too. Tauc plots measured for the tapered NHs showed a slight increase in the bandgap compared to that of planar structure, as shown in Fig. 6b. These results, accompanied by Mott–Schottky measurements, explained that the enhancement in the photocurrent could also be due to the suppression of electron-hole recombination at surface traps. The slight increase in the bandgap of the tapered NHs created a band offset at the interface with planar Si, which formed an electric field by the accumulation of charges. This, in turn, created a passivation effect preventing recombination. Figure 6c compares the probability of recombination in planar Si with the passivation effect that occurred at the interface between planar Si and the tapered Si NHs.

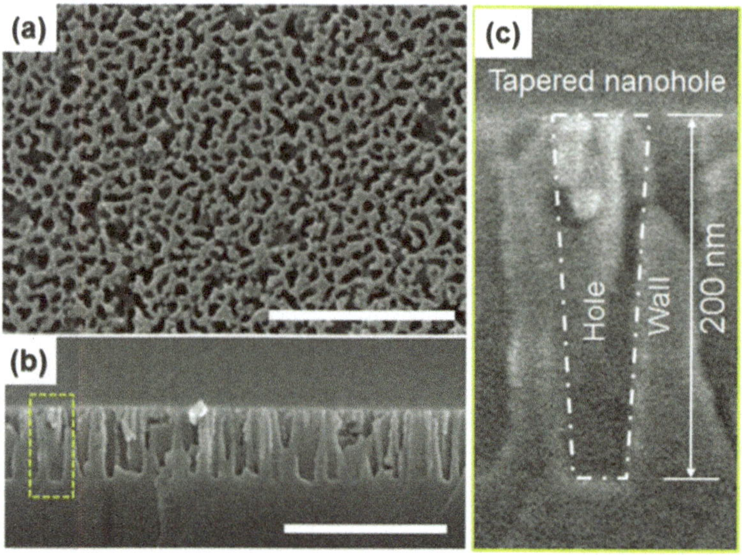


**Fig. 5** (**a**) Top-view and (**b**) cross-sectional SEM images of the Si NH array fabricated by metal-assisted etching method. (**c**) Magnified view of the yellow box in (**a**). All scale bars are 500 nm [30]

## 2.3 Nanobelts

In attempts to attain a higher photocurrent density, Bao et al. [31] used metal-assisted etching process of prepatterned silicon to form Si nanobelt (NB) arrays. The morphology is shown in Fig. 7 panels a–c, in which the NBs are tilted relative to the Si substrate with short porous rods on their top. The measured photocurrent for the Si NBs was higher than that for planar Si and SiNWs fabricated by the same procedure [32]. The authors explained that the enhancement is a result of the belt-like structure that allowed the multireflection of the coupled light absorbed by the top rod structure, as illustrated in the inset schematic of Fig. 7d. The photocurrent enhancement of the Si NBs is shown in comparison with the planar structure in Fig. 7d.

## 2.4 Black Silicon

Another approach for surface treatment is to form black Si structures that have minimum reflectivity in the visible and infrared region. An advantage of these structures is the possibility of obtaining them at the macroscale with cheap, well-established methods. For instance, Ao et al. [33] achieved macroporous Si via

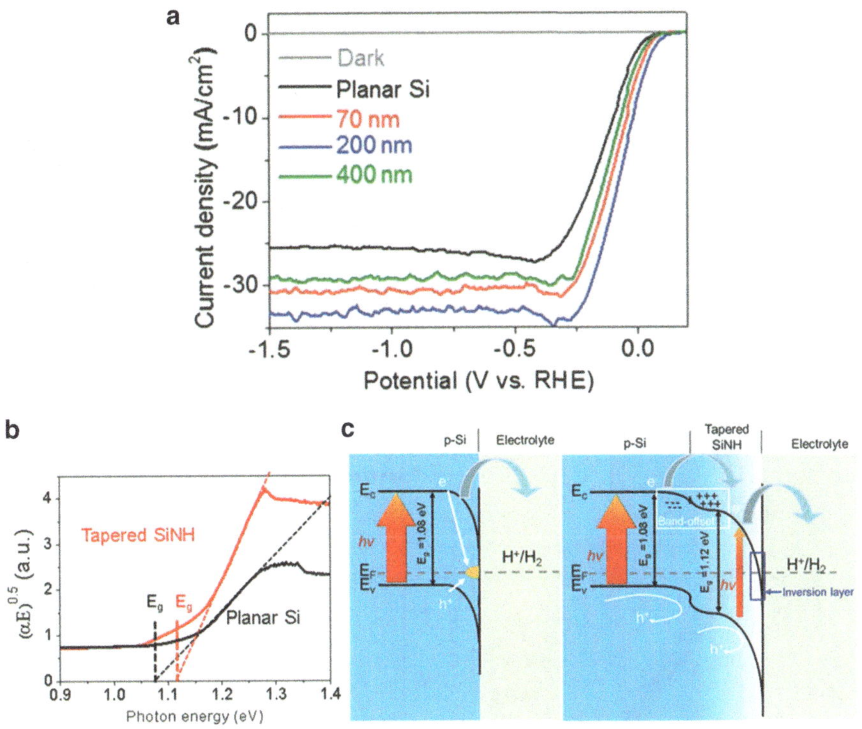


**Fig. 6** (**a**) *J–V* curves of planar Si (black) and Si NH arrays with depths of 70 nm (red), 200 nm (blue), and 400 nm (green) (**b**) *J–V* curves of tapered Si NH arrays obtained with Ag deposition times of 1 s (black), 5 s (red), 10 s (blue), and 30 s (green). The gray line displays the dark current density of Si NHs. (**c**) Tauc plot of planar Si and Si NH arrays showing their optical bandgap. (**d**) Schematic illustration showing the band structure of planar Si and tapered Si NH photocathodes in contact with the electrolyte. The Si/SiMH heterojunction was formed through bandgap widening of Si NHs [30]

electrochemical etching with diluted hydrofluoric acid. The reflectance varied according to the morphology, while the macroporous black structure showed total reflectance that was below 5% over the region from 400 to 100 nm, as shown in Fig. 8a. For constructing the photoanode, the authors used black n-Si and p-Si coated with $TiO_2$. Figure 8b shows the enhancement in the photocurrent achieved by the black structure.

In summary, nanostructuring and surface treatments are necessary to increase the light trapping and surface area of Si photocathodes allowing higher conversion efficiency and reduced overpotential. However, the morphology needs to be controlled, and a thorough understanding of its effects is dire to avoid the negative consequences of the surface defects and the recombination centers in nanostructured films.

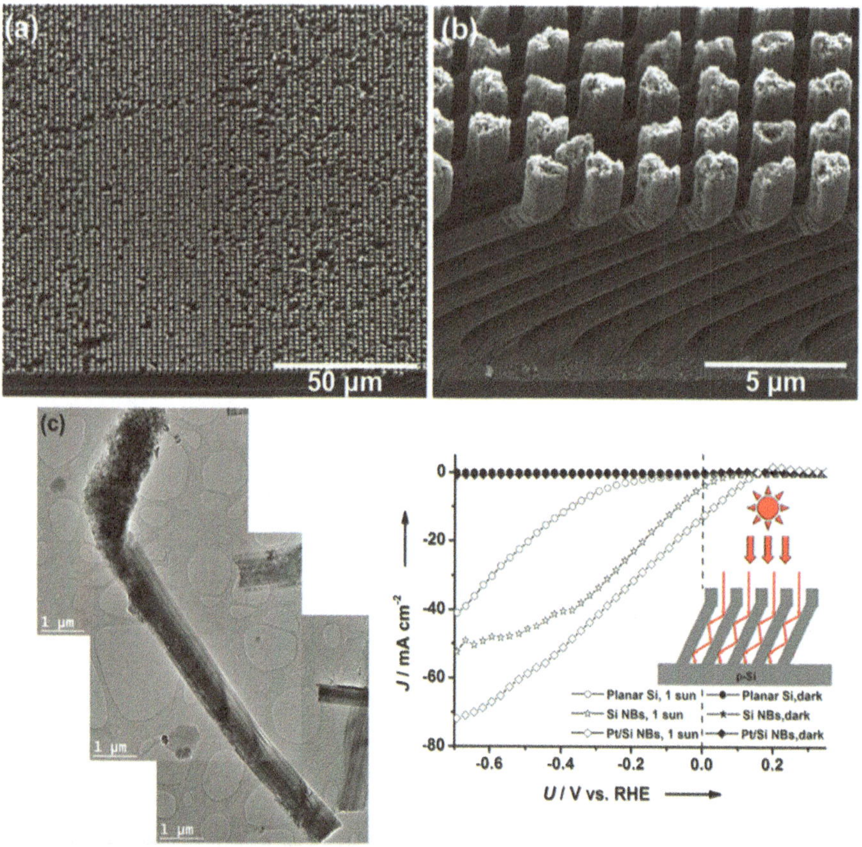


**Fig. 7** Morphology and microstructure of the as-fabricated ordered array of tilted Si NBs. (**a**) and (**b**) Overview and zoomed view SEM micrographs of the array. (**c**) Stitched TEM micrograph of a single NB. (**a**) *J*–*U* characteristics of the planar Si (circles), Si NB (stars), and Pt/Si NB (diamonds) array electrodes measured in 0.5-M $K_2SO_4$ solution under dark (solid symbols) and 100-mW/cm$^2$ illumination (open symbols) conditions. The inset shows schematic illustration of multireflection of the incident light within the tilted NB arrays [31]

# 3 Decorated Silicon

## *3.1 Platinum Nanoparticles*

Indeed, platinum is one of the most used materials for cathodes in water splitting. This fact emerges from the position of its work function, where a very small overpotential would be needed to initiate the water reduction reaction. Thus, decorating Si with Pt is a very promising area of research. In this sense, Warren et al. [34] took out a comparative study between Pt and Ni as two candidate catalysts decorating Si. They made use of the merits of the microwire morphology,

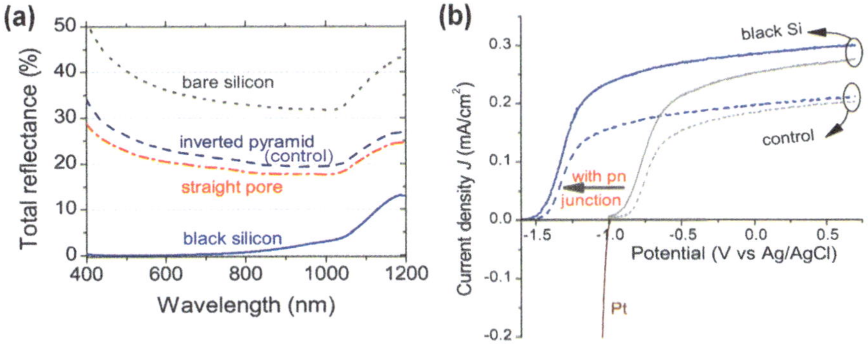


**Fig. 8** (**a**) Total reflectance from bare silicon (dotted gray line), silicon with inverted pyramid-like pits (dotted blue line), silicon with straight pores (dotted red line), and black macroporous silicon (solid blue line). (**b**) Photocurrent density versus the bias potential measured in 1-M KOH electrolyte at light intensity of 85 mW/cm$^2$ for $TiO_2$-coated n-Si (gray lines) and $TiO_2$-coated p–n-Si (blue lines) showing the enhanced photocurrent after black Si etching (solid lines) in comparison with the samples with inverted pyramid-like pits (dashed lines). For reference, the *J*–*V* curve for hydrogen evolution at the Pt electrode is shown (dark red) [33]

such as the high surface area of the rods, to lower the flux of the charge carriers over the rod arrays' surface, which in turn lowers the photocurrent density at the solid–liquid junction, resulting in reducing the demand for electrocatalysis. Thus, Si microwires were fabricated via vapor–liquid–solid mechanism (VLS) to study the effects of decoration with Pt and Ni nanoparticles. Galvanic displacement was used to form separate Pt- and Ni-coated Si arrays. As shown in Fig. 9, coating the microwire arrays with Ni nanoparticles relatively enhanced the photovoltage of p-Si. On the other hand, Pt showed a more significant enhancement in the photovoltage by reducing the kinetic barrier for the charges to pass across the semiconductor/liquid interface.

With Pt proving potential, Oh et al. [32] focus on the effect of nanostructuring on the enhancement witnessed in Pt-decorated Si, where they were one of the first groups to investigate the generation of $H_2$ on p-SiNWs decorated with Pt NPs. As aforementioned, decorated NW electrodes were fabricated via metal-catalyzed electroless etching followed by impregnation of Pt NPs as the electrocatalyst. Figure 10 shows the results of PEC measurements. The enhancement in the $V_{OS}$ was estimated by observing the positive shift in its value. The $V_{OS}$ for SiNW was 0.2 V, which was more positive than that for the planar Si. The Pt/SiNW showed a $V_{OS}$ of 0.42 V, compared to 0.33 V for the Pt/Si planar electrodes. This enhancement in photovoltage was attributed to the much smaller size of the Pt NPs coating the SiNWs, due to its rough surface, as compared to those coating the planar Si substrate. This enhancement in the photovoltage assisted in relieving the burden on the photoanode in the tandem cell configuration. It is notable that the Pt/SiNW structure showed a robust stability of photocurrent at a longer time scale, yet showed a decreased photocurrent density.

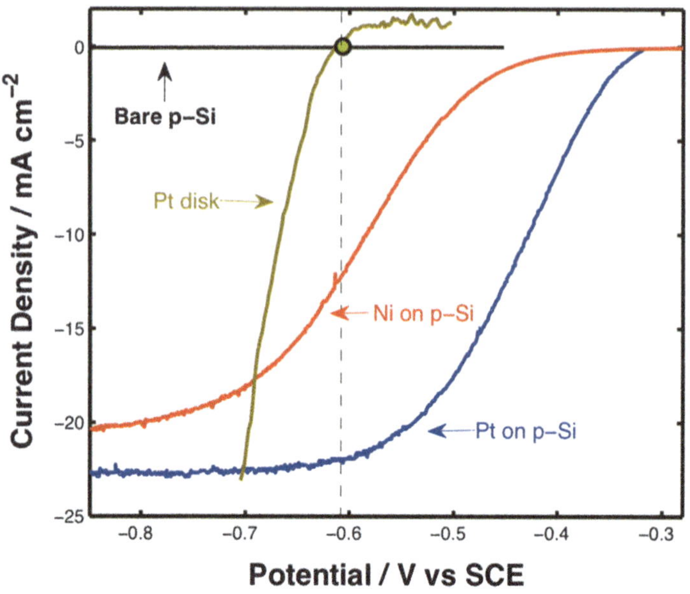


**Fig. 9** Comparison of Ni and Pt nanoparticles on planar Si photocathodes in pH 6.2 phosphate buffer. All samples were illuminated with 100-mW/cm$^2$ ELH illumination, and the system was continuously purged with $H_2$. Both the Ni- and Pt-decorated electrodes passed cathodic currents at potentials positive of the thermodynamic hydrogen potential in the solution (indicated by dashed line) [34]

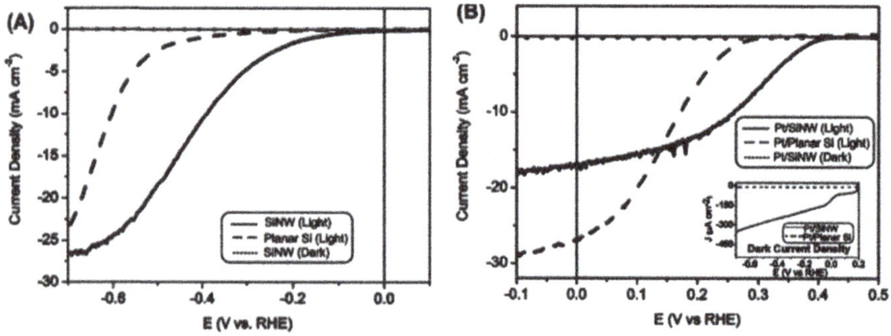


**Fig. 10** Photoelectrochemical $H_2$ generation on p-type SiNW array electrode (both illuminated and dark) in a stirred solution of $H_2SO_4$ + 0.5-M $K_2SO_4$ (pH 1). Same measurements on planar p-type Si are performed for reference. (**a**) Bare SiNW and planar Si photocathodes. While planar Si gives little photocurrent in positive photovoltage region, SiNW shows a dramatic boost in photocurrent of $H_2$ generation. (**b**) When SiNWs are impregnated with Pt nanoparticle by electroless deposition, overpotential for $H_2$ generation is significantly reduced both for SiNWs and planar Si. Photovoltage of SiNWs is ~0.2 V larger than that of planar Si. The inset shows the dark current densities of PtNP-impregnated SiNWs and planar Si electrode in the same electrolyte [32]

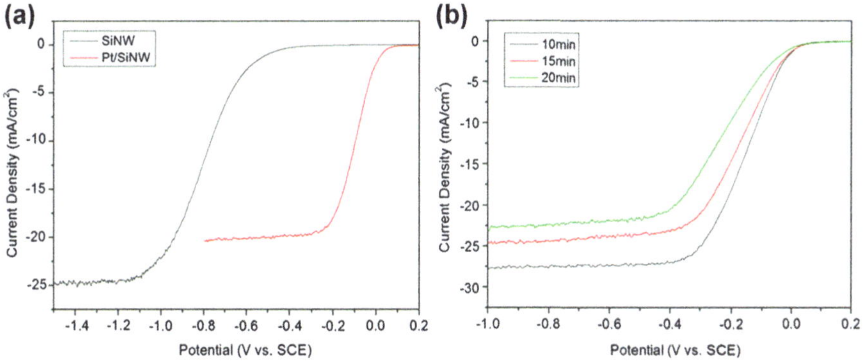


**Fig. 11** (**a**) *J–V* curves of SiNW arrays and Pt/SiNW array electrodes. The curves were measured in a solution of $H_2SO_4$ and 0.5-M $K_2SO_4$ (pH = 1) under illumination of 100 mW/cm$^2$. (**b**) *J–V* curves of Pt/SiNW array electrodes with different Pt NPs electroless deposition time. The curves were measured in a solution of $H_2SO_4$ and 0.5-M $K_2SO_4$ (pH = 1) under illumination of 100 mW/cm$^2$ [35]

Following on the previous results, Ji et al. [35] prepared p-type SiNW arrays decorated with discrete Pt NPs using the same method. As apparent in Fig. 11a, similar results were also obtained, where a large positive shift in $V_{OS}$ was witnessed, which means that a lower bias was required to reduce $H^+$ at the same rate, as well as a decrease in the photocurrent, which meant deterioration in the photocatalytic performance. This reduction in photocurrent was explained in terms of the decrease in the amount of photogenerated carriers due to the reduction in the amount of photons absorbed by the SiNW arrays as a result of the coating NPs, in addition to the recombination of photogenerated carriers at the Pt/Si interface. This adverse effect was overcome by decreasing the Pt deposition time as shown in Fig. 11b.

In an attempt to study the effect of the deposition method, Dasgupta et al. [36] deposited Pt NPs on SiNW arrays via atomic layer deposition (ALD) starting from as little as one cycle. The combination of highly uniform coating and precise control of the NP size and loading in the ALD process proved to be a powerful tool that can take advantage of the benefits of the NW geometry while minimizing the total volume of Pt required. Most notable is that controlling the Pt loading and particle size was easily available by controlling the number of cycles in the ALD process, as presented in Table 2. Furthermore, the stability of the NW arrays against oxidation was enhanced by ALD of $TiO_2$ on the SiNW, forming SiNW/$TiO_2$ core–shell structures, followed by ALD of Pt. It was found that $TiO_2$ thin layer was needed to get a positive photovoltage with Pt catalyst on the Si surface. As the number of Pt ALD cycles increased, the cathodic current also linearly increased. Using this method, the onset voltage was positively shifted up to 0.25 V vs RHE. It is worth noting that Dasgupta et al.[36] also studied the effect of the NW surface area on the catalytic activity by applying the ALD/$TiO_2$/Pt coating on NW arrays fabricated via gas-phase etching and others fabricated via electroless etching, as

**Table 2** Summary of electrochemical and photoelectrochemical performance of Pt catalysts formed by ALD [36]

| No. of Cycles | Surface Pt Loading (ng/cm$^2$) | $j_o$ (μA/cm$^2$) | $J_{sc}$ (mA/cm$^2$)[a] |
|---|---|---|---|
| 1 | 13 ± 3 | 27 ± 4 | 7.1 |
| 2 | 24 ± 5 | 59 ± 9 | 8.3 |
| 3 | 34 ± 7 | 85 ± 10 | 12.5 |
| 10 | 105 ± 21 | 130 ± 20 | 20.7 |

[a] Current density at 0 V *versus* RHE

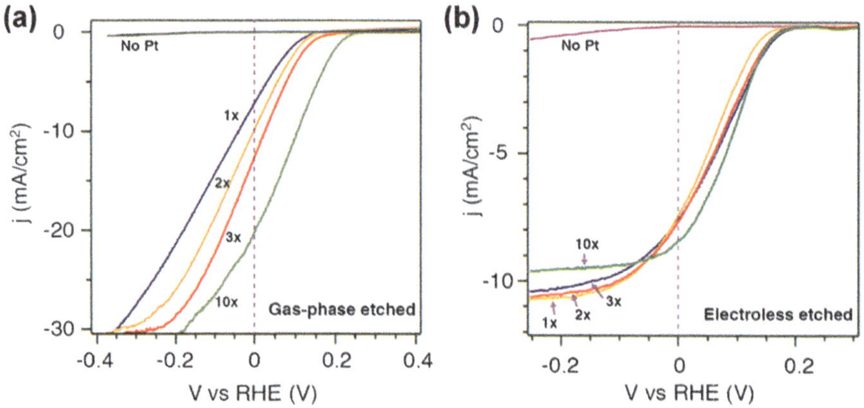


**Fig. 12** Photoelectrochemical performance of Pt cocatalysts loaded via ALD in 0.5-M $H_2SO_4$ electrolyte. (**a**) HER performance (under 1 sun illumination) of SiNW array photocathodes synthesized via gas-phase etching and loaded with ALD Pt cocatalyst. (**b**) HER performance of a SiNW photocathode synthesized via electroless etching under the same condition as in (**a**) [36]

shown in Fig. 12 panels a and b, respectively. The electroless etched NW arrays exhibited a higher surface roughness, leading to a decrease in photocurrent due to surface recombination caused by the large surface area.

## 3.2 Additions/Alternatives to Pt

### 3.2.1 Gold Nanofilms

Kye et al. [37] prepared planar silicon structures, which were decorated by a gold nanostructured film (Au NF) via galvanic displacement technique. After that, a Pt monolayer was deposited on Au NF using galvanic exchange of copper by platinum in order to get Pt/Au NF/Si photocathode, as shown in Fig. 13a. Photoelectrochemical measurements showed a significant enhancement in the photocurrent and the onset

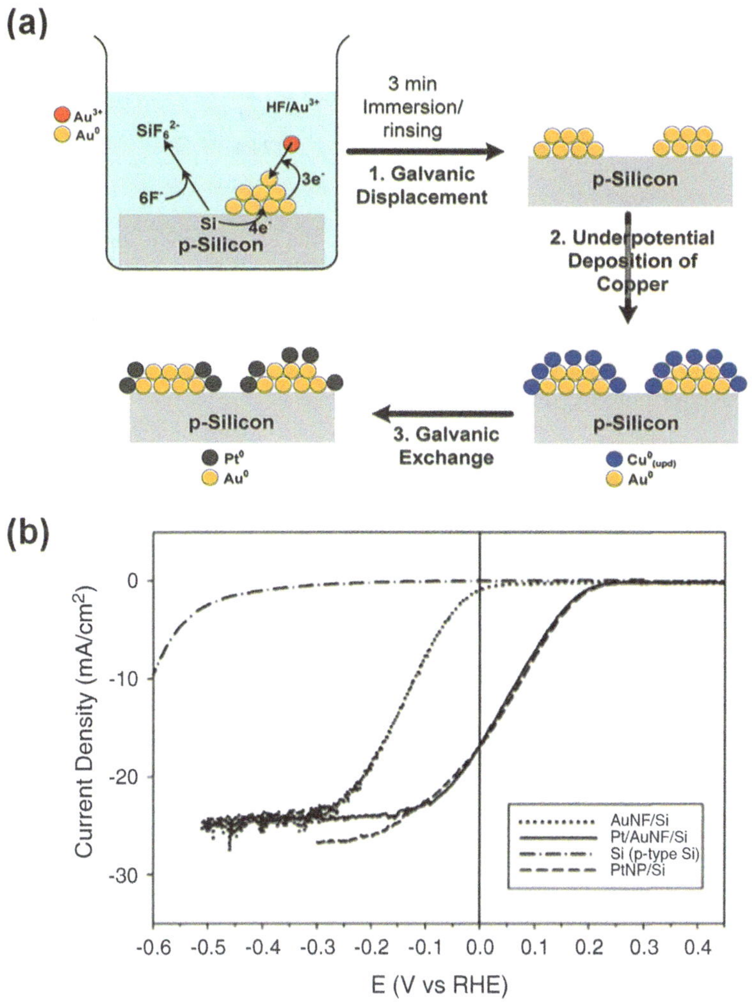


**Fig. 13** (**a**) The fabrication schematic of Pt monolayer-decorated Au nanofilm on p-Si. First, Au deposition occurs by immersion of a Si substrate into a mixture of HF and metal salts, which are reduced with the oxidation of silicon. A monolayer of copper is then underpotentially deposited on the Au nanofilm under illumination. Finally, Pt monolayer is deposited with labile copper monolayer by galvanic exchange. (**b**) Photoelectrochemical $H_2$ generation on planar p-Si electrodes with various catalysts under illumination in a stirred solution of $H_2SO_4$ + 0.5-M $K_2SO_4$ (pH 1). Scan rate was 0.02 V/s [37]

potential in the presence of AuNF on Si due to the catalytic effect of Au clusters. This catalytic effect was further enhanced by the deposited Pt monolayer on the surface of Au NF (Fig. 13b). In order to determine whether the increased activity and the reduction in overpotential in case of Pt/Au NF/Si are either due to geometric effects or electronic effects, both factors were studied separately.

**Table 3** Comparison of electrode parameters for photoelectrochemical $H_2$-evolving photocathodes with various catalysts [37]

| Sample | Au NFs | Pt/Au NFs | Pt NPs |
|---|---|---|---|
| $E_{os}$ (mV) | 137 | 249 | 252 |
| $J_{Ph}$ (mA/ $cm^2$) | 24.5 | 24.4 | 26.6 |

Utilizing the data obtained from estimating the electroactive surface area of the catalysts (Pt/Au NF and Pt nanoparticles – Pt NP) and from analyzing their Tafel plots, as presented in Table 3, it was concluded that the enhancement was due to the nature of the catalyst (electronic effect). Accordingly, Key et al. [37] proposed that the deposition of Pt monoatomic layer on the metal catalyst (Au) could be a good alternative for pure Pt NP alone.

### 3.2.2 Molybdenum Sulfide

With the main aim of studying $MoS_x$ as an alternative catalyst to Pt, Seger et al. [38] firstly sought to solve the problem of the spontaneous oxidation of Si on contact with water. They found that, in order to prevent Si oxidation during $H_2$ evolution, the concentration of the dissolved $O_2$ must be lower than 15 ppb, which is inapplicable in a water-splitting system. The authors also mentioned that electrodeposition of a catalyst at oxidative potentials would not work effectively as the photoabsorber will undergo oxidation during the deposition process. This problem was solved by depositing a conductive and protective thin layer of Ti on Si ($n^+$p-Si) in order to effectively enhance the stability, such that when Ti undergoes oxidation, it will form the semiconducting $TiO_2$ instead of the insulating $SiO_2$. With this problem out of the way, the photoelectrodeposition of the $MoS_x$ catalyst was done on the Ti-protected Si (Ti-$n^+$p-Si), as illustrated in Fig. 14a, in order to form $MoS_x$–Ti-$n^+$p-Si. Table 4 summarizes the enhancements in the onset potential of the novel electrode, also shown in Fig. 14b, as compared to Pt–Ti–$n^+$p-Si. As shown, although the Pt-decorated samples showed a more positive $V_{OS}$, the photocurrent density of the $MoS_x$-decorated samples was higher, which showed promise for this catalyst.

Hou et al. [39] also fabricated planar silicon (100) and structured silicon pillars (100) decorated with $Mo_3S_4$ clusters. These clusters were designed to be soluble in polar organic solvents but insoluble in water. As shown in Fig. 15, $Mo_3S_4$ clusters enhanced the onset potential of both the planar and the pillar structures by shifting the onset potential to more positive values. However, and in contrast to the results obtained by Seger et al. [38], the photocurrent decreased slightly. This was attributed to the adhering of the hydrogen bubbles on the electrode's surface, which resulted in decreasing the active surface area.

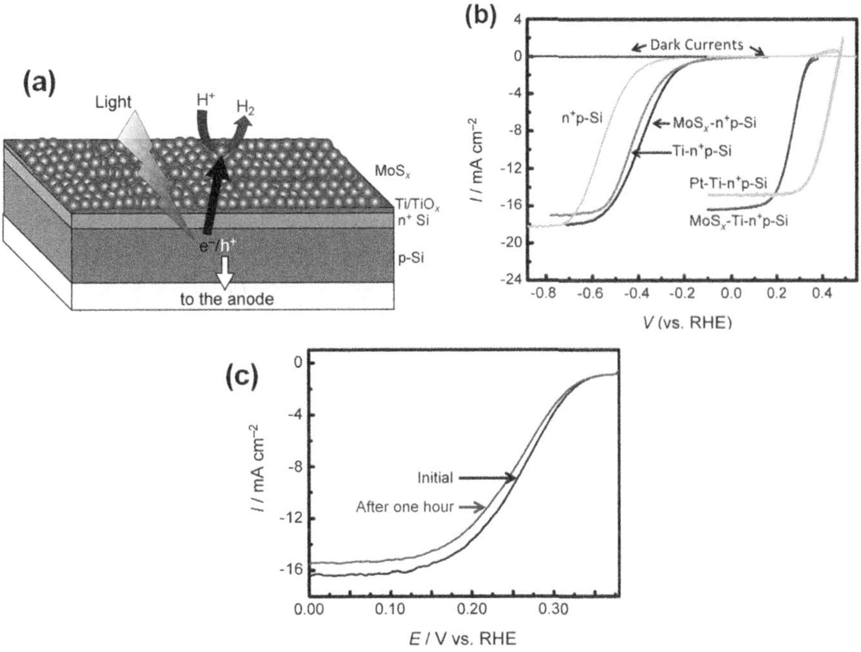


**Fig. 14** (**a**) Scheme of a $MoS_x$–Ti–$n^+$p-Si electrode showing how photo-irradiated electrons react with protons to evolve $H_2$. $e^-/h^+$ refers to the photogenerated electron-hole pair. (**b**) Cyclic voltammograms of the photoelectrocatalytic HER of various $n^+$p-Si electrodes. The samples were irradiated with red light (AM1.5 cutoff < 635 nm, 38.6 mW/cm$^2$) and scanned at 50 mV/s. (**c**) Photocurrent before and after a 1 h chronoamperometry test at +200 mV versus RHE [38]

**Table 4** Comparison of electrodes onset potential for photoelectrochemical $H_2$ evolving photocathodes with various catalysts [38]

| $n^+$p-Si | Ti–$n^+$p-Si | $MoS_x$–$n^+$p-Si | Pt–Ti–$n^+$p-Si | $MoS_x$–Ti–$n^+$p-Si |
|---|---|---|---|---|
| −0.25 V | −0.1 V | ~ −0.1 V | 0.47 V | 0.33 V |

### 3.2.3 Cuprous Oxide

Xiong et al. [40] fabricated SiNW/$Cu_2O$ core–shell structures via metal-assisted electrochemical etching of Si wafers, followed by the electroless deposition of $Cu_2O$ on the SiNWs (Fig. 16a–d). As shown in Fig. 16e, SiNWs/$Cu_2O$ core–shell structure increased the photocatalytic performance compared to SiNWs alone by 50%. Moreover, loading the core–shell structure with Pt NPs via potentiostatic electrodeposition resulted in a significant enhancement in the $H_2$ evolution reaction, as apparent in Fig. 16f. Also shown in Fig. 16f is the superiority of $Cu_2O$ over Pt NPs in terms of hydrogen evolution, which makes this material very promising for the decoration of Si.

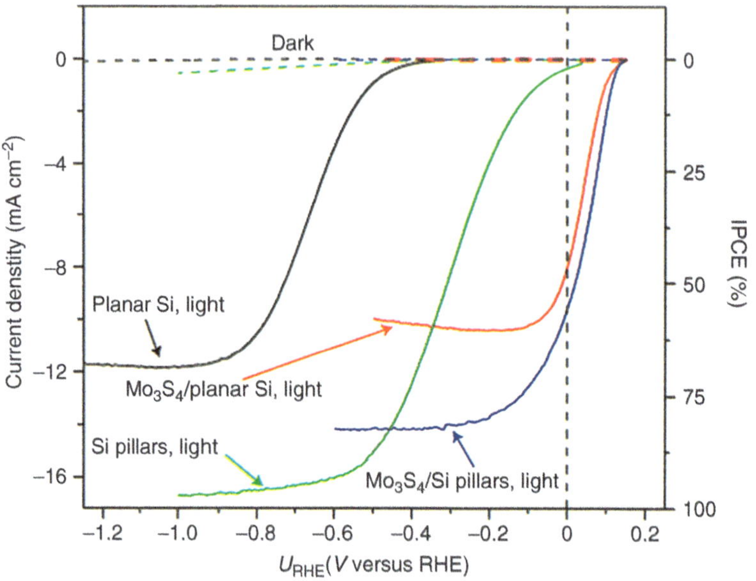


**Fig. 15** Photoelectrocatalytic activity measurements on planar and pillared Si together: the potentiodynamic runs on the photoelectrodes. Steady-state current density–voltage (left axis) is run in aqueous 1.0-mol/l $HClO_4$ solution under red-light irradiation ($\lambda$ >620 nm, 28.3 mW/cm$^2$) with the calculated IPCE shown on the right axis. The almost horizontal dashed lines denote the current measured in darkness (almost 0 mA/cm$^2$) [39]

### 3.2.4 Graphene

Huang et al. [41] studied the effect of decorating SiNWs with graphene. They fabricated the SiNWs via metal-assisted chemical etching and then used spin coating of reduced graphene oxide (rGO) to get a composite of SiNWs/rGO. Coating the SiNWs with rGO enhanced the short-circuit photocurrent density (at 0 V vs RHE) by a factor of four times larger than that of the pristine SiNWs, as shown in Fig. 17a. This enhancement was attributed to the fact that rGO exemplifies a channel that provides fast electron transfer between SiNWs and the electrolyte. Furthermore, the open-circuit potential was positively shifted by 20 mV in the case of the SiNWs/rGO composite as compared to pristine SiNWs. Huang et al. [41] explained the fluctuation in the photocurrent density around −0.4 V (vs RHE) in terms of accumulation of the hydrogen bubbles at the surface of the SiNWs/rGO photoelectrode, which resulted in the reduction of the active area of the surface and/or scattering of the incident illumination through its path in the electrolyte to reach the sample. Figure 17b shows the stability of SiNWs/rGO; it was found that the degradation in the photocurrent density with time can be neglected, which indicates the enhancement done by addition of rGO to SiNWs. To study the interface physics, the flat band potentials and band bending values were calculated

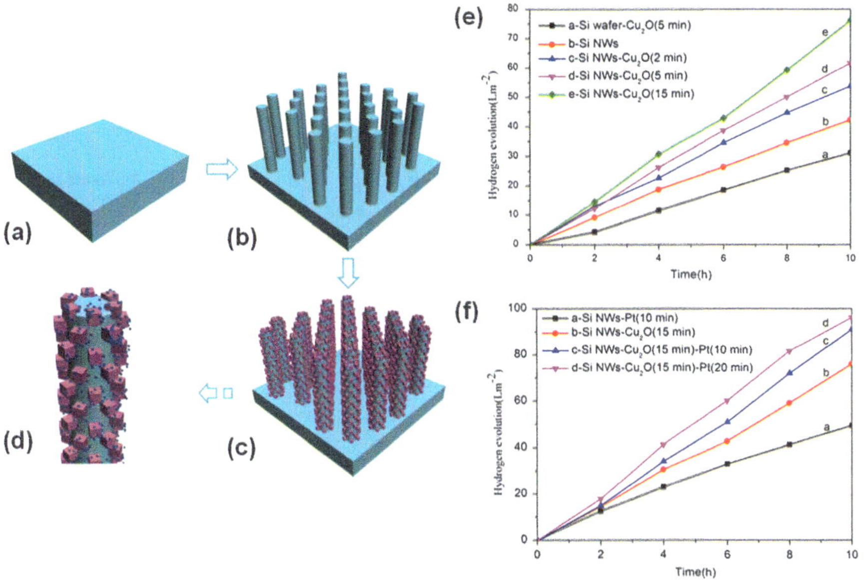


**Fig. 16** Schematic diagram of the fabrication process for the SiNW/$Cu_2O$ core–shell nanosystem: (**a**) Si wafer, (**b**) SiNWs fabricated via metal-assisted chemical etching, (**c**) $Cu_2O$ nanocrystallites deposited on SiNWs via electroless deposition, and (**d**) SiNW/$Cu_2O$ core–shell nanosystem with Pt loading. (**e**) Photocatalytic $H_2$ evolution of SiNWs, planar Si/$Cu_2O$ structure, and SiNW/$Cu_2O$ core–shell nanosystem under simulated sunlight irradiation. (**f**) Photocatalytic $H_2$ evolution of SiNWs and SiNW/$Cu_2O$ core–shell nanosystem with Pt nanoparticles as cocatalyst under simulated sunlight irradiation [40]

using Mott–Schottky plots, shown in Fig. 17c. Indeed, better charge separation requires larger band bending values associated with higher flat band potentials. In the case of the SiNWs/rGO composites, a flat band potential smaller than pristine SiNWs was observed, indicating the negative effect of rGO in charge separation. Moreover, electrochemical impedance spectroscopy (EIS – Fig. 18d) revealed that SiNWs/rGO exhibited smaller-charge transfer double-layer capacitance (around four times smaller than SiNWs) and so a higher HER rate. Using EIS and *J–V* curves, Huang et al. [41] derived the relation between current density, the charge transfer resistance ($R_{ct}$), and the amount of coated rGO, showing that a sample coated with 200 μL of rGO would exhibit the best photocurrent density due to the smallest $R_{ct}$ values (Fig. 17e). Further coating will result in increasing the series resistance of rGO. Table 5 shows the reported short-circuit photocurrent for different modified Si-based composites.

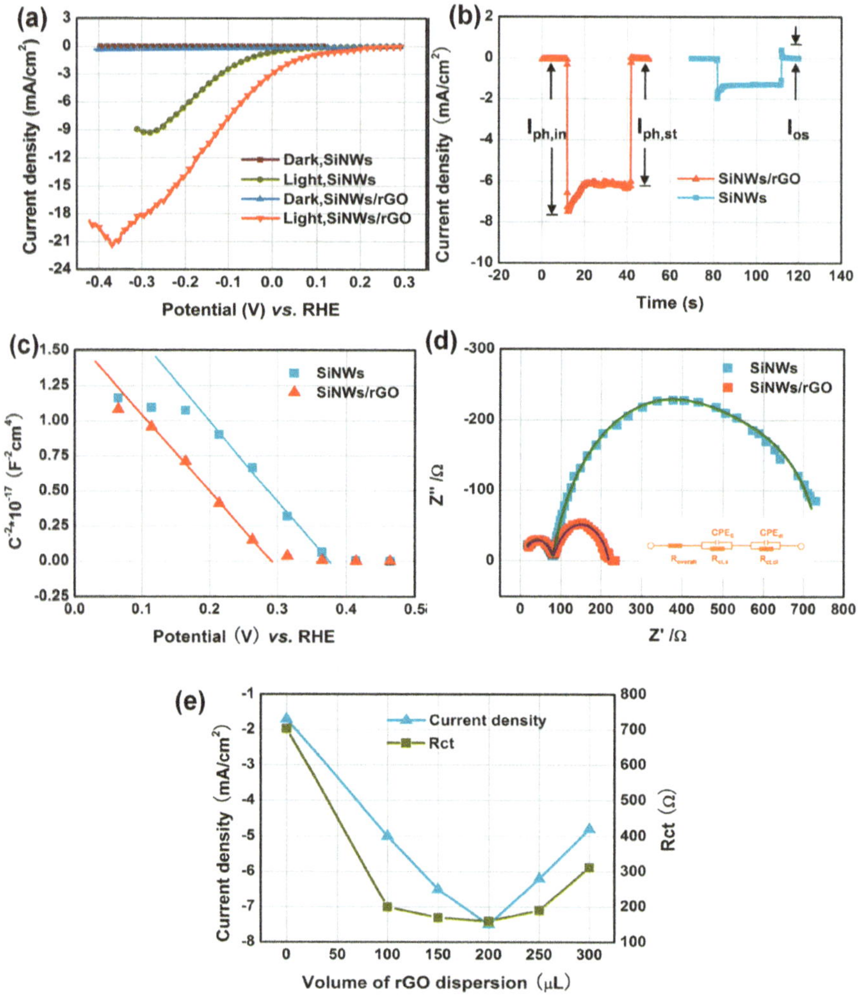


**Fig. 17** (**a**) Photoelectrochemical behavior of SiNWs and SiNWs/rGO samples. (**b**) Transient *J–V* behaviors of SiNWs and SiNWs/rGO samples under chopped illumination at $V_a = 0.09$ V (vs RHE). (**c**) Mott–Schottky plots of SiNWs and SiNWs/rGO samples. (**d**) EIS spectra of SiNWs and SiNWs/rGO samples measured at $V_a = -0.11$ V (vs RHE). (**e**) Variation of $R_{ct,dl}$ and photocurrent density at −0.11 V (vs RHE) with the amount of rGO [41]

# 4 Silicon Carbide

Silicon carbide (SiC) occurs in more than 200 crystal structures [42]. Although all the crystal structures have the same 1:1 ratio of Si to C, each structure offers its own electrical properties. The most common structures of SiC that were studied for electrical applications are 3C-SiC (β phase), 4H-SiC, and 6H-SiC. Table 6 states the

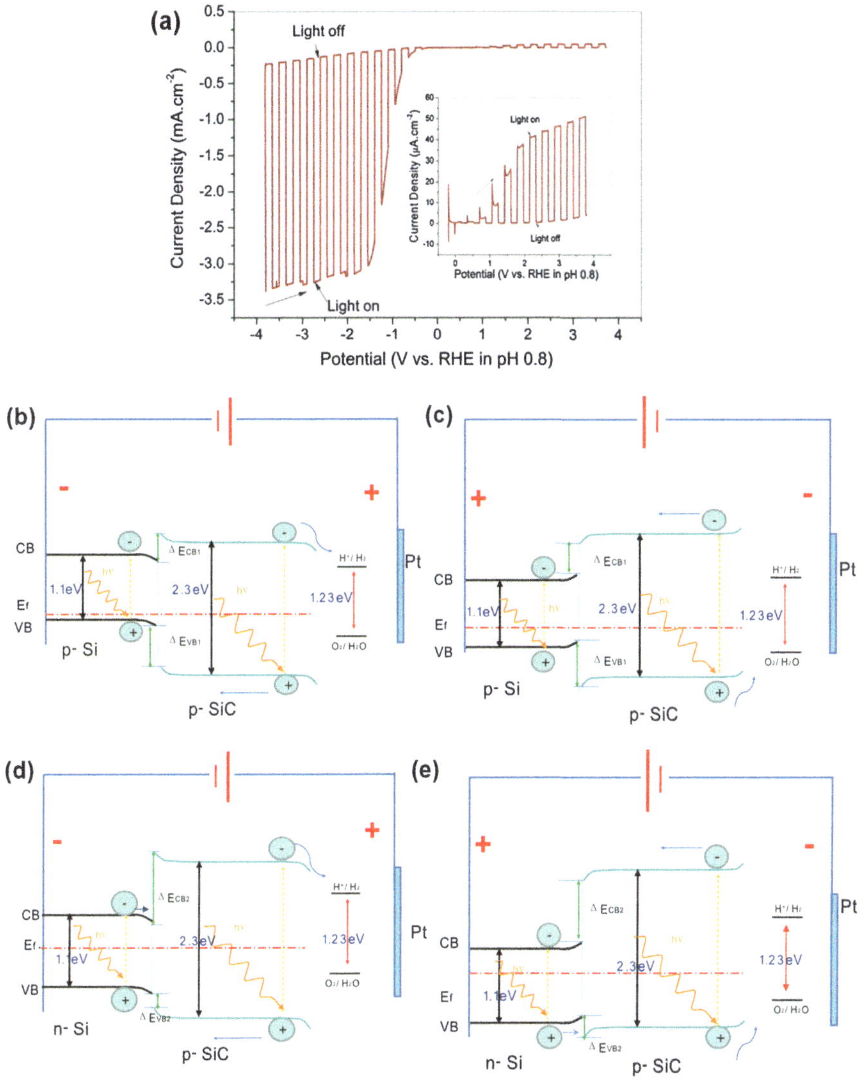


**Fig. 18** (**a**) Photocurrent density–potential curve of the p-SiC film on p-Si substrate under chopped light illumination. The light intensity was set at 100 W/m$^2$. Energy diagrams of PEC components under illumination are also shown: (**b**) photocathode (p-SiC on p-Si) and anode (Pt) with an external negative bias and (**c**) with an external positive bias. (**d**) Photocathode (p-SiC on n-Si) and anode (Pt) with an external negative bias and (**e**) with an external positive bias [46]

electronic properties of these three different structures of SiC compared to Si. These have respective bandgaps of 2.3, 3.2, and 3.0 eV. Compared to 1200 cm$^2$ V$^{-1}$ s$^{-1}$ for Si, the electron mobility (at $N_D = 10^{16}$ cm$^{-3}$) of the 3C-SiC and 4H-SiC structures is lower, having values of 750 and 800 cm$^2$ V$^{-1}$ s$^{-1}$, respectively. The 6H-SiC structure

**Table 5** Reported $J_{sc}$ (photocurrent density at RHE) of photoelectrochemical hydrogen production of SiNWs or Si microwire-based composites as presented in Ref. [41]

| Ref | Semiconductor | Modification | Strategy | $J_{sc}$ (mA/cm$^2$) |
|---|---|---|---|---|
| 1 | CVD-grown silicon microwires, core–shell p–n + junction | Pt | Electrocatalyst | 15 |
| | CVD-grown p-type silicon microwires | | | 7.3 |
| | Planar silicon p–n+ | | | 28 |
| | Planar p-type silicon | | | 23 |
| 2 | P-type silicon micropillar | $Mo_3S_4$ cluster | Electrocatalyst | 10 |
| | P-type planar silicon | | | 8 |
| 3 | P-type SiNWs fabricated by MACC | Pt | Electrocatalyst | 17 |
| | P-type planar Si | | | 27 |
| 4 | CVD-grown silicon microwires, core–shell p–n + junction | Ni–Mo | Electrocatalyst | 9.1 |
| 5 | P-type SiNWs fabricated by MACC | $MoS_2$ | Electrocatalyst | 1 |
| 6 | P-type planar silicon/ZnO nanowires array | Pt | Electrocatalyst | 1 |
| | | Pd | | 0.5 |
| | | Ni | | 0.5 |
| 6 | P-type SiNWs fabricated by MACC | Branched ZnO | Energy band engineering | 0.1 |

**Table 6** A comparison of the band gap and the mobility of electrons and holes in Si as well as the three major polytypes of SiC at 300 K [42]

| Property | Si | 3C-SiC | 4H-SiC | 6H-SiC |
|---|---|---|---|---|
| Bandgap (eV) | 1.1 | 2.3 | 3.2 | 3.0 |
| Electron mobility at $N_D = 10^{16}$ cm$^{-3}$ (cm$^2$ V$^{-1}$ s$^{-1}$) | 1200 | 750 | // c-axis: 800<br>⊥ c-axis: 800 | // c-axis: 60<br>⊥ c-axis: 400 |
| Hole mobility at $N_A = 10^{16}$ cm$^{-3}$ (cm$^2$ V$^{-1}$ s$^{-1}$) | 420 | 40 | 115 | 90 |

has an electron mobility of 60 cm$^2$ V$^{-1}$ s$^{-1}$ parallel to the c-axis and 400 cm$^2$ V$^{-1}$ s$^{-1}$ perpendicular upon the c-axis [42]. Furthermore, Si structures also show higher hole mobility (420 cm$^2$ V$^{-1}$ s$^{-1}$), compared to 3C-SiC (40 cm$^2$ V$^{-1}$ s$^{-1}$), 4H-SiC (115 cm$^2$ V$^{-1}$ s$^{-1}$), and 6H-SiC (90 cm$^2$ V$^{-1}$ s$^{-1}$) [42].

Kato et al. [43] studied the photocatalytic behavior of the three structures; 3C-, 4H-, and 6H-SiC grown by molecular epitaxy for hydrogen production in a two-electrode system without applying any external bias. The highest STH efficiency (0.38%) was measured for the 3C-SiC film due to its low bandgap compared to the other two structures. Hao et al. [44] showed that the surface modification of 3C-SiCNWs by acid oxidation increased the rate of hydrogen production by 76.1% higher than the unmodified NWs due to the increase in the hydrophilicity.

In another study, Hao et al. [45] showed the effect of the morphology on the average hydrogen production rate of 3C-SiC.

Usually SiC and other p-type semiconductors are grown on a substrate which can be p- or n-type. These multilayers work as a whole as a photoelectrode for water splitting. Ma et al. [46] analyzed the photoresponse from p-type 3C-SiC films grown on p-Si and others grown on n-Si. The p-SiC film on p-Si generated a cathodic photocurrent as a photocathode when exposed to negative bias and generated an anodic photocurrent as a photoanode when exposed to positive bias, as shown in Fig. 18a. However, the anodic photocurrent was found to be much lower than the cathodic photocurrent. On the other hand, the p-SiC film on n-Si was shown to generate appreciable anodic photocurrents at positive potentials but very weak cathodic current at negative potentials. Depending on the measurements of incident photon to electron conversion efficiency (IPCE) and gravimetric analysis for the resulting gases, the behavior of p-SiC on p-Si and n-Si was explained upon applying negative and positive bias by the energy diagrams shown in Fig. 18 panels b–e. Upon illumination, both p-SiC and p-Si or n-Si absorb photons generating electrons and holes. For p-SiC on p-Si at negative external potential (Fig. 18b), the electrons from p-SiC are derived to the hydrogen redox potential producing hydrogen. When the applied negative bias increases, electrons from p-Si can pass the conduction band barrier toward p-SiC to undergo the reduction of water, which increases the cathodic photocurrent. At positive external potential (Fig. 18c), the holes from p-Si cannot pass to p-SiC due to the large valence band barrier. Therefore, the anodic photocurrent is weak as it is only generated from the SiC holes that are driven to the oxygen redox potential producing oxygen. For the p-SiC on n-Si at negative external potential (Fig. 18d), the photogenerated electrons from n-Si cannot pass to SiC due to the large barrier between the conduction bands. At positive external potential (Fig. 18e), the photogenerated holes from SiC are driven to oxidize water, and the holes from Si contribute to the anodic photocurrent. In conclusion, the p-SiC on p-Si can be used as a photocathode, and p-SiC on n-Si can be used for photoanodes. These contrasting effects originate from the fact that SiC on Si substrate doesn't act as a single electrode but rather as a heterojunction semiconductor.

In summary, ever since it was originally shown to have promise in water splitting in 1990 by Nariki et al. [47], SiC has not been extensively studied for that purpose. Therefore, further studies are required to understand the effect of the rich chemistry of SiC on its photocatalytic performance.

## 5 Amorphous Si

Crystalline semiconductors including Si and Ge are gaining a lot of attention in the field of photovoltaic and solar energy conversion [48–50]. For c-Si, one atom of Si is covalently bonded to four other atoms in a regular patterned lattice. On the other hand, a-Si lacks the highly ordered pattern, and there instead exist a number of bond angle variations and disorders that disrupt the crystalline structure. Historically, a-Si gained an interest in the middle of the twentieth century, when amorphous Se and Si were first used in xerography; the flexible structure of the amorphous atoms allowed the production of different sizes of the semiconductors. Today, a-Si is widely used in the design of Si solar panels.

Indeed, the structure and properties of a-Si have been widely studied [48, 51]. Unlike c-Si, a-Si is characterized by a large number of dangling bonds that originate from different bonding angles in the structure, which can lead one atom to bond with only three covalent bonds [51]. These dangling bonds increase the resistance and the internal strain throughout a-Si structure limiting its usage in photovoltaic applications [48, 51]. Thus, it is of engineering importance to bond the dangling bonds to dopant atoms in order to improve the electrical properties of a-Si. Indeed, this was achieved by sputtering silane gas over a-Si [52], where hydrogen atoms got attached to dangling bonds resulting in an amorphous structure with comparable electrical properties to its crystalline counterpart. The electrical conductivity enhancement is directly related to the origin of the gases used to react with a-Si [53]. A mixture of gases such as phosphine and silane can be used for enhancing the conduction of electrons (n-type), while diborane with silane is used for enhancing the mobility of holes (p-type) [48]. The most effective method for the efficient doping of hydrogen is plasma-enhanced chemical vapor deposition, where plasma of different gases is heated at 450 °C to allow doping of a-Si [48, 51]. This type of a-Si bonded to hydrogen through its dangling bonds is called hydrogenated silicon (a-Si:H), and it is the mostly used form nowadays.

## 5.1 Properties

As a photocathode, a-Si:H is considered to have advantageous properties when compared to c-Si. Firstly, a-Si is characterized by having a larger bandgap than c-Si due to the fact that the Si–H bond has a higher binding energy than Si–Si bond [48]. Furthermore, because binding energies vary directly with bandgaps, a-Si has a bandgap ranging from 1.6 to 1.8 eV [54], while that of c-Si is about 1.1 eV [48]. This increase in the bandgap offers a greater opportunity for a-Si, such that the conduction band of a-Si will exceed the hydrogen reduction band in water splitting [55], decreasing the amount of external bias needed. In addition, c-Si is known to have an indirect bandgap due to its rigid, highly ordered structure that conserves momentum K of both valence and conduction bands [51]. With nanocrystalline Si in particular (especially quantum dots), there is a debate between the direct and indirect bandgap mechanisms, as the electrons and holes under strong quantum confinement reduce the phonon relaxation mechanisms [56]. But for a-Si, the highly ordered structure is absent, which leads to breaking the K conservation momentum rule that originally renders a-Si a direct bandgap semiconductor. This, in turn, leads to great enhancements in the absorption coefficient if compared to c-Si. Figure 19 shows the difference in absorbance coefficients in

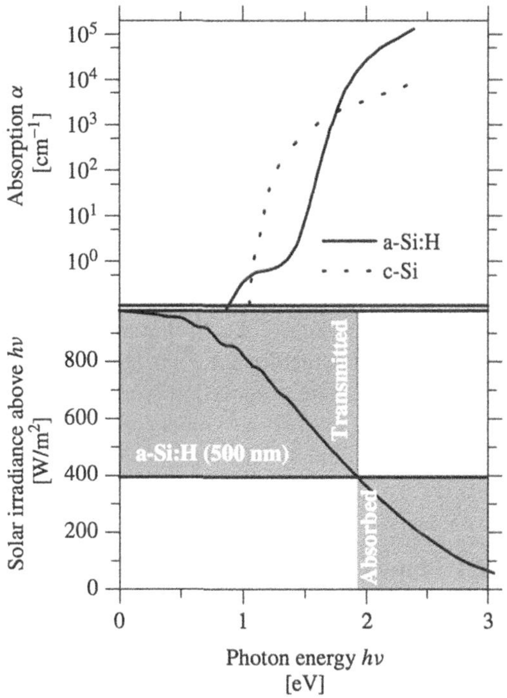


**Fig. 19** The difference in absorbance coefficient between a-Si and c-Si. It is clear that a-Si can absorb most of the solar irradiance above 1.9 eV [48]

both a-Si and c-Si [48]; it is clear that a-Si can absorb most of the solar irradiance above 1.9 eV. Thus, in order to make c-Si absorb the same amount of solar irradiance as a-Si, the thickness of c-Si must be increased, which means higher cost. This is why a-Si:H is advantageous, in several aspects, as a photocathode.

## 5.2 *Staebler–Wronski Effect*

The main problem arising with the usage of a-Si as photocathodes is its stability, as it was found by operation that the photo and dark conductivities decline after 1000 h of illumination, so-called the Staebler–Wronski effect [54]. This effect arises mainly due to several factors, which are represented by various theories [57–59]. Although those theories are different, they are all based on a similar background. The introduction of new defect states in the bandgap of a-Si results in an increase in the recombination rate, which leads to breaking of the Si–Si bonds. Most models assume that weak bonds that are present at the band tail, which mainly originate from configurational disturbances in the ground state, are the main source of the dangling bonds. Those configurational disturbances are less likely to occur in c-Si with its four-bonded structure; due to the highly ordered structure, very high temperature (~1000 °C) is required to create a defect in Si–Si bond. On the other hand, for amorphous Si, defects are more likely to occur due to the flexible structure. As a result, the bonding disorder increases in a-Si, increasing the strain on the weak band tail bonds, which in turn leads to their breakage and formation of dangling ones. Thus, besides the low-energy barrier potential for dislocation in the amorphous structure, weak band tail bonds are considered to be the precursors of the dangling bonds.

Staebler and Wronski [54] noted that this photodegradation is attributed to the introduction of the metastable defects in the mid-gap of a-Si after prolonged illumination. They also found that annealing the a-Si above 150 °C reverses this process and returns the structure to its prior photoconductivity. Inspired by the different morphologies of a-Si that are represented by the highly disordered structure difference in the dopant concentration (hydrogen) and impurity concentration, several models were proposed in order to properly understand this process. Because such models have a direct effect on the performance of a-Si in photoelectrochemical cells, these models will be briefly presented herein.

### 5.2.1 Impurity-Related Model

Ishii et al. [60, 61], who proposed the impurity-related model, suggested that the produced electrons by photo-illumination would be captured by an incorporated positively charged oxygen ion (acting as an impurity) that is coordinated by three Si atoms. This absorbed electron will neutralize the oxygen atom and will result in forming neutral, threefold-coordinated Si dangling bonds. However, this model showed contrast with experimental results performed later in 1996, whereby in incorporating a very low impurity concentration (smaller than the density of the produced dangling bonds), the density of dangling bond increased, which meant that there was no direct relation between the impurities present and dangling bond formation.

### 5.2.2 Change in the Charge State

Another model [62–64] suggested that, depending on the change in the charge state, the Si–H bonds are randomly distributed. Thus, given the ability of the hydrogen-rich regions to capture electrons, it was proposed that the absorbed electrons would neutralize the dangling bonds. However, this model was also disapproved when the charged dangling bonds concentration was measured in a photodegraded sample and after it annealing (where annealing removes the dangling bonds produced by illumination); it was found that the concentration didn't change [65]. This showed that there is no relation between the charged dangling bonds and the photodegradation.

### 5.2.3 Breaking Si–Si Bond

Stutzman et al. proposed a new model depending on the breaking of the weak Si–Si bond [66, 67]. Upon illumination, electrons and holes are produced; some electrons will bind non-radiatively with the dangling bonds in the mid-band, producing phonons with sufficient energy to break the weak Si–Si bond in the band tail. The deduced relation assumed that the rate of neutral dangling bonds formation, which results from breaking of weak Si–Si bonds, should be proportional to the electron and hole densities, i.e.,

$$\frac{dN_s}{dt} \alpha\, dt\, \alpha\, np\, \alpha \frac{G}{N^2}, \tag{3}$$

where $N_s$ is the density of neutral dangling bonds, $t$ is the illumination time, $np$ is the holes density, and $G$, which is proportional to the generation rate, is the light intensity. Once bond breaking takes place, Si–Si should recombine again. However, Stutzman et al. proposed that, if there was a hydrogen atom within the surrounding 0.4 nm, it would stabilize the system, and then an exchange will take place between Si–H and the Si dangling bonds. It was also supposed that bond switching can take place with neighboring atoms. This model was highly appreciated till 1998, where Branz produced a new model, known as the hydrogen collision model. Figure 20 shows the mechanism of the Si–Si breaking and formation of dangling bonds.

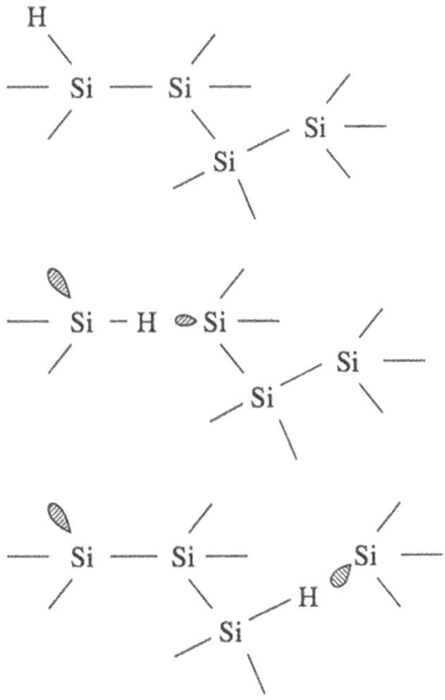


**Fig. 20** The mechanism of the Si–Si breaking and formation of dangling bonds [59]

### 5.2.4 Hydrogen Collision Model

The hydrogen collision model [67–69] solved the problems of the Stutzman model, by assuming that the neighboring hydrogen atom is not the only case for this mechanism to take place. Instead, Branz proposed that the excited electrons and holes (biexcitons) could produce mobile hydrogen atoms (low-barrier potential associated with biexciton absorption), which can bind to different dangling bonds originally present on the surface of a-Si. A lower concentration of H atoms, however, can bind to other H atoms resulting in forming a metastable state of two Si–H atoms. By this, the hydrogen collision model solved the problem of the Si–Si breakage model. Furthermore, another point where both models differed was the dependence of the excitation rate of hydrogen on $N_s$. Instead, Branz assumed that it depended on $G$.

$$\frac{dN_{\mathrm{m}}}{dt} = K_{\mathrm{H}}N_{\mathrm{H}}G - K_{\mathrm{s}}N_{\mathrm{m}}N_{\mathrm{s}} - 2K_{\mathrm{c}}N_{\mathrm{m}}^{2}, \tag{4}$$

where $N_{\mathrm{m}}$ is the density of mobile hydrogen, $N_{\mathrm{H}}$ is the density of Si–H bond, $K_{\mathrm{H}}$ is a constant responsible for creating mobile hydrogen, and $K_{\mathrm{s}}$ is the rate of trapping mobile hydrogen by a dangling bond. Although this model resolved different unclear issues such as the mechanism of breaking bonds and creation of dangling bonds at low temperature (4.2 K), it dismissed a direct relation between $G$ and $N_{\mathrm{s}}$, disapproving different theories that were proposed by Zhang et al. and Stutzman et al.

### 5.2.5 Floating Bond Model

The floating bond model [67, 70, 71] aimed to solve some of the raised problems against H collision model. Simply, a floating bond is a fivefold coordinated Si atom, surrounded by Si neighbors. This over-coordinated Si atom is characterized by having lower energy than a dangling bond and thus can be mobile and mediate the dangling bond – same as H atoms in the hydrogen collision model. This process continues by bond switching, where the floating bond can lose one Si atom to form a dangling bond. This model depended on Stutzman's rate of generation rule and refuted the relations in Branz's model [49, 67, 72, 73]. Figure 21 shows an illustration of the floating band model.

### 5.2.6 Microvoid-Induced Dangling Bonds

In 1986, Carlson introduced a model that relied on what was called hydrogenated microvoids [74]. Although this model was not popular as other models, toward the end of 2014, a group of scientists at Helmholtz Institute Jena were able to prove this model experimentally [75]. Near microvoids, where weak Si–Si bonds reside,

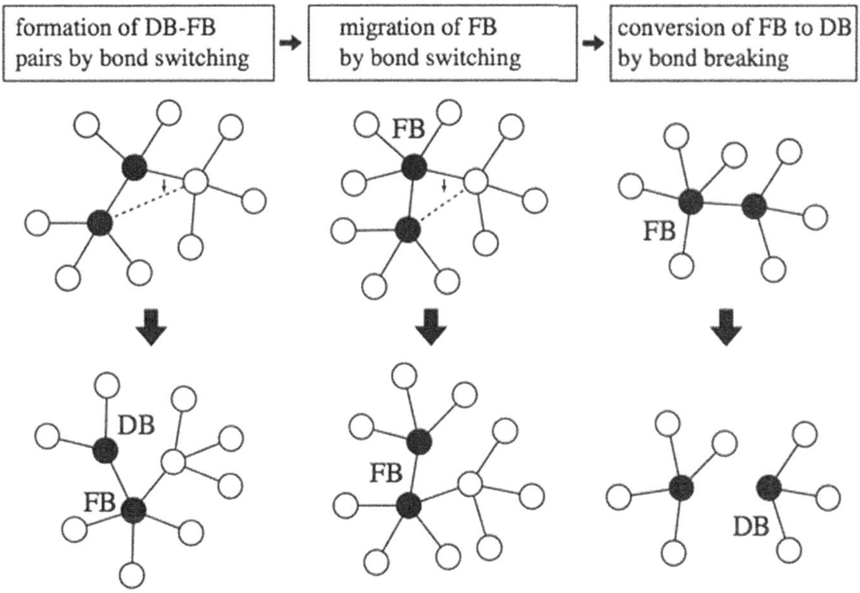


**Fig. 21** A detailed mechanism for formation of the dangling bonds in the proposed floating band model [59]

photoexcited holes become trapped, which results in the release of hydrogen atoms. This release forms a dangling bond that can act on other Si–Si bonds leading to an increase in the density of the dangling bonds. By annealing, hydrogen atoms return to their normal energy positions. Figure 22 illustrates the microvoid effect in absorbing holes.

## 5.3 *Band Edge and Band Tail*

For a crystalline semiconductor [59, 61], the positions of the valence and conduction bands can be sharply determined. In the case of amorphous structures, there is an associated tail known as Urbach tail [2, 51, 76] that can be expressed by an exponential distribution of band tail states. For a valence band tail, it can be calculated from the following equation [48]:

$$g(E) = E_{\mathrm{v}} \exp\left[-\frac{E - E_{\mathrm{v}}}{\Delta E_{\mathrm{v}}}\right], \tag{5}$$

where $g(E)$ is the density of the electronic states and $E_{\mathrm{v}}$ and $E_{\mathrm{c}}$ are the valence and conduction band energies, respectively. In the case of a-Si, the typical value for $\Delta E_{\mathrm{v}}$ is about $5 \times 10^{-3}$ eV and for $\Delta E_{\mathrm{c}}$ about $22 \times 10^{-3}$ eV.

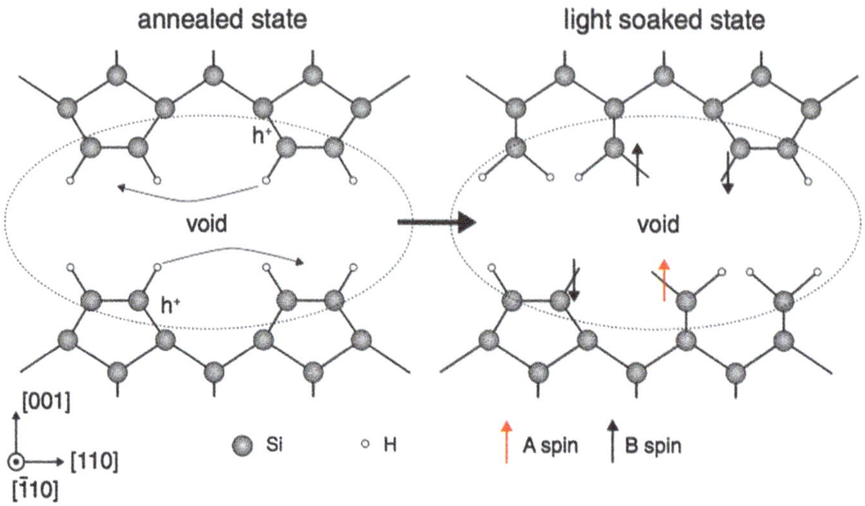


**Fig. 22** A schematic illustration showing the role of microvoids in a-Si [75]

Because the presence of band tails is assured, the discussion of band edges would be inappropriate. However, a considerable number of scientists consider that band edges as the energy bands that separate between discrete and localized energy levels. Indeed, there are several ways to determine band edges in semiconductors, the most popular of which is by the well-known Tauc procedure [56].

## *5.4 Band Defects*

a-Si:H is characterized by having dangling bonds. These dangling bonds are amphoteric in nature; they are classified into positive, neutral, and negative [48, 59–61]. Those three types can be divided into two groups, from negative to neutral and from neutral to positive. The positive zero level roughly lies below the negative zero level by 0.3 eV, which is called correlation energy of d centers [59, 60]. Doping is an efficient technique to enhance photoelectrical properties, as it can directly reduce the bandgap and either changes the semiconductor to p-type or n-type. Doping c-Si is accompanied with a number of problems [48]. First, the accommodation of phosphorous, for instance, is problematic; it will induce de-bonding to allow the presence of dangling bonds in the structure in order to accommodate its additional electron [57, 76]. These dangling bonds result in charge recombination and reduce the effectiveness of the doping process. In a- Si, phosphorus atoms can form three bonds in the p-level with the threefold Si, and the additional two electrons become stuck to s-level and do not participate in bonding [58, 60, 62]. This is more favorable for a-Si with its less rigid structure. Street et al.

[64] proposed a mechanism for that, where fourfold phosphorus is formed independently with a dangling bond. Thus, no additional electrons are contributed in this case to a-Si, rendering doping obsolete when it comes to enhancement.

## *5.5 Application in Water Splitting*

Multijunction solar cells (tandem cells) have recently been gaining an increasing interest [48, 55, 77, 78], as they can be tuned in order to absorb most of the visible spectrum. The idea of using tandem solar cells can be extended to photoelectrodes. Relying on tandem photocathodes in water splitting applications can eliminate the need of the applied external bias required to reduce the protons into hydrogen gas [55]. However, there are a lot of problems that arise with tandem solar cells such as high cost and crystalline mismatch, which can lead to higher recombination rates [77].

Being a cheap, Earth-abundant, nontoxic, and an amorphous material (where the crystalline mismatch problem is absent), a-Si proves promising in the field of photoelectrodes. In addition, well known for its high absorption coefficient, a very thin film (less than 1 μm) of a-Si can suffice for efficiently capturing light across the solar spectrum [48, 77, 78].

In that sense, Javari et al. [78] produced a novel technique to use a-Si tandem photocathode p-i-n, with a protection layer of $TiO_2$, as to stabilize the photocathode from corrosion or various reductions in photocurrent. They used ZnO as a conductive substrate and Ni–Mo as a catalyst for hydrogen evolution reaction together with platinum. The final form of the tandem photocathodes, shown in Fig. 23, was ZnO/p-

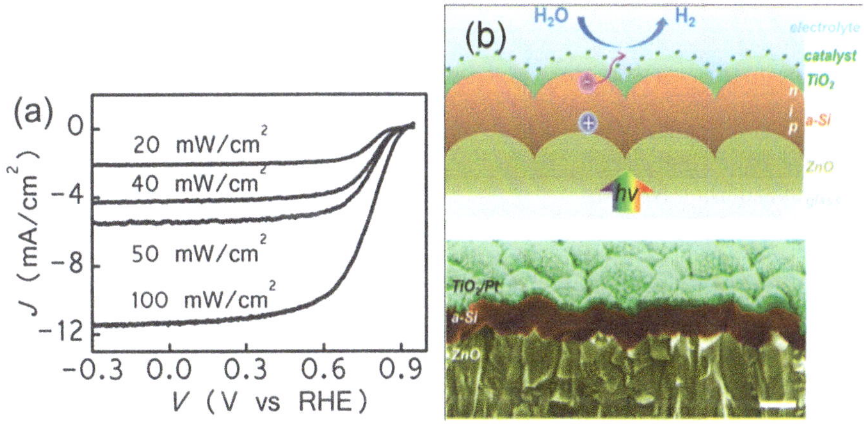


**Fig. 23** (**a**) Photocurrent density transients for a-Si photocathodes. (**b**) Design of the photocathode with ZnO as a conducting substrate, a-Si as light absorber, $TiO_2$ as a protective layer, and hydrogen evolution cocatalyst (either Pt or Ni–Mo). The cathode is illuminated through the ZnO side. Also shown is an SEM image of the cross-section of the photocathode system [78]

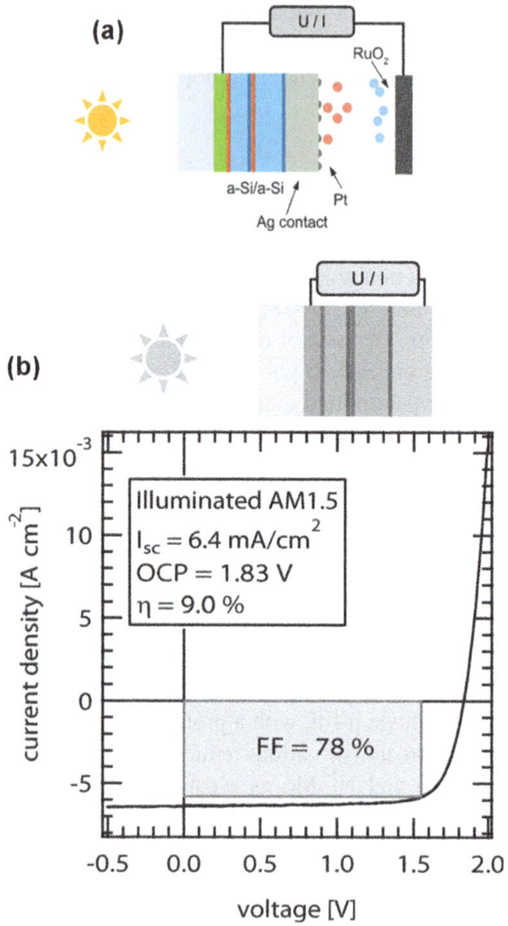


**Fig. 24** (**a**) a-Si a-Si photocathode tandem cell. (**b**) *J*–*V* plots of tandem cell [79]

i-n a-Si/$TiO_2$/Pt. They were able to produce the highest recorded photocurrent at positive bias. Photocurrent obtained was compared between using platinum as a catalyst and using Ni–Mo.

Finger et al. [79] introduced a new system to get rid of the high recombination rates in a-Si photocathode systems occurring at the electrode/electrolyte interface. The charge transfer from the electrode to the electrolyte was directed instead to the inside of the semiconductor, which is made out of a-Si/a-Si tandem cell (homojunction); see Fig. 24. His group used silver as a back contact, platinum nanoparticles as a catalyst, and ruthenium dioxide as a counter electrode. It was reported that this system produced STH 5.5% without any external bias.

Although the aforementioned a-Si tandem cell photocathode showed superior properties, Finger et al. [80] introduced another mixture by using a-Si:H with microcrystalline Si as a tandem photocathode (configuration: a-Si:H(n)/a-Si:H(i)/

μc-Si:H(i)) so as to provide higher efficiency and lower light degradation that is characteristic of a-Si photocathodes. This tandem system produced over 1.5 V as well as a photocurrent efficiency over 11% for a thickness less than 1 μm.

## 6 Silicon-Based P-Type Photocathodes for Carbon Dioxide Reduction

$CO_2$ reduction into useful fuels is another approach for the photoelectrochemical conversion of solar energy. Kumar et al. [81] studied the performance of planar p-type silicon with hydrogen termination in the presence of Re(bipy-Bu$^t$)$(CO)_3Cl$ as an electrocatalyst for the selective photoreduction of $CO_2$ to CO. Although the conduction band edge of the Re catalyst is above that of the H–Si as shown in Fig. 25a, reduction occurred due to Fermi-level pinning/unpinning of band edges as shown in Fig. 25b. The junction achieved quantum efficiency of 61% for light-to-chemical energy conversion at short-circuit condition with good stability [81]. In another study, iron porphyrin simple electrocatalyst was used with hydrogen-terminated p-type Si for the photoconversion of $CO_2$ to CO [82].

LaTempa et al. [83] presented a p–n junction photoelectrochemical cell that was composed of p-type Si nanowires as photocathode and n-type $TiO_2$ nanotubes as photoanode, as illustrated in Fig. 26a, for $CO_2$ conversion into hydrocarbon fuels. To evaluate the performance of the p–n junctions, the formation rates of the different products were compared to that obtained by an electrolysis cell composed of copper cathode and platinum anode. Figure 26b reveals that the formation of methane is higher for the Si–$TiO_2$ photoelectrochemical cell. The rate of formation of methane was also increased by the electrodeposition of copper nanoparticles on the Si nanowires photocathode for up to 30 sec [83]. One alternative approach for the direct reduction of $CO_2$ into fuels is the photofixation of $CO_2$ and using it as a carbon source to form useful organic compounds, which is mimicked from natural photosynthesis. In that sense, Liu et al. [84, 85] proposed the use of p-type Si nanowire photocathodes for $CO_2$ photofixation as illustrated in Fig. 27.

## 7 Conclusions

The use of silicon in the production of solar fuels is a global aim that has been the focus of many researchers throughout the past four decades due to its Earth abundance, as well as its readily available fabrication technology. The most important research efforts aiming at modifying the electronic/chemical/optical properties of Si photoelectrodes have been briefly reviewed in this chapter. It can be concluded

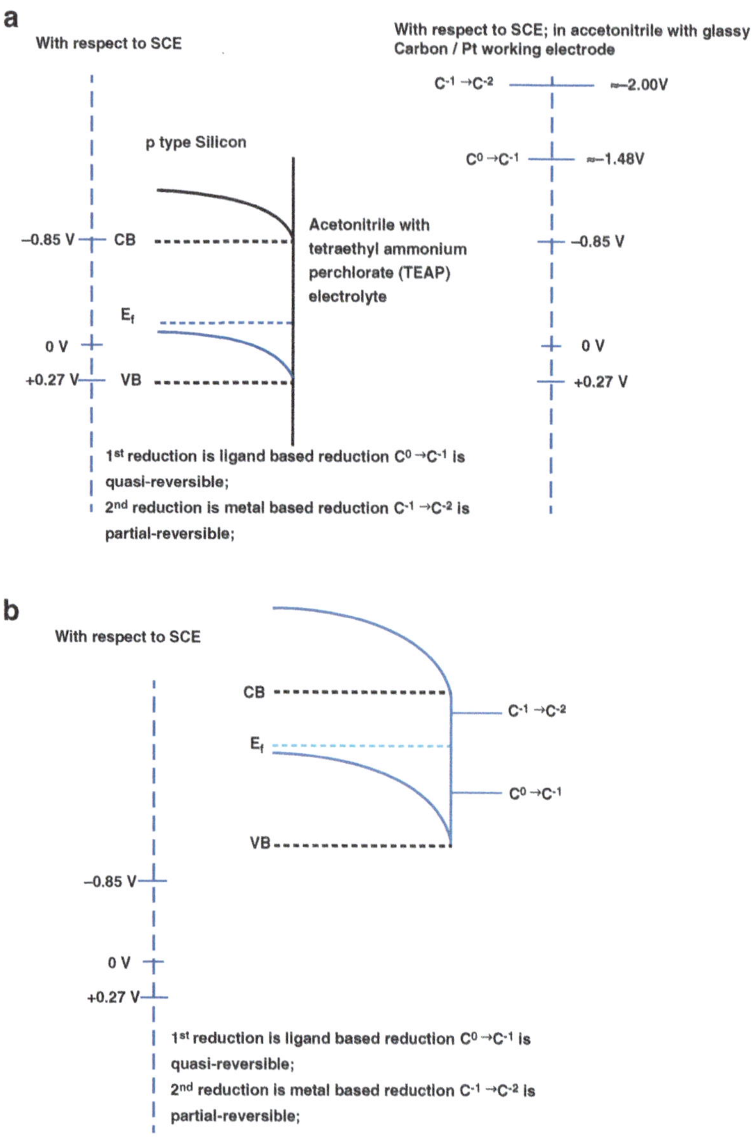


**Fig. 25** (**a**) Band edge position of conduction band (CB) and valence band (VB) of p-type Si and redox potentials of Re catalyst before contact. (**b**) The band edge positions after coming into contact with respect to saturated calomel electrode (SCE); where $E_f$ is the Fermi level of p-type Si [81]

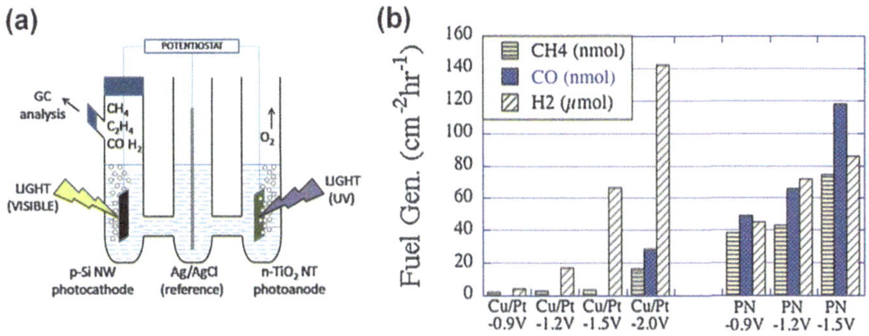


**Fig. 26** (**a**) The photoelectrochemical cell composed of p-Si nanowire photocathode illuminated with simulated AM 1.5G sunlight and n-$TiO_2$ nanotube photoanode illuminated with a UV light source in $CO_2$-saturated 1-M $NaHCO_3$ electrolyte. (**b**) Product generation rates for a Cu cathode/Pt anode electrolysis experiment compared with p-Si nanowire/n-$TiO_2$ nanotube photoelectrochemical cell under varying potential illuminated for 30 min [83]

that, when it comes to plain p-Si, nanostructuring is always a good strategy to obtain higher efficiencies, with nanobelts showing the highest photocurrent attainable. On the other hand, decorating Si with platinum, with the minimum overpotential, results in photocathodes that emerge as the most efficient and the most stable in the tested electrolytes today. However, with Pt being relatively expensive, other cheaper materials, such as molybdenum sulfide and graphene, show promise as decorating materials for Si as well. However, it can be inferred from the literature that further research on the modification of such decorating materials is required if these materials are to compete with or outperform Pt. Likewise, silicon carbide is a relatively cheap form of Si that may replace Si in the near future, again given further research is conducted.

Amorphous Si is indeed preferred over its crystalline counterpart due to its larger, variable bandgap, which renders its band positions more preferable over c-Si, especially in water decomposition systems. Also, its absorption coefficient is much better than that of c-Si. However, the Staebler–Wronski effect makes its use in industrial systems inefficient. It can be concluded from the results reviewed herein that passivation of dangling bonds may be the most promising strategy to elevate a-Si to industrial standards, just like its crystalline form.

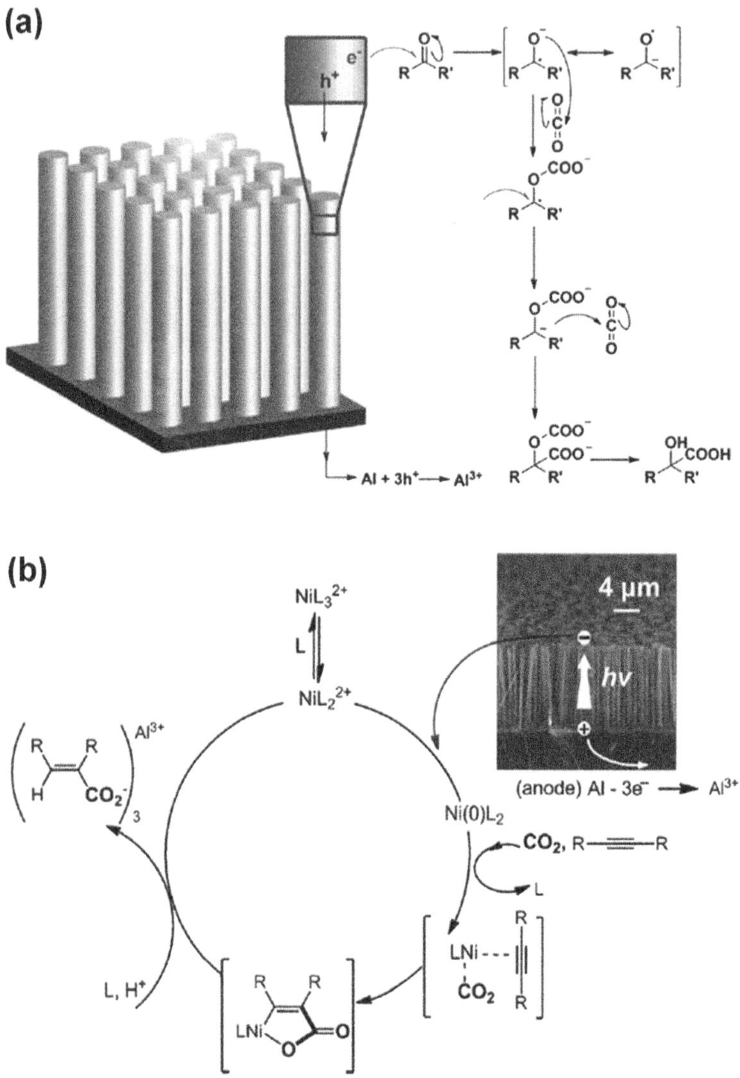


**Fig. 27** The use of p-type Si nanowires for $CO_2$ photofixation in (**a**) carboxylic reactions [84] and (**b**) catalyzed process [85]

## References


1. S. Dunn, Hydrogen futures: toward a sustainable energy system. Int. J. Hydrog. Energy **27**, 235 (2002)
2. K. Sivula, Chimia **67**, 155 (2013)
3. A. Fujishima, K. Honda, Nature **238**, 37 (1972)
4. P.V. Kamat, J. Bisquert, J. Phys. Chem. C **117**, 14873 (2013)
5. H. Wang, T. Deutsch, J. Turner, Electrochem Soc **155**, F91 (2008)
6. N. Serpone, A.V. Emeline, J. Phys. Chem. Lett. **3**, 673 (2012)
7. J. Nozik, Appl. Phys. Lett. **96**, 150 (1976)
8. J. Oh, T.G. Deutsch, H.C. Yuan, H.M. Branz, Energy Environ. Sci. **4**, 1690 (2011)
9. A. Paracchino, V. Laporte, K. Sivula, M. Grätzel, E. Thimsen, Nat. Mater. **10**, 456 (2011)
10. A. Paracchino, N. Mathews, T. Hisatomi, M. Stefik, S.D. Tilley, M. Grätzel, Energy Environ. Sci. **5**, 8673 (2012)
11. S.D. Tilley, M. Schreier, J. Azevedo, M. Stefik, M. Grätzel, Adv. Funct. Mater. **24**, 303 (2014)
12. C.Y. Lin, Y.H. Lai, D. Mersch, E. Reisner, Chem. Sci. **3**, 3482 (2012)
13. N.K. Awad, E.A. Ashour, N.K. Allam, J Renew Sustainable Energy **6**, 022702 (2014)
14. H. Tang, M.A. Matin, H. Kang, S. Sudhakar, L. Chen, M.M. Al-Jassim, Y. Yan, J. Electron. Mater. **40**, 3062 (2012)
15. Y. Tachibana, L. Vayssieres, J.R. Durrant, Nat. Photonics **6**, 511 (2012)
16. Y. Mao, J. He, X. Sun, W. Li, X. Lu, J. Gan, Z. Liu, L. Gong, J. Chen, P. Liu, Y. Tong, Electrochim. Acta **62**, 1 (2012)
17. T.M. McCormick, B.D. Calitree, A. Orchard, N.D. Kraut, F.V. Bright, M.R. Detty, R. Eisenberg, J. Am. Chem. Soc. **132**, 15480 (2010)
18. C.A.N. Fernando, L.L.A. De Silva, R.M. Mehra, K. Takahashi, Semicond. Sci. Technol. **16**, 433 (2001)
19. A. Kay, I. Cesar, M. Grätzel, J. Am. Chem. Soc. **128**, 15714 (2006)
20. M.G. Walter, E.L. Warren, J.R. Mckone, S.W. Boettcher, Q. Mi, E.A. Santori, N.S. Lewis, Chem. Rev. **110**, 6446 (2010)
21. Z. Zhang, P. Wang, J. Mater. Chem. **22**, 2456 (2012)
22. I. Cesar, K. Sivula, A. Kay, R. Zborik, M. Grätzel, J. Phys. Chem. C **113**, 772 (2009)
23. W. Schockley, H.J. Queisser, J. Appl. Phys. **32**, 510 (1961)
24. T. Sjodin, H. Petek, H.L. Dai, Phys. Rev. Lett. **81**, 5664 (1998)
25. S. Cattarin, L.M. Peter, J. Phys. Chem. B **101**, 3961 (1997)
26. N.P. Dasgupta, P. Yang, Front. Phys. **8**, 1 (2013)
27. U. Sim, H.Y. Jeong, T.Y. Yang, T.N. Ki, J. Mater. Chem. A **1**, 5414 (2013)
28. M.D. Kelzenberg, S.W. Boettcher, J.A. Petykiewicz, D.B. Turner-Evans, M.C. Putnam, E.L. Warren, J.M. Spurgeon, R.M. Briggs, N.S. Lewis, H.A. Atwater, Nat. Mater. **9**, 239 (2010)
29. J. Jia, H. Zhang, Y. Qiu, L. Wang, Y. Wang, L. Hu, Appl. Surf. Sci. **292**, 86 (2014)
30. J.Y. Jung, M.J. Choi, K. Zhou, X. Li, S.W. Jee, H.D. Um, M.J. Park, K.T. Park, J.H. Bang, J.H. Lee, J. Mater. Chem. A **2**, 833 (2014)
31. X.Q. Bao, R. Ferreira, E. Paz, D.C. Leitao, A. Silva, S. Cardoso, P.P. Freitas, L. Liu, Nano **6**, 2097 (2014)
32. I. Oh, J. Kye, S. Hwang, Nano Lett. **12**, 298 (2012)
33. X. Ao, X. Tong, D.S. Kim, L. Zhang, M. Knez, F. Müller, S. He, V. Schmidt, Appl. Phys. Lett. **101**, 111901 (2012)
34. E.L. Warren, S.W. Boettcher, J.R. McKone, N.S. Lewis, in *Proceeding of SPIE7770, Fifth Conference on Solar Hydrogen and Nanotechnology* ed. by H. Idriss, H. Wang (Wiley, California, 2010)
35. J. Ji, H. Zhang, Y. Qiu, Y. Wang, Y. Luo, L. Hu, J. Mater. Sci. Mater. Electron. **24**, 4433 (2013)
36. N.P. Dasgupta, C. Liu, S. Andrews, F.B. Prinz, P. Yang, J. Am. Chem. Soc. **135**, 12932 (2013)

37. J. Kye, M. Shin, B. Lim, J.W. Jang, I. Oh, S. Hwang, ACS Nano **7**, 6017 (2013)
38. B. Segar, A.B. Laursen, P.C. Vesborg, T. Pedersen, O. Hansen, S. Dahl, I. Chorkendorff, Angew. Chem. **124**, 9262 (2012)
39. Y. Hou, B.L. Abrams, P.C.K. Vesborg, M.K. Björketun, K. Herbst, L. Bech, A.M. Setti, C.D. Damsgaard, T. Pedersen, O. Hansen, J. Rossmeisl, S. Dahl, J.K. Norskov, I. Chorkendorf, Nat. Mater. **10**, 434 (2011)
40. Z. Xiong, M. Zheng, S. Liu, S. Ma, W. Shen, Nanotechnology **24**, 265402 (2013)
41. Z. Huang, P. Zhong, C. Wang, X. Zhang, C. Zhang, ACS Appl. Mater. Interfaces **5**, 1961 (2013)
42. P.G. Neudeck, in *The VLSI Handbook*, ed. by W.K. Chen (CRC Press & IEEE Press, Boca Raton, 1999) p. 6–1
43. M. Kato, T. Yasuda, K. Miyake, M. Ichimura, T. Hatayama, Int. J. Hydrog. Energy **39**, 4845 (2014)
44. J.Y. Hao, Y.Y. Wang, X.L. Tong, G.Q. Jin, X.Y. Guo, Int J Hydrog Energy **37**, 15038 (2012)
45. J.Y. Hao, Y.Y. Wang, X.L. Tong, G.Q. Jin, X.Y. Guo, Catal. Today **212**, 220 (2013)
46. Q.B. Ma, B. Kaiser, W. Jaegermann, J. Power Sources **253**, 41 (2014)
47. Y. Nariki, Y. Inoue, K. Tanaka, J. Mater. Sci. **25**, 3101 (1990)
48. X. Deng, E.A. Schiff, in *Handbook of Photovoltaic Science and Engineering*, ed. by A. Luque, S. Hegedus (Wiley , Chichester, 2003), p. 504
49. J.I. Pankove, D.E. Carlson, Ann. Rev. Mater. Sci. **10**, 43 (1980)
50. Y. Abdulraheem, I. Gordon, T. Bearda, H. Meddeb, J. Poortmans, AIP Adv. **4**, 057122–057121 (2014)
51. D. Kovalev, H. Heckler, M. Ben-Chorin, G. Polisski, M. Schwartzkopff, F. Koch, Phys. Rev. Lett. **81**, 2803 (1998)
52. H. Mannsperger, S. Kalbitzer, Appl. Phys. A Mater. Sci. Process. **41**, 253 (1986)
53. Y. Tawada, K. Tsuge, M. Kondo, H. Okamoto, Y. Hamakawa, J. Appl. Phys. **53**, 5273 (1982)
54. C.R. Wronski, J.M. Pearce, R.J. Koval, A.S. Ferlauto, R.W. Collins, Progress in amorphous silicon based solar cell technology. Paper presented at the world climate & energy event. Rio **67**, 6–11 (2002)
55. H. Zhang, S. Huang, G. Conibeer, Energy Procedia **22**, 10 (2012)
56. A. Madan, P.G. Le Comber, W.E. Spear, J. Non-cryst, Soldiers **20**, 239 (1976)
57. J. Geissbühler, S. De Wolf, B. Demaurex, J.P. Seif, D.T.L. Alexander, L. Barraud, C. Ballif, Appl. Phys. Lett. **102**, 231604 (2013)
58. A. Kolodziej, Opto-electron. Revue **12**, 21 (2004)
59. T. Shimizu, Jap J App Phys **43**, 3257 (2004)
60. N. Ishii, M. Kumeda, T. Shimizu, Jpn. J. Appl. Phys. **24**, L244 (1981)
61. T. Kamei, N. Hata, A. Matsuda, T. Uchiyama, S. Amano, K. Tsukamoto, Y. Yoshioka, T. Hirao, Appl. Phys. Lett. **68**, 2380 (1996)
62. D. Adler, Sol. Cell **9**, 133 (1983)
63. D. Adler, J. Phys, Collogues **42**, C4–C3 (1981)
64. R.A. Street, N.F. Mott, Phys. Rev. Lett. **35**, 1293 (1975)
65. Q. Zhang, M. Kumeda, T. Shimizu, Jpn. J. Appl. Phys. **32**, L371 (1993)
66. H.M. Branz HM, Phys. Rev. B **42**, 7420 (1990)
67. H.M. Branz, Sol. Energy Mater. Sol. Cells **78**, 425 (2003)
68. H.M. Branz, Phys. Rev. B **59**, 5498 (1999)
69. H.R. Biswas, B.C. Pan, Sol. Energy Mater. Sol. Cells **78**, 447 (2003)
70. S.T. Pantelides, Phys. Rev. Lett. **57**, 2979 (1986)
71. T. Shimizu, R. Durny, M. Kumeda. in *Materials Research Society Symposium Proceedings*, ed. by M. Hack, A. Matsuda, E.A. Schiff, R. Schropp, S. Wagner MRS Spring Meeting Symposium A, California, April 1996 (Cambridge University Press, Cambridge, 1996) p. 553
72. T. Shimizu, M. Kumeda, Jpn. J. Appl. Phys. **38**, L911 (1999)
73. F. Kohler, T. Zimmermann, S. Muthmann, A. Gordijn, R. Carius, IEEE J Photovolt **4**, 4 (2014)
74. D.E. Carlson, Appl. Phys. A Mater. Sci. Process. **41**, 305 (1986)

75. M. Fehr, A. Schnegg, B. Rech, O. Astakhov, F. Finger, R. Bittl, C. Teutloff, K. Lips, Phys. Rev. Lett. **112**, 066403–066401 (2014)
76. J.H. Yoon, J. Appl. Phys. **74**, 1838 (1993)
77. F. Zhu, J. Hu, I. Matulionis, T. Deutsch, N. Gaillard, A. Kunrath, E. Miller, A. Madan, Philos. Mag. **89**, 2723 (2009)
78. Y. Lin, C. Battaglia, M. Boccard, M. Hettick, Z. Yu, C. Ballif, J.W. Ager, A. Javey, Nano Lett. **13**, 5615 (2013)
79. J. Ziegler, B. Kaiser, W. Jaegermann, F. Urbain, J.P. Becker, V. Smirnov, F. Finger, Chem. Phys. Chem. **15**, 4026 (2014)
80. F. Urbain, V. Smirnov, J.P. Becker, U. Rau, F. Finger, j. Ziegler, B. Kaiser, W. Jaegermann, J. Mater. Res. **29**, 2605 (2014)
81. B. Kumar, J.M. Smieja, C.P. Kubiak, J. Phys. Chem. C **114**, 14220 (2010)
82. K. Alenezi, S.K. Ibrahim, P. Li, C.J. Pickett, Chem. Eur. J. **19**, 13522 (2013)
83. T.J. LaTempa, S. Rani, N. Bao, C.A. Grimes, Nano **4**, 2245 (2012)
84. R. Liu, G. Yuan, C.L. Joe, T.E. Lightburn, K.L. Tan, D. Wang, Angew. Chem. **124**, 6813 (2012)
85. R. Liu, C. Stephani, J.J. Han, K.L. Tan, D. Wang, Angew. Chem. **125**, 4319 (2013)

# Silicon Nanowire Solar Cells

**Guijun Li and Hoi-Sing Kwok**

**Abstract** Over the past decade, silicon nanowire solar cells have been intensively explored as potential platforms for the next-generation photovoltaic (PV) technologies with high power conversion efficiency and low production cost. This chapter discusses the details of the silicon nanowire solar cells in terms of their device structures, fabrication and characterization, electrical and optical properties benefited from the nanowire geometry. These benefits are not only expected to increase the power conversion efficiency, but also considered to reduce the requirement for the material quantity and quality, allowing for potential efficiency improvements and substantial cost reductions.

## 1 Introduction

Over the past few decades, great effort has been expended to develop photovoltaic technologies with high power conversion efficiency and low production cost. Today, photovoltaics is the third most important renewable energy after hydro and wind power. In 2014, worldwide installation of photovoltaics has increased to supply 1% of global electricity demands. In many countries, photovoltaic (PV) electricity has levelized its cost to "grid parity" so that it can now compete with traditional electricity.

In order to sustain the extremely fast growth of the PV industry and be more successful competing against traditional energy, substantial further power conversion efficiency improvement and production cost reduction are needed, both of which are key levers for the entrance of the multi-terawatt electricity markets. Specifically, improving the solar cell conversion efficiency is preferred, because

G. Li (✉)
College of Electronic Science and Technology, Shenzhen University, Shenzhen, China
e-mail: gliad@connect.ust.hk

H.-S. Kwok
Department of Electronic & Computer Engineering,The Hong Kong University of Science and Technology, Hong Kong, China

 
S. Ikhmayies (ed.), *Advances in Silicon Solar Cells*,
https://doi.org/10.1007/978-3-319-69703-1_10

high efficiency devices can not only help to reduce the module price but also lower the price of balance of system (BOS), especially in the cases of expensive land or building roofs. In principle, for high efficiency devices, solar cells have to be thick enough to absorb sunlight sufficiently while thin enough to separate and collect the photo-generated carrier effectively. In traditional wafer based solar cells, light absorption can be easily resolved by using a thick absorbing layer. However, the thick absorber deteriorates the charge carrier separation and collection because the direction of the separation/collection process and light absorption is the same. The simultaneous requirement of needing to be optically thick and electrically thin is a paradox in the design of high conversion efficiency solar cells. In thin film solar cells, which adopt an ultra-thin absorber, their ultra-thin thickness assures the requirement of being electrically thin, but additional advanced light trapping strategies are required to make them optically thick, which absolutely increase the production cost and complexity.

Despite a great amount of progress that has been achieved to increase solar cell efficiency so far, the efficiency of conventional solar cells has been limited to the Shockley Queisser limit [1]. Recently, a number of unconventional strategies which may promise either significantly higher efficiencies or extremely lower production cost have emerged. Of particular interest are the PV technologies based on the nanostructures or nanomaterials, which may hold great promise for third-generation photovoltaics and for powering nanoelectronics [2, 3]. Semiconductor nanowire, which is a structure with a diameter in the order of nanometers and a length upwards of a few micrometers [4], is a prototypical example, from a broad number of novel photonic and electronic device, primarily due to its unique electrical, optical, magnetic and mechanical properties.

For a solar cell, which is an optoelectronic device, the nanowire geometry provides advantages because it allows strong light absorption owing to its nanoscale structure and efficient carrier separation and collection due to the direct short path for charge transport [5]. Indeed, one of the most important key attributes of the nanowire solar cell is that, in some structures (i.e., radial junction), it decouples the light absorption from the carrier collection, which means the light absorption occurs in one direction while the carrier collection occurs in another direction. Subsequently, a solar cell can be made optically thick and electrically thin simultaneously. In addition to the decoupling, the use of nanowires as the building blocks of solar cells present other advantages: (a) Nanowire solar cells lower the requirement for the material quality, and as a result, a wide variety of materials including silicon [6], germanium [7], a-Si [8], μc-Si [9], copper zinc tin sulfide/selenide (CZTS) [10], $Cu_2S/CdS$ [11], cadmium telluride [12], cadmium selenide [13], copper oxide [14, 15], and many polymers [16, 17] can be used to make nanowire solar cells with the potential for high efficiency and low production cost, helping to drive large-scale implementation with versatility. (b) Nanowire solar cells are expected to favor the lattice strain relaxation, giving a great degree of freedom to the design of solar cells with lattice-mismatched materials [18]. (c) The seamless integration of nanowire solar cells with electronic and photonic devices also affirms them as power suppliers for these devices. (d) The bottom-up and top-down designs allow the rational control of some PV parameters, such as the dopant profile/composition, junction shape/size, and interface gradient. (e) The studies of solar cell properties in the nanoscale geometries provide a platform for exploring physical limits and novel device concepts for solar cells [19].

Among different types of PV materials, silicon (Si) is the most widely used, not only because it is Earth-abundant and non-toxic, but also due to the well-established Si manufacturing technology in today's semiconductor industry. Furthermore, Si can be made with different forms such as amorphous, nanocrystalline, microcrystalline, polycrystalline and single crystalline, providing a great freedom for the electronic and photonic devices. An Si nanowire solar cell was initiated with the theoretical demonstration of the benefit of such a configuration by Kayes, et al. in 2005 [5]. However, it wasn't until the year 2007 that a single coaxial Si nanowire solar cell was demonstrated by Tian et al. with a power conversion efficiency of up to 3.4% [3]. Almost in the same year, the first attempt at a vertical Si nanowire array solar cell was reported by Tsakalakos et al. using a conformal plasma–enhanced chemical vapor deposited n type amorphous silicon (a-Si) and a p-core Si nanowire grown by means of a vapor-liquid-solid (VLS) process on stainless steel foil [6]. After that, Si nanowire solar cells with different structures and different fabrication methods were widely reported. But, even with these efforts, the resulting Si nanowire solar cells showed very low efficiency, primarily limited by their low open-circuit voltage ($V_{oc}$) and poor fill factor (FF) [20]. In the year 2010, nanowire solar cells with an efficiency above 5% were realized by Garnett et al. using a process combining nanosphere lithography, deep reactive ion etching and high temperature dopant diffusion [21]. Ultra-high density silicon nanowire arrays of 10.8% were fabricated by parallel electron lithography afterwards [22]. In addition to c-Si, a-Si nanowire solar cells were also demonstrated with the silicon or ZnO acting as the cores, with an efficiency of around 6% [8].

Although a power conversion efficiency up to 11% can be achieved in Si solar cells with a microwire radial junction structure and 13% in hybrid organic/silicon nanowire solar cells [23, 24], the conversion efficiencies of Si nanowire solar cells still are not able to compete with that of planar Si solar cells. In this chapter, we will start with the introduction of basic aspects of Si nanowire solar cells, including devices structures, fabrication and characterization methods, and optical and electrical properties. In addition, the challenges and future works of silicon nanowire solar cells are also addressed in the last part of this chapter.

## 2 Device Structures

Device structure can be related to its geometrical and physical aspects. According to the geometry, nanowire can be classified into three type of structures: radial junction, axial junction and substrate junction, as shown in Fig. 1. Additionally, the junction can be made with a homo-junction or hetero-junction, with p/n or p/i/n structures, etc.

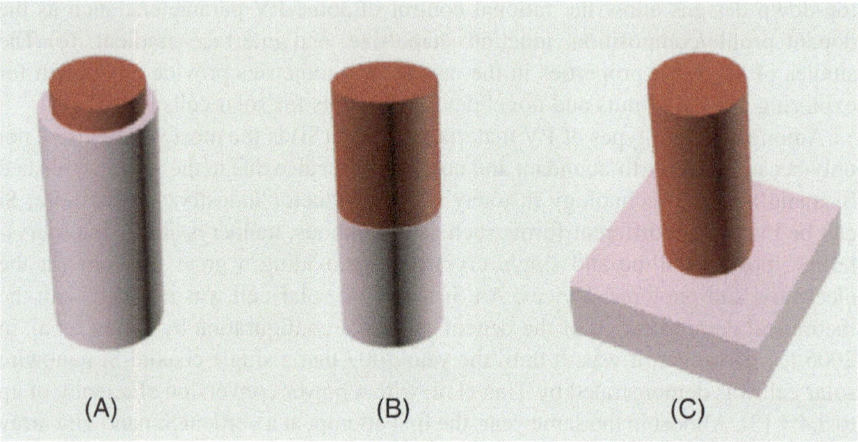


**Fig. 1** Device structure of nanowire solar cells. (**a**) Radial junction, also called core-shell structure, has a core wire inside and a shell outside. The junction is formed along the core/shell interface in the axial direction. Nanowire arrays with radial junctions have the advantages of enhanced light absorption and efficient charge collection. (**b**) Axial junction is formed in a similar way to the planar structure solar cell with a stacking sequence of the p-type and the n-type segments. The junction is formed at the core/shell interface in the radial direction. Axial junction solar cells lose the advantages of lateral charge separation and collection but keep the advantages of improved light absorption. (**c**) Substrate junction is formed between the substrate and the nanowire. Substrate junction cells lose the charge separation and collection advantages, but have the advantages of light trapping and lattice relaxation

## *2.1 Radial Junction*

In a solar cell, an incident photon creates an e-h pair; the e-h pair is separated by the build-in-field, and collected by electrodes, respectively. A comparison of conventional planar p/n junction solar cell and a radial junction nanowire solar cell is shown in Fig. 2. In a conventional planar structure, the semiconductor thickness $L$ should be larger than the optical thickness $L_\alpha$ to maximize the light absorption, otherwise, the light absorption will be insufficient, decreasing the photocurrent of the solar cell:

$$L > L_\alpha. \quad (1)$$

The optical thickness $L_\alpha$ is generally defined as $1/\alpha$, where $\alpha$ is the integration of the wavelength-dependent absorption coefficient over all wavelengths. Alternatively, the optical thickness can also be defined as the thickness of the material required to absorb 90% of the incident photons with an energy above the band-gap energy. With this definition, silicon has an optical thickness of 125 μm.

In terms of the carrier collection, the minority carrier diffusion length $L_n$ must also be greater than the device thickness $L$ to make sure the carrier can be collected effectively by the electrodes:

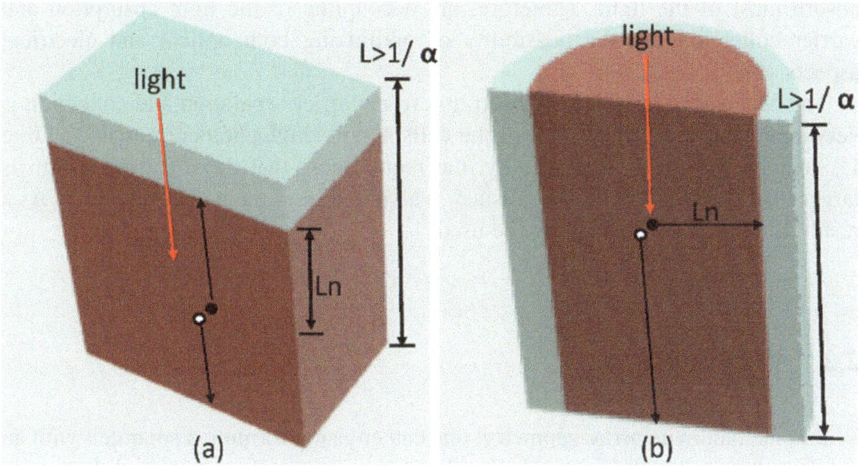


**Fig. 2** Schematic of optical and electrical behavior in solar cells with (**a**) a conventional planar structure; and (**b**) a radial junction structure. $L$ is the thickness of semiconductor, $L_\alpha$ is the optical thickness equals to $1/\alpha$ and $L_n$ is the minority diffusion length. In principle, $L$ should be larger than $L_\alpha$ to fully absorb the light, while $L_n$ must be longer than $L$ to ensure the photo-generated carrier can reach the junction before recombination

$$L_n > L. \tag{2}$$

In a solar cell with p type silicon as the absorbing layer, $L_n$ is defined as the electron diffusion length. Si has an electron mobility of 1350 $cm^2/Vs$ and carrier life time of 1 μs, the electron diffusion length is therefore calculated to be 53 μm, which is smaller than the optical thickness. A possible way to overcome the above dilemma is by orthogonalizing the directions of the light absorption and carrier collection. Fortunately, in nanowire solar cells with radial junction, as shown in Fig. 2b, incident light is absorbed axially, while photo-generated minority carriers are collected in the radial direction, through this way, the light absorption and carrier extraction are decoupled into orthogonal spatial directions.

The radial junction structure consists of a densely packed silicon wire array grown directly on a substrate. Unlike the planar structure, the p/n junction in a radial structure is formed in a core-shell geometry, and extends along the length of the nanowire. In general, the core is the absorber and the shell is the emitter. In order to reduce the parasitic optical loss in the emitter, the emitter should be as thin as possible. When light incidents on the device, electrons and holes will be generated in the core of the wire, in the direction of the nanowire. Meanwhile, photo-generated carrier separation and collection takes place in the radial versus the longer axial direction. The maximum length for the minority carrier diffusion is thus the radius of the wire, generally, quite a lot shorter than the length of the wire (with radius from tens of nanometers to hundreds of micrometers). Because light absorption occurs in the direction of the nanowire, it can be as long as needed to absorb most of the light. Therefore, the decoupling of the light absorption and carrier collection has the possibility of optimizing both optical and electrical aspects of solar cells.

The decoupling of the light absorption from carrier separation and collection is very interesting and important for solar cells. Firstly, high photocurrent, and a large $V_{oc}$ and FF are expected; secondly, the requirement that the diffusion length is larger than the radius is easily satisfied, which increases the defect tolerance. As a result, low-quality materials can be used.

## 2.2 Axial Junction

Within the nanowire array geometry, one can envision forming a solar cell with an axial junction which has a similar stacking sequence to the p-type and the n-type segments as in the planar structure solar cell, as shown in Fig. 1b. In an axial junction structure, e-h pairs are generated throughout the device upon absorption of light and swept in the direction of the electric field, namely, the direction of the nanowire. So, it loses the advantage of the radial junction that light absorption and carrier collection are spatially orthogonalized. Compared with the planar structure, an axial junction holds the enhanced light trapping properties as in the radial junction.

The main drawback of the axial junction is the additional surface recombination and junction loss, both of which are much worse than those in the radial junction. In the radial junction, the depletion region is embedded within the material and the minority carrier cannot see the nanowire outside surface and can only be swept across the core. However, in the axial junction, the depletion region is exposed outside, and the minority carrier in both the p and n sides can see the outside surface and be possibly recombined or trapped due to the high surface defect density of nanowires. In this sense, the material requirement and surface passivation for the axial junction are extremely high. Additionally, in order to suppress the recombination, the p and n type regions should be made arbitrarily short in length and large in diameter. Going to a small diameter nanowire has no benefits to its electrical performance, provided the surface passivation can be well done or light trapping effects dominate the cell performance.

The above results suggest that high efficiency solar cells can be obtained with a radial junction, which is true for silicon solar cells. In other cases, if one is able to grow nanowire materials of high quality in a simple or inexpensive way, or passivation of the wire surface is easier than the formation of a high quality p/n junction in the radial ways, axial junction solar cells may be of particular importance. In practice, axial junction solar cells play an important role in nanowire solar cells, such as GaAs and InP. For instance, a p–i–n InP axial junction nanowire solar cell with an efficiency of up to 13.8% was reported [25]. In 2016, a GaAs nanowire array solar cell with an independently verified solar energy conversion efficiency of

15.3% and open-circuit voltage of 0.906 V was fabricated, which is the best result on nanowire based solar cells so far [26].

With respect to the III-V semiconductor nanowire solar cells, there are seldom reports of silicon nanowire solar cells with axial junction, primarily due to unwanted impurities, defects and stacking faults, which are quite easily introduced at the core-shell interface during the nanowire growth [27]. These factors also limit the application of tandem or multi-junction structures with a radial junction configuration, since an axial junction is more favorable for tandem or multi-junction solar cells.

## 2.3 Substrate Junction

When the epitaxy of a thin film layer on a substrate is difficult, substrate junction is an option. The small diameter of nanowires makes it possible to grow large lattice mismatch material on top of a substrate. The schematic of the substrate junction solar cell is shown in Fig. 1c, where a p/n junction is formed between the substrate and the wire. Similar with the other nanowire structures, a substrate junction also shows its advantages of enhanced light trapping properties, and a tolerance of the stress and strain relaxation.

In addition to the geometrical junction, physically, solar cells can be formed with a p/n junction or a p/i/n junction. Compared with the p/n junction, a p/i/n junction exhibits high performance in terms of $V_{oc}$, FF and $J_{sc}$ [28]. As shown in Fig. 3a, the insertion of an intrinsic layer causes 40–140 mV improvements in $V_{oc}$. Furthermore, the p/i/n device exhibits a substantially high fill factor (FF) of 73%. The improvement in $V_{oc}$ is directly related to the dramatic reduction in leakage current (Fig. 3b).

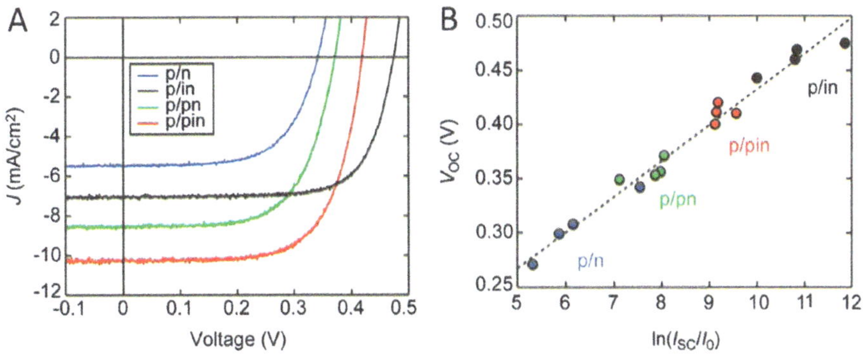


**Fig. 3** (**a**) J-V curves for single-nanowire solar cells with four distinct diode geometries. (**b**) $V_{oc}$ versus the logarithm of the ratio of $I_{sc}$ to $I_0$ for different diode structures. $I_{sc}$ is short-circuit current and $I_0$ is the dark saturation current (Reprinted with permission from Ref. [28]. Copyright © 2012, National Academy of Sciences)

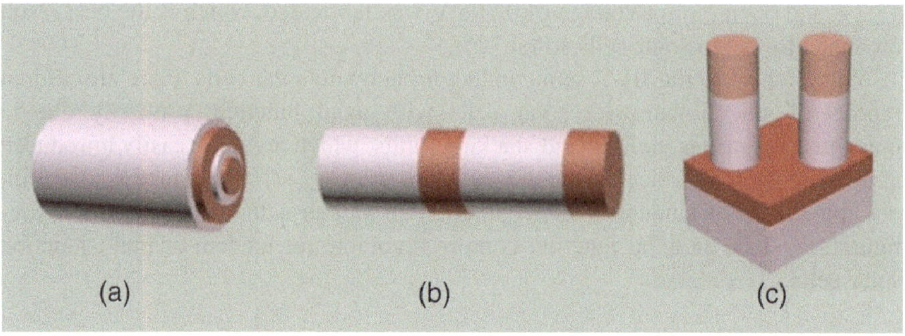


**Fig. 4** Multi-junction nanowire solar cells with (**a**) a multi-core-shell structure; (**b**) a stack axial structure; and (**c**) a nanowire/planar structure

Compared with the homo-junction, a hetero-junction is another widely used structure in nanowire solar cells. A-Si, poly-Si and organic materials can be used with c-Si to make a hetero-junction device. In some cases, a hetero-junction is a good choice because it can provide surface passivation to the nanowire and device. For instance, a core–shell hetero-junction solar cells with an efficiency of up to 7.29% was obtained [29]; using a single crystalline n-core/p-a-Si/TCO structure; a hybrid nanowire solar cell with Si/PEDOT:PSS structure was also reported with a high efficiency of 13% [24].

Multi-junction solar cells provide us with a viable approach to achieve efficiencies higher than the Shockley–Queisser limit. Nanowire can also be made with tandem or multi-junction structures. In practice, nanowire multi-junction solar cells can be made in a multi-core-shell structure, a stack axial structure or a nanowire/planar structure, as shown in Fig. 4. Nanowire solar cells with a multi-junction structure were reported previously, with GaAs nanowire as the top cell and Si as the bottom cell. The cell had an efficiency of 11.4% [30]. An axially connected nanowire core-shell p-n junction, which combines the radial and axial junction geometry, was also proposed as a composite structure for high-efficiency solar cells [31].

## 3 Fabrication and Characterization

The development of nanowires as building blocks of photovoltaic devices began with the demonstration of controlled growth of nanowire with morphologies, composition and doping variation [32–34]. In general, nanowire solar cells fabrication consists of three primary steps: nanowire synthesis, junction formation, and contacting.

## 3.1 Nanowire Fabrication

There is an overwhelming amount of literature describing the nanowire synthesis over the last few years. Generally, nanowire fabrication can be classified into two basic techniques. They are the bottom-up and top-down approaches. The bottom-up approach is a process to grow silicon nanowire from a precursor with a silicon element, while the top-down approach is a process to prepare silicon nanowire by etching bulk silicon with nanowire geometry.

### 3.1.1 Bottom-Up Approach

Bottom-up fabrication is an additive process that starts with precursor atoms or molecules to build up the designed objects such as nanostructures and nanomaterials. In the beginning, most of the nanowire fabrication methods were based on the bottom-up approach that nanowires are epitaxially grown, or formed, on a substrate via the reaction between the substrate and precursors. From the point view of the growth mechanism, bottom-up approaches contain vapor phase deposition (i.e., chemical-vapor deposition (CVD), atomic layer deposition (ALD), molecular beam epitaxy (MBE), metal-organic chemical vapor deposition (MOCVD)), liquid phase deposition, and electrodeposition, etc.

The most extensively explored bottom-up approach is the vapor phase deposition. In vapor phase synthesis, the substrate is exposed to the reaction chamber flowing with a chemical precursor, mostly in the form of vapor. The precursor reacts and/or decomposes on the substrate and the desired film can be deposited. In order to promote one-dimensional nanowire growth instead of uniform thin film deposition, the deposition condition must be regulated to allow the nanowire growth mechanism to predominate and to suppress secondary nucleation [35]. The mechanism includes: growing an intrinsically anisotropic crystal structure; directing the formation of a one-dimensional nanostructure with templates; introducing a liquid/solid interface to reduce the seed symmetry; and controlling the growth rates of various facets of a seed using an appropriate capping reagent.

One of the most successfully employed vapor phase deposition techniques of making silicon nanowire is the Vapor-liquid-solid (VLS) method. This process was originally suggested and investigated more than 40 years ago by Wagner and Ellis, and was developed to explain the growth of Si whiskers [36]. Then the thermodynamics and kinetics were justified, and reexamined by Lieber, Yang and other researchers [32, 34, 37]. Now VLS has been demonstrated to be able to synthesize a wide range of nanowire materials, such as GaAs, InP, GaP, ZnO and other nitride compounds. As shown in Fig. 5, the VLS process starts with introducing a metal catalyst deposited on the wafer substrate. Upon decomposition and dissolution of gaseous reactant at a temperature higher than the metal-Si eutectic point, metal-Si alloy droplets are formed on the substrate by absorbing the vapor component. The liquid droplet is a preferred site to limit the lateral growth of an individual wire. It

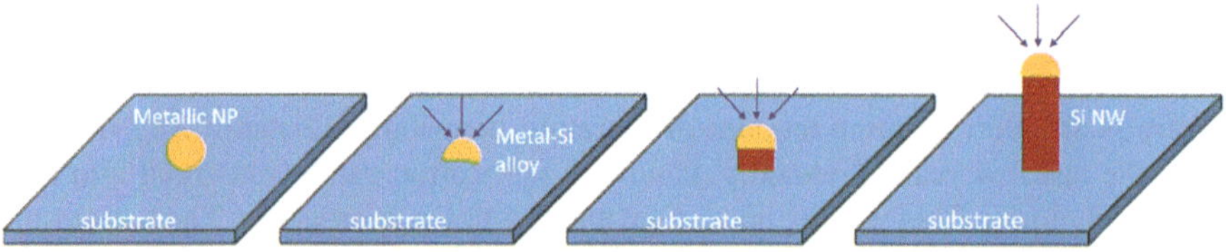


**Fig. 5** Schematic of Si nanowire growth by VLS method. (**a**) The deposition of metallic nanoparticles on the substrate; (**b**) metal-Si alloy droplets formation; (**c**) silicon deposited preferentially at the nanoparticle-substrate interface; and (**d**) silicon nanowire growth

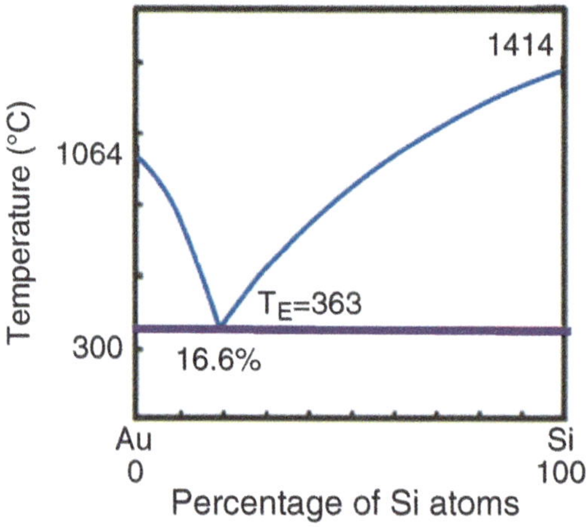


**Fig. 6** Phase diagram of a gold (Au)-silicon (Si) system. The eutectic point temperature is 363 °C, along with a 16.6% of Si in the alloy. The melting points of pure gold and pure silicon are 1064 °C and 1414 °C, respectively

easily becomes supersaturated and overcomes the nucleation barrier to precipitate and lead to nanowire growth at the solid-liquid interface. With continued exposure to the silicon vapor precursor, a wire tends to grow underneath the metal with the continuous precipitation of silicon.

A phase diagram is usually used to illustrate the thermodynamic process of nanowire growth in the VLS method. A typical phase-diagram of an Au-Si system is shown in Fig. 6. From the diagram we can see that the eutectic point temperature of a Au-Si system is 363 °C, somewhat lower than the melting points of pure gold, 1064 °C, or pure silicon, 1414 °C. At the beginning, the process temperature is set above the eutectic temperature, and silicon from the substrate or from the silicon precursor will diffuse into the gold nanoparticle. As the percentage of the silicon in the Au-Si alloy increase to a certain level, a liquid Au-Si alloy will be produced. Because of the surface and interface energy, Au-Si alloy tends to be distributed as droplets. With continued absorption of the silicon atoms from the gaseous reactant, the liquid droplet becomes supersaturated, leading to the nucleation of the solid silicon.

There are two interfaces during the nanowire growth [38]. The solid-liquid interface acts as a sink for the growth of nanowire in the wire direction and the gas-solid interface favors the vapor-solid growth in the radial direction. These two interfaces are competing during the nanowire growth. Precipitation through the solid-liquid interface results in axial growth, while adsorption on the gas-solid interface results in growth in the radial direction. Varying the process conditions can therefore be used to control the shape of the nanowire, and of multi-core-shell structures. For example, tapered silicon nanowire can be grown by controlling the simultaneous growth in both the axial and radial directions [39]. Uniform axial nanowire can be grown with the introduction of $H_2$ as the carrier gas to suppress either the adsorption of the reactants by terminating the Si surface or the dissociation of silane [40–42].

With control of the substrate, precursor, temperature, catalyst, and concentration, silicon nanowire can be grown with rational precise control of chemical composition, structure, size and morphology. In particular, nanowire heterojunctions can be achieved in a rational way.

Beside the VLS method, other techniques are also proposed to grow silicon nanowire. These include vapor-solid (VS), vapor-solid-solid (VSS) etc. The vapor-solid-solid (VSS) process is similar to the VLS process, except that the metal catalyst remains a solid instead of forming an eutectic liquid droplet.

### 3.1.2 Top-Down Approach

Top-down techniques provide a different path for fabricating high density, uniform distributed and vertically aligned Si nanowire arrays. Compared with the bottom-up approach, top-down fabrication is a subtractive process in which macroscopic (bulk) material is removed to produce features of a controlled shape and size. Bottom-up grown semiconductor nanowires typically lack the periodic ordering and placement for large-scale devices, while the top-down method is capable of producing well-aligned semiconductor nanowires with large-scale application.

In a top-down approach, an aligned pattern is often created on the substrate at the beginning, by using photoresist (PR) or other polymer materials. Lithography is one such method to make periodic patterning. The most used lithography techniques are those developed in the microelectronics industry, such as photolithography and electron-beam lithography. However, new areas of research in soft lithography, imprint lithography, and various types of self-assembly methods have also emerged as useful tools for realistic implementations in wide-ranging classes of applications. For example, nanoimprinting is a relatively low cost technology that allows high-resolution, high-throughput and large-area patterning in industry. Nanosphere lithography is another widely used technique for fabricating precisely controlled patterns by using self-assembled monolayers of spheres (i.e., polystyrene) as masks.

After the formation of the pattern, an etching step is needed to remove the unwanted silicon. Reactive ion etching (RIE) is a standard dry etching process used in microfabrication. It uses a strong electromagnetic field to generate chemically reactive plasma to attach and react with materials deposited on a wafer and then remove them. Combing RIE with different lithography techniques, ordered arrays of silicon nanowire can be easily produced with controlled size, length, shape and density. However, the dry etching process often induces defects and impurities, which can seriously degrade the optical and electrical performance of the device. An additional oxidization step is thus needed to remove these surface defects and impurities.

Wet chemical etching is another way to remove the silicon. It is a process that uses liquid chemicals to remove Si not protected by masks on a wafer. Generally, there are three basic steps during wet etching: (a) the liquid etchant diffusion to the structure, (b) the reaction between Si and the etchant, and (c) diffusion of the byproduct from the reacted surface. Depending on the chemical used, wet etching can be classified into isotropic and anisotropic etching. In practice, isotropic etching can be achieved with a mixture of hydrofluoric acid, nitric acid, and acetic acid (HNA), while anisotropic etching can be achieved with potassium hydroxide (KOH), ethylenediamine pyrocatechol (EDP), or tetramethylammonium hydroxide (TMAH). Metal-assisted chemical etching of Si is a typical, simple, low-cost and useful method in patterning Si nanowire arrays using wet etching. In a typical process, a Si substrate covered with a noble metal pattern is subjected to a chemical solution containing HF and an oxidizing agent such as $H_2O_2$. Because Si beneath the noble metal has a high dissolution rate, pores will be generated at the site of the metal. As a result, Si nanostructures such as nanowires, nanoholes, and nanopillars can be obtained [43, 44].

## 3.2 Junction Formation

After the formation of the nanowire array, the second step is to form a junction to separate charge carriers and promote carriers collection. The junction formation process is slightly different in bottom-up and top-down processed nanowires.

In the VLS method, the formation of a junction can be achieved by rationally controlling the deposition conditions during the nanowire growth. Generally, when the vapor reactant is introduced preferentially for forming a junction at the same nanocluster catalyst, crystalline grows along the axial direction and an axial junction can be obtained. On the other hand, when the new reactant uniformly reacts with the wire on the side, a shell will grow on the original nanowire surface. Therefore, alternating reactants can produce super-lattice (axial) or multi-core-shell structures (radial). In both junction processes, low temperature deposition is preferred, because the metal catalyst has the potential to be incorporated into the silicon as an impurity at high temperature.

In practice, a p-i-n axial heterojunction can be obtained via the procedures described in Ref. [45]. Au nanoparticles are firstly deposited on a substrate with a preferred pattern, and then the substrate is transferred to a vacuum chamber with the pressure and temperature kept at 40 Torr and 450 °C, respectively. Precursor gases such as diborane ($B_2H_6$), silane ($SiH_4$), and phosphine ($PH_3$) are introduced as appropriate to form p-type, intrinsic, and n-type wires. With the same temperature and pressure, axial p-i-n wire can be grown axially by varying the precursor gases with a sequencing of $B_2H_6$ (5 sccm), $SiH_4$ (2 sccm), and $PH_3$ (1 sccm). Tandem structures or multi-junction structures which have two or more p-i-n nanowires connected in series can also be formed easily in this way. During the process, the doping concentration and profile can be well controlled by adjusting the gas flow rate or the doping gas. However, in these sequential deposition processes, cross contamination is possible. Eliminating this cross contamination will lead to high performance devices.

The strategies for forming a radial junction are based upon the control of radial versus axial growth. We know that the silicon core is obtained when the reactant activation and deposition condition occurs at the solid-liquid interface (catalyst site) and not on the gas/solid interface (nanowire surface); correspondingly, the shell is grown by altering the condition to favor homogeneous vapor-phase deposition on the nanowire surface, analogous to the layered growth of planar heterostructures. For example, in the growth of a p-i-n radial junction silicon nanowire, the p-type core is deposited at a relatively low temperature to favor the growth at the solid-liquid interface, while the deposition of the i and n shells is at a high temperature and lower pressure region to drive the deposition on the p-type core surface. The high temperature and low pressure used here help to inhibit axial growth along the silicon nanowire during the shell deposition.

Junction formation in a top-down approach is simpler than that in the bottom-up method. Typically, in an axial junction, the p-n junction can be formed as in the planar structure before the lithography process. After the etching step, an axial junction is obtained; while in a radial junction, after the formation of the nanowire array, high temperature dopant diffusion or implantation processes are usually used.

After the junction formation, a sequential passivation step is essential to passivate the extrinsic surface to prevent the current shunting. As in conventional Si solar cell technologies, $SiO_2$, $SiN_x$ and a-Si are widely used as passivation layers. The high-aspect-ratio of the nanowire geometry requires controlling the process to favor the conformal deposition or increasing the thickness of the shell to eliminate or migrate the shunting problem.

Doped a-Si can be used as the shell to form a hetero-junction structure with the c-Si core. Because of the low temperature processing of a-Si, this type of heterojunction solar cell has a high degree of freedom in the device design. In addition, organic functional material can also be used to passivate the Si surface and make a hybrid junction with a c-Si core. The organic functional material shows superior surface passivation for the c-Si core, and it is a low temperature process for polymer deposition, making this type of solar cell fabrication compatible with low-cost substrates such as glass, plastic and aluminum foil.

## 3.3 Contact Formation

Contact formation is the last step to finish the nanowire solar cell fabrication. Methods of making the contact in nanowire solar cells is similar to that in planar solar cells, and some strategies are the same, such as heavy doping, interfacial engineering, and transparent conductive coating. However, in nanowire solar cells, multiple lithography and etching steps are usually needed to make the pattern to the side of nanowires. In addition, conformal deposition is preferred; because contact formation on nanowires can be challenging due to the high-aspect-ratio of nanowires. A thicker contact layer or a suitable deposition process (i.e., sputtering, or electrodeposition) are commonly suggested.

In some cases of single silicon nanowire devices, the nanowire has a small radius, and the metal contact defining by the lithography is much larger than the nanowire. An interconnection method is suggested through the selective transformation of silicon nanowires into metallic metal silicide (NiSi) nanowires [46].

After the nanowire synthesis, junction formation and contacting, nanowire solar cells fabrication is finished. A typical process for radial nanowire solar cells using the top-down approach is shown in Fig. 7 [47]. First, a $SiO_2$ layer is thermally grown on top of a p-type Si wafer, then photoresist is spin coated on top of $SiO_2$, followed by the photolithography patterning and deep reactive ion etching. After the removal of $SiO_2$, an oxidizing step is used to passivate surface defects and impurities. The oxidizing step also creates a layer of oxide on the back side of wafer. In order to protect the back side of the wafer, prior to stripping off the oxide by HF, a photoresist layer is needed on the back side of wafer. N-type shells are formed by the high temperature diffusion using $POCl_3$ as the doping gas at a temperature around 900 °C. The junction depth can be controlled by the diffusion temperature

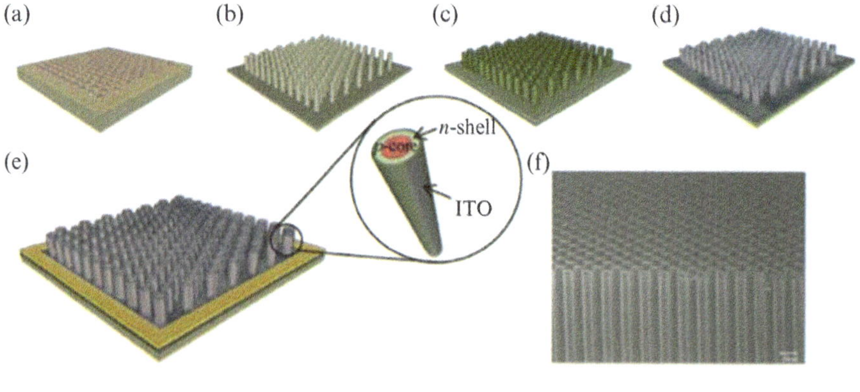


**Fig. 7** Schematic of the formation of silicon nanowire solar cells by top-down approach. (**a**) Photolithographic patterning; (**b**) deep reactive ion etching; (**c**) high temperature $POCl_3$ diffusion; (**d**) front and rear sides contacting; (**e**) Ti/Au evaporation; and (**f**) SEM image of a fabricated nanowire array (Reprinted with permission from Ref. [47]. Copyright © 2012, Wiley-VCH )

and process time. Right after the dopant diffusion, oxide grown during the diffusion process and native oxide are removed by buffered HF. Finally, transparent conductive oxide (i.e. indium tin oxide (ITO), aluminum doped zinc oxide (ZnO:Al)) and aluminum are sputtered on the front and back sides, respectively.

### 3.4 Characterization

In device-scale nanowire arrays, the characterization techniques such as the I-V measurement, light absorption, and external quantum efficiency (EQE) are the same as in planar cells. In single nanowire solar cells, because of the nanoscale geometry, traditional characterization methods cannot be implemented with the resolution necessary to obtain meaningful results.

In order to extract the carrier concentration profile and surface state density in a radial silicon nanowire device, capacitance-voltage (C-V) measurements [48], local electrode atom probe (LEAP) microscopy, [49] Kelvin probe force microscopy (KPFM) [50], and high-angle annular dark-field (HAADF) imaging [51] have shown their ability. In addition, E-beam-induced current (EBIC) measurements provide the information of recombination by determining the minority carrier diffusion length.

## 4 Optical and Electrical Properties

The potential advantages of silicon nanowire solar cells are primarily their light absorption, defect tolerance and strain relaxation. This section describes their optical and electrical properties.

### 4.1 Optical Properties

Light absorption in nanowire device obeys Beer's law that light intensity is attenuated exponentially as a function of material thickness.

$$I_{out} = (1 - R)I_{in}\exp(-\alpha d), \tag{3}$$

where $R$ is the surface reflectivity, $I_{out}$ is the light intensity after it passes through the sample, $I_{in}$ is the initial light intensity, $d$ is the material thickness, and $\alpha$ is the absorption coefficient. The total absorption is then given by the following equation:

$$I_{abs} = I_{in} - I_{out} = I_{in}[1 - (1 - R)\exp(-\alpha d)], \tag{4}$$

where $I_{abs}$ is the light absorbed by the material. The total absorption of a piece of nanowire is then dependent on the surface reflectivity $R$ and the thickness $d$. In order to enhance the light absorption in the device, one has to suppress the light reflection at surface and increase the thickness or the light path length.

As is well known, light reflection arises from a difference in the refractive index between two media. In conventional technologies of bulk silicon and thin film solar cells, surface reflection is addressed by using an antireflection coating. The refractive index of the antireflection coating is between that of the semiconductor layer and the air, with the optimum value given by the geometric mean of the two surrounding indices. For example, in a bulk silicon solar cell, $SiO_2$ or $Si_3N_4$ are widely used as an antireflection coating. In some cases, an antireflection coating with multilayers is used to reduce reflection over a broadband wavelength. Even in that, the effect is still limited to a specific wavelength. Another way to achieve the broadband antireflection is using a texturing surface. If the wavelength is greater than the feature size of the texture, the texture provides a gradual index change to reduce reflection. Effective medium approximations can be used to calculate the reflection. If the wavelength is smaller than the texturing size, rays will be multiply reflected, as a result, light reflection is reduced, which can be estimated with the help of geometric optics (i.e., the ray tracing method). For example, in bulk silicon solar cells, the use of microscale pyramids (or inverted pyramids) is proven to be a very successful approach; while in thin film solar cells, nanoscale structures such as nanoporous, nanocones, nanopyramids, and nanopillars were reported to be very effective to suppress the light reflection [52–55].

The nanostructured Si solar cells include nanoporous, nanopyramides, nanopillars, nanocones, nanowires, et al., and most of the nanostructures are formed on the top of Si surface, which is slightly different from the nanowire structure. Because of the nanoscale feature of nanowire, it can help to reduce surface light reflection. The antireflection effect of nanowire solar cells is also the first observed light trapping property at the beginning of the development of nanowire solar cells [6]. However, the antireflection effect in nanowire is not significant compared with the other nanostructures such as nanoporous, nanocones or nanopyradmids, most probably due to the steep change of its geometry, which cannot provide a soft gradual change of refractive index given by the weighted average of the material and air [56]. Therefore, nanowires with a controlled shape of the tip (i.e., tapered) were proposed to greatly reduce the reflection [57]. Silicon solar cells with controlled nanowires or other nanostructures at the surface have also been developed over the last few years. These solar cells are called "black-silicon solar cells" since the silicon surface is modified with very low reflectivity and correspondingly high absorption of visible (and infrared) light. In the year 2012, an independently confirmed 18.2%-efficient nanostructured 'black-silicon' solar cell was reported. In the year 2015, black silicon solar cells with interdigitated back-contacts achieved 22.1% efficiency, demonstrating the ability of reducing the surface reflection of the nanowire arrays [58]. The surface morphology and light reflection are shown in Fig. 8. As can be seen, extremely low reflection could be achieved in the 300–1000 nm wavelength range.

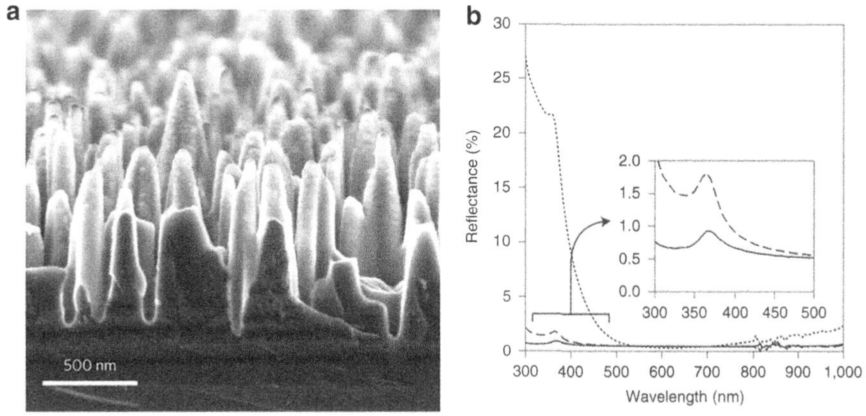


**Fig. 8** (**a**) Scanning electron microscopy image of a black silicon surface with controlled nanowire structure; (**b**) measured light reflectance spectra in the 300–1000 nm wavelength range (Reprinted with permission from Ref. [58]. Copyright © 2015, Nature Publishing Group)

Another way to overcome the light absorption loss is to increase the thickness of the absorbing layer. However, this is not a straightforward way because of the limitation of the carrier diffusion length, which is required to be larger than the physical thickness of the absorber to assure efficient carrier collection. A common method is to increase the light path length without increasing the physical thickness by using light trapping schemes. Theoretically, with a perfectly implemented Lambertian light trapping scheme, a maximum light path length can be enhanced with a factor of $4n^2$ [59].

The light path length enhancement in nanowire solar cells is primarily achieved through the guided resonant modes which are excited when incoming light interacts with nanowire structures. Semiconductor nanowire geometry is a subwavelength optical cavity that can support resonant modes. The possible resonance modes are shown in Fig. 9. In the axial junction nanowire, when the condition $m\lambda_{\text{eff}} = 2\pi r$ is satisfied, Mie resonance is excited. Because the Mie resonance is only dependent on the diameter, it can occur in both radial and axial junction nanowires. In radial junction solar cells, when the refractive index of outside and inside media is different, plasmon resonance is possible. Furthermore, Fabry-Perot mode can also be excited in radial junction structure solar cells.

Through numerical full-field electromagnetic simulations and measurements, light absorption behavior in nanowire can be clearly understood. For example, the Mie-type resonances were observed in random nanowire arrays by using broadband enhanced backscattering spectroscopy, and the resonance can be controlled over the entire visible spectrum [60]. The Fabry-Perot and whispering-gallery resonant absorption modes were identified within the hexagonal Si nanowire structure. The experimental and simulated external quantum efficiencies (Fig. 10a, b) show several peaks with identical peak positions [28]. The 1, 2, 4 and 5 peaks can be assigned to wavelength dependent Fabry-Perot resonance, while 3 and 6 peaks are

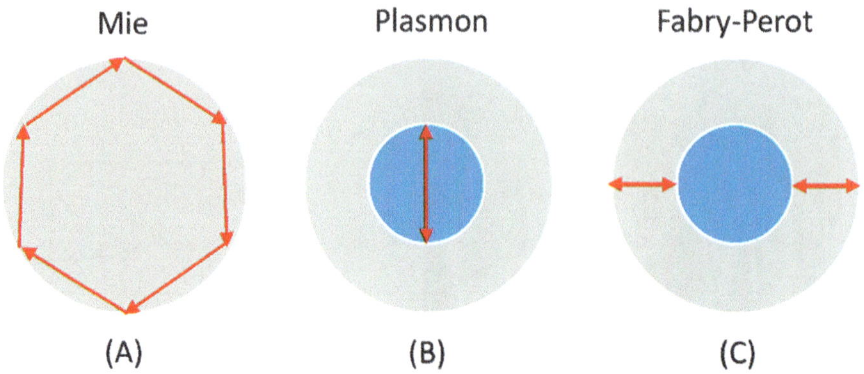


**Fig. 9** Possible guided resonant modes in nanowire solar cells. (**a**) Mie resonance; Mie resonance occurs in both axial and radial junction nanowires. Its resonant properties depend on the nanowire diameter. (**b**) Plasmon resonance; Plasmon resonance occurs only in core-shell nanowires. Its resonant properties depend on the refractive index of the core and shell. (**c**) Fabry-Perot modes. Fabry-Perot resonance occurs only in core-shell nanowires. Its resonant properties depend on the shell thickness

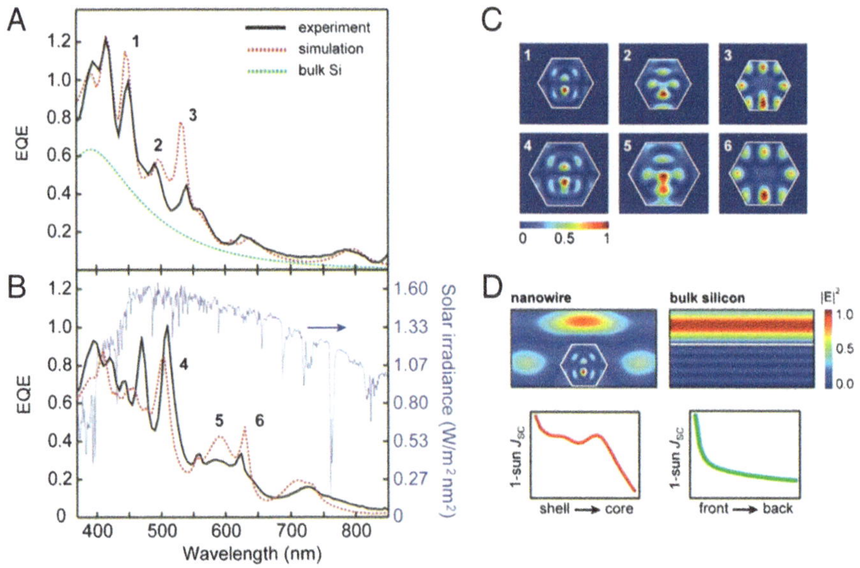


**Fig. 10** (**a**) Experimental and simulated external quantum efficiency (EQE) for a p/in nanowire solar cell. (**b**) Experimental and simulated external quantum efficiency (EQE) for a p/pin nanowire solar cell. (**c**) FDTD simulations of resonant mode profiles for p/in (profiles 1–3) and p/pin (profiles 4–6) structures. (**d**) Electric field distribution for nanowire and bulk silicon at a wavelength of 445 nm (Reprinted with permission from Ref. [28]. Copyright © 2012, National Academy of Sciences)

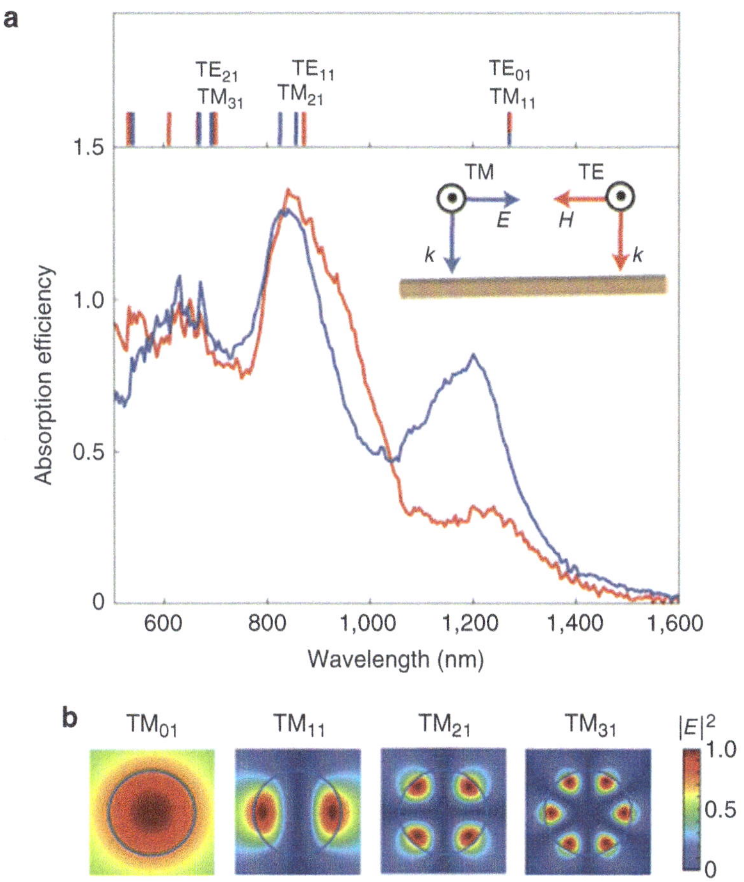


**Fig. 11** (**a**) Absorption efficiency of a 110 nm germanium nanowire under linearly polarized transverse-electric (TE) and transverse-magnetic (TM) light; and (**b**) simulated electric filed intensity distribution for typical TM leaky modes (Reprinted with permission from Ref. [61]. Copyright © 2009, Nature Publishing Group)

related to whispering-gallery resonance. These resonant information and electrical field distribution are further illustrated in Fig. 10c,d by the FDTD simulation. With the Lorentz-Mie scattering calculations and finite difference time domain simulations, Cao, et al. found that the photocurrent shows strong enhancement peaks in the absorption spectra (Fig. 11) [61]. The absorbed incident photons beyond its physical cross section at a specific wavelength is because of an optical antenna effect caused by the subwavelength diameter of the nanowire, the same as the Mie-type resonance. By coating the wires with conformal dielectric shells of $SiN_x$ or $SiO_x$, the optical antenna effect in NW can be dramatically enhanced, along with a $\sim$ 80% increase in short-circuit current density in Si photovoltaics [62].

In addition to the scattering and resonance, the diffraction effect from a periodic array of nanowires can also contribute to the light absorption enhancement. For instance, by measuring the photocurrent of solar cells made with different silicon nanowire length, Gernett et al. found a strong correlation between path length enhancement and roughness factor (proportional to nanowire length), as well as path length enhancement factors exceeding that of the randomized case (Fig. 12) [21]. Furthermore, waveguiding modes were also identified in a periodic array [63].

Because most of the resonant modes are dependent on nanowire geometry, the morphology of nanowire (i.e., size, cross-section, length, filling ratio) can also influence the light absorption characteristics. By varying the diameter from 170 nm to 280 nm and 380 nm, the number of measured peaks increases and, absorption modes shift to longer wavelength [64]. The over integrated photocurrent is also increased. In addition to the diameter, the shape-dependent absorption behavior has also been studied. For example, FDTD simulation shows that rectangular NW has high absorption compared with that of a hexagonal NW [64].

## *4.2 Electrical Properties*

In addition to good light absorption performance, a solar cell must separate and collect carriers to generate power. As discussed above, the nanowire with a radial junction geometry is especially beneficial for the charge separation and carrier collection, due to the fact that the direction of light absorption and carrier collection are decoupled into orthogonal directions. Therefore, in single nanowire radial junction solar cells, the efficiency shows an increase of approximately a factor of five compared with the axial junction cells [3, 45]. In addition, in a previous report, simulation using density functional theory has predicted that a spontaneous potential gradient can be generated without doping in nanowires, primarily due to difference of degree of quantum confinement along the wire [65]. Varying strain along the nanowire can also be used to separate charges in a similar manner [66]. However, these carrier separation mechanisms have not yet been proven experimentally.

In order to gain a quantitative understanding of the electrical performance, Kayes et al., has developed a model to characterize carrier collection in a cylindrical coordinated system [5]. Figure 13 shows the schematic of a nanowire radial junction geometry and its corresponding energy band diagram. The junction is formed between the nanowire core and shell. The diffusion length occurs in the radial direction, which is set equal to the nanowire length. Although the carrier collection is independent on the nanowire length, it should be noted that the nanowire length has an impact on the junction area, and as a result, the electrical performance is also dependent on the nanowire length. In silicon nanowire solar cells, the depletion region recombination is one of the most significant loss factors. For example, as shown in Fig. 13, in the radial p-n junction, the $V_{oc}$ decreases as the cell thickness increases because the junction area increases, so the trap density near

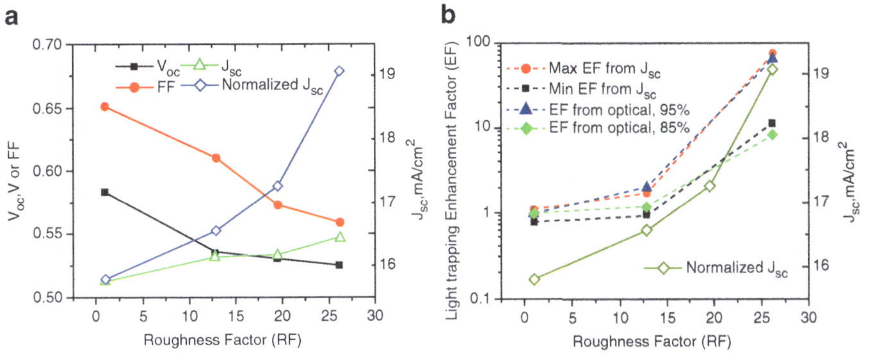


**Fig. 12** (**a**) The $V_{oc}$, FF, and $J_{sc}$ of periodic silicon nanowire array solar cells fabricated from an 8 μm thick silicon absorber with different roughness factors (RF); and (**b**) light-trapping path length enhancement factor (EF) as a function of roughness factor (RF). The EF is derived by comparing the $J_{sc}$ from ordered silicon nanowire arrays with the same RF but different absorber thicknesses (Reprinted with permission from Ref. [21]. Copyright © 2009, American Chemical Society)

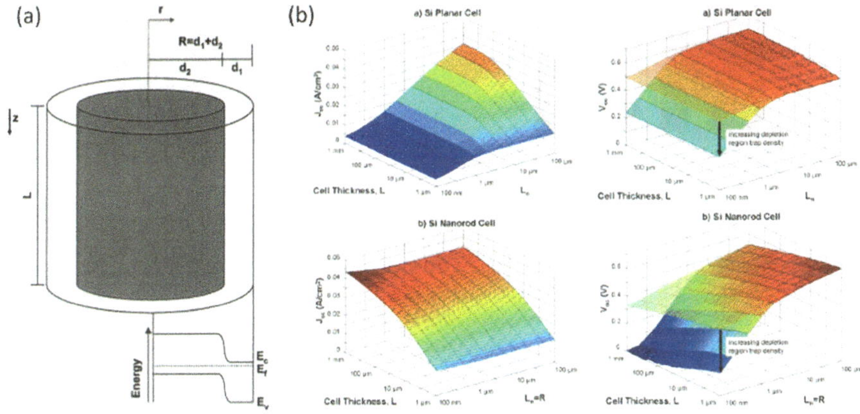


**Fig. 13** (**a**) Schematic of a single rod from the radial p-n junction nanorod cell and its corresponding energy band diagram; (**b**) short-circuit current density $J_{sc}$, open-circuit voltage $V_{oc}$ vs cell thickness $L$ and minority-electron diffusion length $L_n$ for a conventional planar p-n junction silicon cell and a radial p-n junction nanorod silicon cell (Reprinted with permission from Ref. [5]. Copyright © 2005, American Institute of Physics)

the core-shell interface should be low enough that depletion region recombination does not dominate, otherwise large $V_{oc}$ losses are expected. Simulation results also show that a silicon solar cell with a radial junction has a high tolerance of the bulk defect so that when the minority carrier diffusion length is reduced to the physical length, the efficiency degrades slightly with decreasing lifetime, whereas planar junction solar cells shown severe degradation. In terms of the photocurrent, $J_{sc}$ of

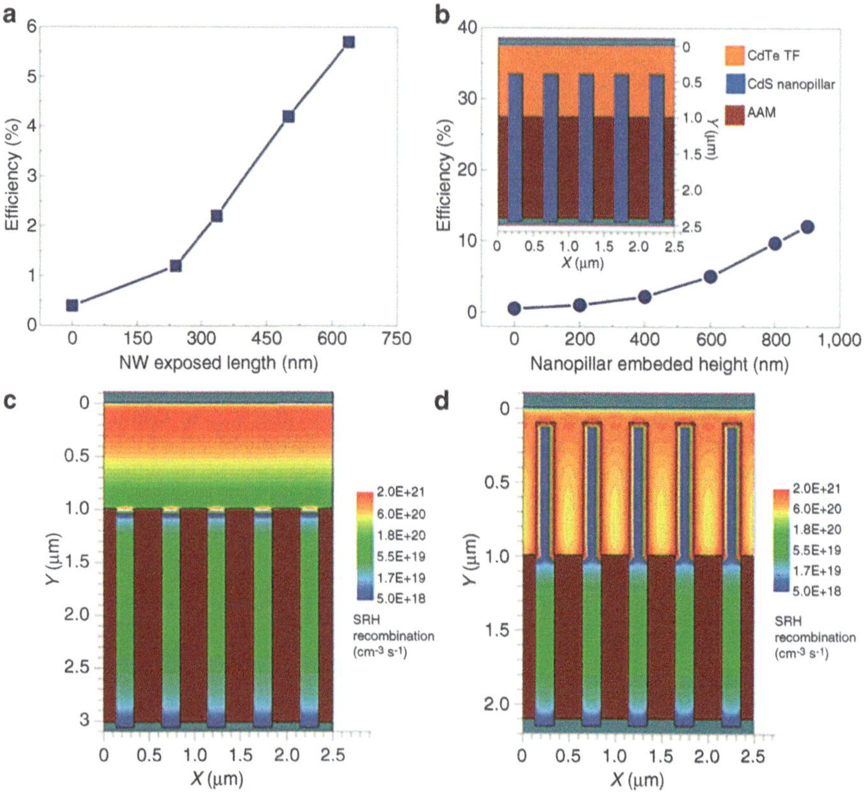


**Fig. 14** (**a**) experimentally measured efficiency of solar cells as a function of the embedded nanopillar height; (**b**) theoretical simulation of the solar cell efficiency as a function of nanopillar height; (**c**) Shockley-Read-Hall (SRH) recombination in solar cells with nanopillar height = 0 nm; and (**d**) Shockley-Read-Hall (SRH) recombination in solar cells with nanopillar height = 900 nm (Reprinted with permission from Ref. [67]. Copyright © 2009, Nature Publishing Group)

nanowire solar cells is essentially independent of the trap density, in striking contrast to the planar one. Kayes' results indicated a guideline in designing high performance nanowire solar cells.

Device physics simulations by Fan shown that the space charge and carrier collection region is significantly enlarged in a radial nanowire geometry [67]. Figure 14 shows that when the minority diffusion length is smaller than the absorber, the collected charges increase dramatically with nanowire length. However, longer nanopillars will reduce the efficiency as a result of the increased interface recombination, which agrees with Kayes' results.

In order to overcome the large leakage current induced by the large high-aspect-ratios of nanowire solar cells, surface passivation is a critical step. However, nanowire surface passivation is a challenging task due to their small size and the fact that multiple facets are exposed. In radial junction nanowire solar cells, surface passivation quality can be quantified with the measurement of the surface recombination velocity $S$, which can be obtained by the following equation:

$$S = \frac{D}{4}\left(\frac{1}{\tau_{\text{eff}}} - \frac{1}{\tau_{\text{b}}}\right), \tag{5}$$

where $\tau_{\text{eff}}$ is the effective carrier lifetime,$\tau_{\text{b}}$ is the carrier lifetime in bulk silicon with the same impurity concentration, and $D$ is the nanowire diameter.

A-Si is often used to make a heterojunction with crystalline silicon. It can provide good passivation for the single crystalline nanowire cores. Experiment using near-field scanning photocurrent microscopy reveals a $\sim$ 100-fold reduction in surface recombination using a thin layer of amorphous silicon (a-Si) coated on a single-crystalline silicon nanowire [68]. In addition to a-Si, a-$SiN_x$ can further reduce the surface recombination velocity to 70 cm/s compared to a value of 450 cm/s provided by a-Si [69]. In recent years, organic such as PEDOT:PSS and organic-inorganic perovskite were also used to make hybrid silicon nanowire solar cells [24, 70], and an efficiency as high as 13% was able to obtained in these types of solar cells partly due to the excellent surface passivation.

The enhanced recombination in nanowire may also result from the metal catalyst used in nanowire growth, especially those metals which can introduce mid-band gap trap state in silicon. However, this possibility was excluded by using Al catalyst to replace the Au catalyst [71]. Totally removing the Au catalyst did not help to further improve the $V_{oc}$. These results affirmed that the quality of the core/multi-shell structure is the most important driver of good electrical performance.

Because of the largely reduced requirement of the minority diffusion length, solar cell materials with a low carrier diffusion length are preferable to be used in nanowire geometry. For example, Zinc oxide and titanium oxide nanowire/nanotube are widely reported in dye-sensitized solar cells. The nanowire/nanotube dye-sensitized solar cells often show higher carrier collection efficiency and reduced surface recombination rates compared with the nanoparticle case.

### *4.3 Ultimate Efficiency*

The ultimate efficiency of nanowire solar cells can be studied with the Shockley-Queisser detailed balance analysis. N. Anttu calculated the Shockley-Queisser efficiency limit for an InP nanowire array solar cell through electromagnetic modeling. It was shown that nanowires longer than 4 μm can produce an efficiency limit of up to 32.5%, despite producing a lower short-circuit current, where the conventional planar InP solar cell shows a Shockley-Queisser efficiency limit of 31% [72]. The same process can be applied to calculate the Shockley-Queisser efficiency limit for a Si nanowire array solar cell. The improvement of the efficiency, as also indicated by the report in [73], results from an increase in the open-circuit voltage, instead of the light absorption. However, the ultimate efficiency of the nanowire solar cell with radial and axial junctions is not expected to exceed that of the Shockley-Queisser limit of thin film design [74, 75].

In order to clarify the point that the intrinsic nanowire properties allow them to have higher $V_{oc}$ than that of planar cells, we have to go back to the thermodynamic limitation of the $V_{oc}$, which is given by [76]

$$qV_{oc} = kT_c\left(\frac{\int A(E)\varphi_{sun}(E)dE}{A(E)\varphi_{bb}(E)dE} \times \frac{\varepsilon_{in}}{\varepsilon_{out}} \times \frac{\varepsilon_{out}A(E)\varphi_{bb}(E)dE}{S_{cell}w(1-p_r)R_{rad}} \times \frac{(1-p_r)R_{rad}}{R_{nrad}+(1-p_r)R_{rad}}\right). \quad (6)$$

The detailed definition of parameters above can be found in [76]. The first term of Eq. (6) accounts for the different temperatures of the incoming and the outgoing photons. It relates to a radiative recombination limited maximum $V_{oc}$ obtained for ideal solar cells when considering the imperfect absorption photons and thermalization losses. The second term is related to the étendue expansion between the incident and emitted photons, in other words, it describes the energy loss caused by the angular dependent photons emission. The third term of Eq. (6) is the ratio between the outgoing flux of photons and the radiative recombination within the solar cell. It describes light concentration by an external lens. The last term describes the energy loss due to the non-radiative recombination process.

In planar solar cells, the emission occurs in an isotropic way, the loss in the second term contributes to about 300 mV of the $V_{oc}$ loss. On the other hand, nanowires have intrinsic wave guiding properties, the photon emission is angularly restricted, which is believed to be of great benefit to high $V_{oc}$. Fig. 15 shows the calculated photon emission rate of nanowire arrays as a function of the emission angle with a nanowire diameter of 170 nm and pitch of 400 nm. More than a 10% enhancement of the $V_{oc}$ could be obtained with a nanowire length smaller than 4 μm and nanowire diameters from 100 nm to 200 nm, for a nanowire solar cell.

## 5 Applications

Single nanowire solar cells are an ideal platform for the fundamental research of optical, electrical and other aspects of the photovoltaic device. Individual and interconnected silicon nanowire photovoltaic elements can serve as robust power sources to drive functional nanoscale electronics, such as sensors, logic gates and other nanoelectronics. As similar to planar solar cells, nanowire array solar cells of a large-scale can be used in generating electrical power from light. For example, they can be used in large-scale territorial application, providing their conversion efficiency can be competitive with commercial technology.

At present, although the nanowire geometry promises to lower the cost by relaxing the material purification requirements, expanding the available materials,

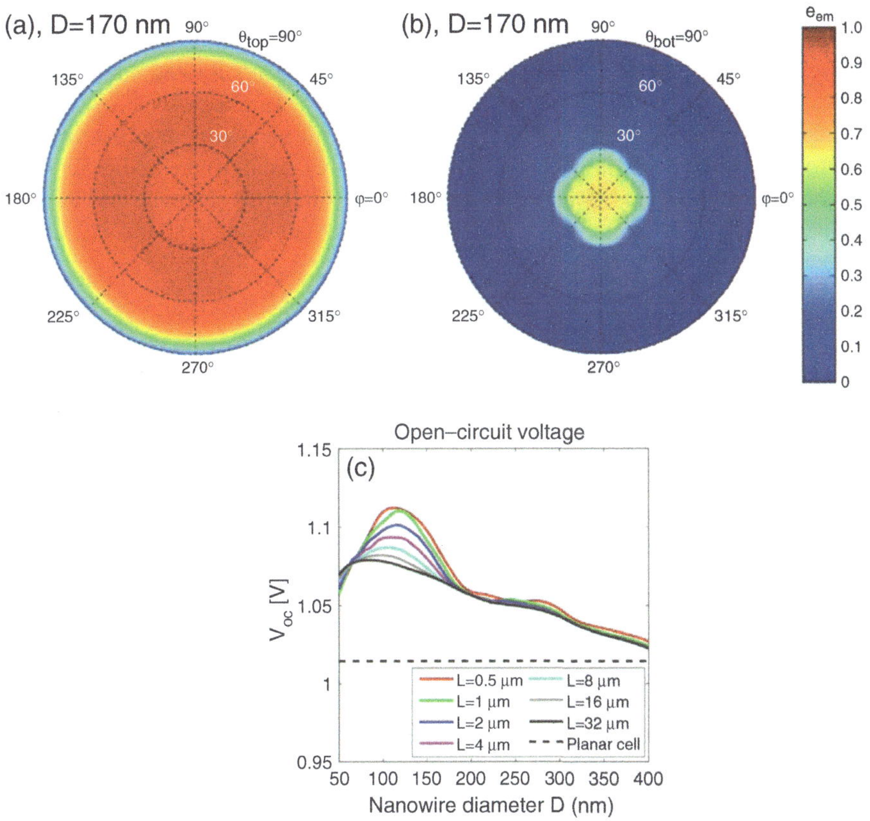


**Fig. 15** Emissivity of a nanowire array with a period of 400 nm and length of 4 μm into (**a**) air region with $n = 1$; and (**b**) substrate region with $n = 3.5$. (**c**) Calculated $V_{oc}$ for a nanowire solar cell as compared to a planar cell with $V_{oc} = 1.01$ V (Reprinted with permission from Ref. [72]. Copyright © 2015, American Chemical Society )

reducing the material usage, or simplifying the fabrication process, nanowire solar cells are still far from practical application due to their low conversion efficiency and the complicated fabrication process involved, which limits the commercialization of nanowire solar cells.

## 6 Challenges and Outlook

To compete with conventional technology, solar cells should be made with an extremely high conversion efficiency and extremely low production cost. In terms of the low production cost, silicon nanowire opens up the possibility to use low quality materials for efficienct solar cells and to grow high single crystalline silicon on inexpensive substrates such as glass and stainless steel. In addition, nanowire solar cells promise the potential of superior light absorption and excellent electrical properties. The present achievement cannot make it be realized commercially, however, some daunting challenges have been addressed for the future; these include (a) the development of robust methods for large-scale assembly; (b) improvement in the large-scale uniformity of nanowire devices; (c) the approach to reduce the surface and interface recombination; (d) the rational control of the morphology, and doping profile; and (e) addressing nanowire mechanical properties. In addition, particle issues such as modules integration and device packaging have also been seriously explored.

## References


1. W. Shockley, H.J. Queisser, Detailed balance limit of efficiency of p-n junction solar cells. J. Appl. Phys. **32**, 510–519 (1961). https://doi.org/10.1063/1.1736034
2. M. Law, L.E. Greene, J.C. Johnson, R. Saykally, P. Yang, Nanowire dye-sensitized solar cells. Nat. Mater. **4**, 455–459 (2005). https://doi.org/10.1038/nmat1387
3. B. Tian, X. Zheng, T.J. Kempa, Y. Fang, N. Yu, G. Yu, J. Huang, C.M. Lieber, Coaxial silicon nanowires as solar cells and nanoelectronic power sources. Nature **449**, 885–889 (2007). https://doi.org/10.1038/nature06181
4. R. Yan, D. Gargas, P. Yang, Nanowire photonics. Nat. Photonics. **3**, 569–576 (2009). https://doi.org/10.1038/nphoton.2009.184
5. B.M. Kayes, H.A. Atwater, N.S. Lewis, Comparison of the device physics principles of planar and radial p-n junction nanorod solar cells. J. Appl. Phys. **97**, 114302 (2005). https://doi.org/10.1063/1.1901835
6. L. Tsakalakos, J. Balch, J. Fronheiser, B.A. Korevaar, O. Sulima, J. Rand, Silicon nanowire solar cells. Appl. Phys. Lett. **91**, 233117 (2007). https://doi.org/10.1063/1.2821113
7. J.-H. Yun, Y.C. Park, J. Kim, H.-J. Lee, W.A. Anderson, J. Park, Solution-processed germanium nanowire-positioned Schottky solar cells. Nanoscale Res. Lett. **6**, 1–5 (2011). https://doi.org/10.1186/1556-276X-6-287
8. M.M. Adachi, M.P. Anantram, K.S. Karim, Core-shell silicon nanowire solar cells. Sci RepSci Rep. **3**, 1546 (2013). https://doi.org/10.1038/srep01546
9. X. Xie, X. Zeng, P. Yang, H. Li, J. Li, X. Zhang, Q. Wang, Radial n-i-p structure SiNW-based microcrystalline silicon thin-film solar cells on flexible stainless steel. Nanoscale Res. Lett. **7**, 1–6 (2012). https://doi.org/10.1186/1556-276X-7-621
10. W. Sun, M. Brozak, J.C. Armstrong, J. Cui, Solar cell structures based on ZnO/CdS core-shell nanowire arrays embedded in Cu2ZnSnS4 light absorber, in *2013 I.E. 39th Photovoltaic. Specialists Conference PVSC*, (2013), pp. 2042–2046
11. J. Tang, Z. Huo, S. Brittman, H. Gao, P. Yang, Solution-processed core–shell nanowires for efficient photovoltaic cells. Nat. Nanotechnol. **6**, 568–572 (2011). https://doi.org/10.1038/nnano.2011.139
12. R. Salazar, A. Delamoreanu, C. Levy-Clement, V. Ivanova, ZnO/CdTe and ZnO/CdS core-shell nanowire arrays for extremely thin absorber solar cells. Energy Procedia **10**, 122–127 (2011). https://doi.org/10.1016/j.egypro.2011.10.164
13. D. Caselli, C.Z. Ning, CdSe nanowire solar cells, in *IEEE 39th Photovoltaic Specialists Conference PVSC 2013*, (2013), pp. 0268–0270
14. S. Brittman, Y. Yoo, N.P. Dasgupta, S. Kim, B. Kim, P. Yang, Epitaxially aligned cuprous oxide nanowires for all-oxide, single-wire solar cells. Nano Lett. **14**, 4665–4670 (2014). https://doi.org/10.1021/nl501750h

15. B.D. Yuhas, P. Yang, Nanowire-based all-oxide solar cells. J. Am. Chem. Soc. **131**, 3756–3761 (2009). https://doi.org/10.1021/ja8095575
16. S.S. Williams, M.J. Hampton, V. Gowrishankar, I.-K. Ding, J.L. Templeton, E.T. Samulski, J.M. DeSimone, M.D. McGehee, Nanostructured Titania–polymer photovoltaic devices made using PFPE-based Nanomolding techniques. Chem. Mater. **20**, 5229–5234 (2008). https://doi.org/10.1021/cm800729q
17. K. Takanezawa, K. Tajima, K. Hashimoto, Efficiency enhancement of polymer photovoltaic devices hybridized with ZnO nanorod arrays by the introduction of a vanadium oxide buffer layer. Appl. Phys. Lett. **93**, 63308 (2008). https://doi.org/10.1063/1.2972113
18. F. Glas, Critical dimensions for the plastic relaxation of strained axial heterostructures in free-standing nanowires. Phys. Rev. B **74**, 121302 (2006). https://doi.org/10.1103/PhysRevB.74.121302
19. B. Tian, T.J. Kempa, C.M. Lieber, Single nanowire photovoltaics. Chem. Soc. Rev. **38**, 16–24 (2009). https://doi.org/10.1039/B718703N
20. E.C. Garnett, P. Yang, Silicon nanowire radial p–n junction solar cells. J. Am. Chem. Soc. **130**, 9224–9225 (2008). https://doi.org/10.1021/ja8032907
21. E. Garnett, P. Yang, Light trapping in silicon nanowire solar cells. Nano Lett. **10**, 1082–1087 (2010). https://doi.org/10.1021/nl100161z
22. Y. Lu, A. Lal, High-efficiency ordered silicon Nano-conical-frustum Array solar cells by self-powered parallel electron lithography. Nano Lett. **10**, 4651–4656 (2010). https://doi.org/10.1021/nl102867a
23. D.R. Kim, C.H. Lee, P.M. Rao, I.S. Cho, X. Zheng, Hybrid Si microwire and planar solar cells: passivation and characterization. Nano Lett. **11**, 2704–2708 (2011). https://doi.org/10.1021/nl2009636
24. X. Yu, X. Shen, X. Mu, J. Zhang, B. Sun, L. Zeng, L. Yang, Y. Wu, H. He, D. Yang, High efficiency organic/silicon-nanowire hybrid solar cells: significance of strong inversion layer. Sci. Rep. **5**, 17371 (2015). https://doi.org/10.1038/srep17371
25. InP Nanowire Array Solar Cells Achieving 13.8% Efficiency by Exceeding the Ray Optics Limit | Science. http://science.sciencemag.org/content/339/6123/1057. Accessed 11 Apr 2016
26. I. Åberg, G. Vescovi, D. Asoli, U. Naseem, J.P. Gilboy, C. Sundvall, A. Dahlgren, K.E. Svensson, N. Anttu, M.T. Björk, L. Samuelson, A GaAs nanowire Array solar cell with 15.3 #x0025; efficiency at 1 sun. IEEE J Photovolt. **6**, 185–190 (2016). https://doi.org/10.1109/JPHOTOV.2015.2484967
27. Y. Cui, J. Wang, S.R. Plissard, A. Cavalli, V. TTT, R.P.J. van Veldhoven, L. Gao, M. Trainor, M.A. Verheijen, J.E.M. Haverkort, E.P.A.M. Bakkers, Efficiency enhancement of InP nanowire solar cells by surface cleaning. Nano Lett. **13**, 4113–4117 (2013). https://doi.org/10.1021/nl4016182
28. T.J. Kempa, J.F. Cahoon, S.-K. Kim, R.W. Day, D.C. Bell, H.-G. Park, C.M. Lieber, Coaxial multishell nanowires with high-quality electronic interfaces and tunable optical cavities for ultrathin photovoltaics. Proc. Natl. Acad. Sci. **109**, 1407–1412 (2012)
29. G. Jia, M. Steglich, I. Sill, F. Falk, Core–shell heterojunction solar cells on silicon nanowire arrays. Sol. Energy Mater. Sol. Cells **96**, 226–230 (2012). https://doi.org/10.1016/j.solmat.2011.09.062
30. M. Yao, S. Cong, S. Arab, N. Huang, M.L. Povinelli, S.B. Cronin, P.D. Dapkus, C. Zhou, Tandem solar cells using GaAs nanowires on Si: design, fabrication, and observation of voltage addition. Nano Lett. **15**, 7217–7224 (2015). https://doi.org/10.1021/acs.nanolett.5b03890
31. S. Wang, X. Yan, X. Zhang, J. Li, X. Ren, Axially connected nanowire core-shell p-n junctions: a composite structure for high-efficiency solar cells. Nanoscale Res LettNanoscale Res Lett. **10**, 269 (2015). https://doi.org/10.1186/s11671-015-0744-3
32. L.J. Lauhon, M.S. Gudiksen, D. Wang, C.M. Lieber, Epitaxial core–shell and core–multishell nanowire heterostructures. Nature **420**, 57–61 (2002)

33. M.S. Gudiksen, L.J. Lauhon, J. Wang, D.C. Smith, C.M. Lieber, Growth of nanowire superlattice structures for nanoscale photonics and electronics. Nature **415**, 617–620 (2002). https://doi.org/10.1038/415617a
34. J. Goldberger, R. He, Y. Zhang, S. Lee, H. Yan, H.-J. Choi, P. Yang, Single-crystal gallium nitride nanotubes. Nature **422**, 599–602 (2003). https://doi.org/10.1038/nature01551
35. M. Law, J. Goldberger, P. Yang, Semiconductor nanowires and nanotubes. Annu. Rev. Mater. Res. **34**, 83–122 (2004). https://doi.org/10.1146/annurev.matsci.34.040203.112300
36. R.S. Wagner, W.C. Ellis, Vapor-liquid-solid mechanism of single Crystal growth. Appl. Phys. Lett. **4**, 89 (1964). https://doi.org/10.1063/1.1753975
37. Y. Xia, P. Yang, Y. Sun, Y. Wu, B. Mayers, B. Gates, Y. Yin, F. Kim, H. Yan, One-dimensional nanostructures: synthesis, characterization, and applications. Adv. Mater. **15**, 353–389 (2003). https://doi.org/10.1002/adma.200390087
38. W. Lu, C.M. Lieber, Semiconductor nanowires. J Phys. Appl. Phys. **39**, R387–R406 (2006). https://doi.org/10.1088/0022-3727/39/21/R01
39. L. Cao, B. Garipcan, J.S. Atchison, C. Ni, B. Nabet, J.E. Spanier, Instability and transport of metal catalyst in the growth of tapered silicon nanowires. Nano Lett. **6**, 1852–1857 (2006). https://doi.org/10.1021/nl060533r
40. Y. Wu, Y. Cui, L. Huynh, C.J. Barrelet, D.C. Bell, C.M. Lieber, Controlled growth and structures of molecular-scale silicon nanowires. Nano Lett. **4**, 433–436 (2004). https://doi.org/10.1021/nl035162i
41. A.B. Greytak, L.J. Lauhon, M.S. Gudiksen, C.M. Lieber, Growth and transport properties of complementary germanium nanowire field-effect transistors. Appl. Phys. Lett. **84**, 4176–4178 (2004). https://doi.org/10.1063/1.1755846
42. D. Wang, Q. Wang, A. Javey, R. Tu, H. Dai, H. Kim, P.C. McIntyre, T. Krishnamohan, K.C. Saraswat, Germanium nanowire field-effect transistors with SiO2 and high-κ HfO2 gate dielectrics. Appl. Phys. Lett. **83**, 2432–2434 (2003). https://doi.org/10.1063/1.1611644
43. Z. Huang, N. Geyer, P. Werner, J. de Boor, U. Gösele, Metal-assisted chemical etching of silicon: a review: in memory of prof. Ulrich Gösele. Adv. Mater. **23**, 285–308 (2011). https://doi.org/10.1002/adma.201001784
44. C. Chartier, S. Bastide, C. Lévy-Clément, Metal-assisted chemical etching of silicon in HF–H2O2. Electrochim Acta. **53**, 5509–5516 (2008). https://doi.org/10.1016/j.electacta.2008.03.009
45. T.J. Kempa, B. Tian, D.R. Kim, J. Hu, X. Zheng, C.M. Lieber, Single and tandem axial p-i-n nanowire photovoltaic devices. Nano Lett. **8**, 3456–3460 (2008). https://doi.org/10.1021/nl8023438
46. Y. Wu, J. Xiang, C. Yang, W. Lu, C.M. Lieber, Single-crystal metallic nanowires and metal/semiconductor nanowire heterostructures. Nature **430**, 61–65 (2004). https://doi.org/10.1038/nature02674
47. A. Dalmau Mallorquí, F.M. Epple, D. Fan, O. Demichel, i. Fontcuberta, A. Morral, Effect of the pn junction engineering on Si microwire-array solar cells. Phys Status Solidi A. **209**, 1588–1591 (2012). https://doi.org/10.1002/pssa.201228165
48. E.C. Garnett, Y.-C. Tseng, D.R. Khanal, J. Wu, J. Bokor, P. Yang, Dopant profiling and surface analysis of silicon nanowires using capacitance–voltage measurements. Nat. Nanotechnol. **4**, 311–314 (2009). https://doi.org/10.1038/nnano.2009.43
49. D.E. Perea, E.R. Hemesath, E.J. Schwalbach, J.L. Lensch-Falk, P.W. Voorhees, L.J. Lauhon, Direct measurement of dopant distribution in an individual vapour–liquid–solid nanowire. Nat. Nanotechnol. **4**, 315–319 (2009). https://doi.org/10.1038/nnano.2009.51
50. E. Koren, N. Berkovitch, Y. Rosenwaks, Measurement of active dopant distribution and diffusion in individual silicon nanowires. Nano Lett. **10**, 1163–1167 (2010). https://doi.org/10.1021/nl9033158
51. J.E. Allen, E.R. Hemesath, D.E. Perea, J.L. Lensch-Falk, Z.Y. Li, F. Yin, M.H. Gass, P. Wang, A.L. Bleloch, R.E. Palmer, L.J. Lauhon, High-resolution detection of au catalyst atoms in Si nanowires. Nat. Nanotechnol. **3**, 168–173 (2008). https://doi.org/10.1038/nnano.2008.5

52. G. Li, H. Li, J.Y.L. Ho, M. Wong, H.S. Kwok, Nanopyramid structure for ultrathin c-Si tandem solar cells. Nano Lett. **14**, 2563–2568 (2014). https://doi.org/10.1021/nl500366c
53. C.-M. Hsu, C. Battaglia, C. Pahud, Z. Ruan, F.-J. Haug, S. Fan, C. Ballif, Y. Cui, High-efficiency amorphous silicon solar cell on a periodic Nanocone back reflector. Adv. Energy Mater. **2**, 628–633 (2012). https://doi.org/10.1002/aenm.201100514
54. Z. Fan, D.J. Ruebusch, A.A. Rathore, R. Kapadia, O. Ergen, P.W. Leu, A. Javey, Challenges and prospects of nanopillar-based solar cells. Nano Res. **2**, 829–843 (2009). https://doi.org/10.1007/s12274-009-9091-y
55. G. Li, J.Y.L. Ho, H. Li, H.-S. Kwok, Diffractive intermediate layer enables broadband light trapping for high efficiency ultrathin c-Si tandem cells. Appl. Phys. Lett. **104**, 231113 (2014). https://doi.org/10.1063/1.4883496
56. J. Zhu, Z. Yu, G.F. Burkhard, C.-M. Hsu, S.T. Connor, Y. Xu, Q. Wang, M. McGehee, S. Fan, Y. Cui, Optical absorption enhancement in amorphous silicon nanowire and Nanocone arrays. Nano Lett. **9**, 279–282 (2009). https://doi.org/10.1021/nl802886y
57. J.-Y. Jung, Z. Guo, S.-W. Jee, H.-D. Um, K.-T. Park, J.-H. Lee, A strong antireflective solar cell prepared by tapering silicon nanowires. Opt. Express. **18**, A286–A292 (2010)
58. H. Savin, P. Repo, G. von Gastrow, P. Ortega, E. Calle, M. Garín, R. Alcubilla, Black silicon solar cells with interdigitated back-contacts achieve 22.1% efficiency. Nat. Nanotechnol. **10**, 624–628 (2015). https://doi.org/10.1038/nnano.2015.89
59. E. Yablonovitch, G.D. Cody, Intensity enhancement in textured optical sheets for solar cells. IEEE Trans Electron Devices **29**, 300–305 (1982). https://doi.org/10.1109/T-ED.1982.20700
60. O.L. Muskens, S.L. Diedenhofen, B.C. Kaas, R.E. Algra, E.P.A.M. Bakkers, J. Gómez Rivas, A. Lagendijk, Large photonic strength of highly tunable resonant nanowire materials. Nano Lett. **9**, 930–934 (2009). https://doi.org/10.1021/nl802580r
61. L. Cao, J.S. White, J.-S. Park, J.A. Schuller, B.M. Clemens, M.L. Brongersma, Engineering light absorption in semiconductor nanowire devices. Nat. Mater. **8**, 643–647 (2009). https://doi.org/10.1038/nmat2477
62. S.-K. Kim, X. Zhang, D.J. Hill, K.-D. Song, J.-S. Park, H.-G. Park, J.F. Cahoon, Doubling absorption in nanowire solar cells with dielectric Shell optical antennas. Nano Lett. **15**, 753–758 (2015). https://doi.org/10.1021/nl504462e
63. T. Voss, G.T. Svacha, E. Mazur, S. Müller, C. Ronning, D. Konjhodzic, F. Marlow, High-order waveguide modes in ZnO nanowires. Nano Lett. **7**, 3675–3680 (2007). https://doi.org/10.1021/nl071958w
64. S.-K. Kim, R.W. Day, J.F. Cahoon, T.J. Kempa, K.-D. Song, H.-G. Park, C.M. Lieber, Tuning light absorption in Core/Shell silicon nanowire photovoltaic devices through morphological design. Nano Lett. **12**, 4971–4976 (2012). https://doi.org/10.1021/nl302578z
65. Z. Wu, J.B. Neaton, J.C. Grossman, Quantum confinement and electronic properties of tapered silicon nanowires. Phys. Rev. Lett. **100**, 246804 (2008). https://doi.org/10.1103/PhysRevLett.100.246804
66. Z. Wu, J.B. Neaton, J.C. Grossman, Charge separation via strain in silicon nanowires. Nano Lett. **9**, 2418–2422 (2009). https://doi.org/10.1021/nl9010854
67. Z. Fan, H. Razavi, J. Do, A. Moriwaki, O. Ergen, Y.-L. Chueh, P.W. Leu, J.C. Ho, T. Takahashi, L.A. Reichertz, S. Neale, K. Yu, M. Wu, J.W. Ager, A. Javey, Three-dimensional nanopillar-array photovoltaics on low-cost and flexible substrates. Nat. Mater. **8**, 648–653 (2009). https://doi.org/10.1038/nmat2493
68. Y. Dan, K. Seo, K. Takei, J.H. Meza, A. Javey, K.B. Crozier, Dramatic reduction of surface recombination by in situ surface passivation of silicon nanowires. Nano Lett. **11**, 2527–2532 (2011). https://doi.org/10.1021/nl201179n
69. M.D. Kelzenberg, D.B. Turner-Evans, M.C. Putnam, S.W. Boettcher, R.M. Briggs, J.Y. Baek, N.S. Lewis, H.A. Atwater, High-performance Si microwire photovoltaics. Energy Environ. Sci. **4**, 866 (2011). https://doi.org/10.1039/c0ee00549e

70. X. Yan, C. Zhang, J. Wang, X. Zhang, X. Ren, A high-efficiency Si nanowire Array/Perovskite hybrid solar cell. Nanoscale Res. Lett. **12**, 1.14 (2017). https://doi.org/10.1186/s11671-016-1785-y
71. Y. Ke, X. Weng, J.M. Redwing, C.M. Eichfeld, T.R. Swisher, S.E. Mohney, Y.M. Habib, Fabrication and electrical properties of Si nanowires synthesized by al catalyzed vapor–liquid–solid growth. Nano Lett. **9**, 4494–4499 (2009). https://doi.org/10.1021/nl902808r
72. N. Anttu, Shockley–Queisser detailed balance efficiency limit for nanowire solar cells. ACS Photonics. **2**, 446–453 (2015). https://doi.org/10.1021/ph5004835
73. Y. Xu, T. Gong, J.N. Munday, The generalized Shockley-Queisser limit for nanostructured solar cells. Sci. Rep. **5**, 13536 (2015). https://doi.org/10.1038/srep13536
74. X. Wang, M.R. Khan, M. Lundstrom, P. Bermel, Performance-limiting factors for GaAs-based single nanowire photovoltaics. Opt. Express **22**, A344 (2014). https://doi.org/10.1364/OE.22.00A344
75. X. Zhai, S. Wu, A. Shang, X. Li, Limiting efficiency calculation of silicon single-nanowire solar cells with considering auger recombination. Appl. Phys. Lett. **106**, 63904 (2015). https://doi.org/10.1063/1.4908294
76. U. Rau, U.W. Paetzold, T. Kirchartz, Thermodynamics of light management in photovoltaic devices. Phys. Rev. B. **90**, 35211 (2014). https://doi.org/10.1103/PhysRevB.90.035211

# Si Nanowire Solar Cells: Principles, Device Types, Future Aspects, and Challenges

**Mrinal Dutta, Lavanya Thirugnanam, and Naoki Fukata**

**Abstract** Dream of low-cost and high-efficiency Si solar cells is going to see daylight through the advent of Si nanowire (SiNW)-based solar cells due to the unique three-dimensional structures over planar solar cells. SiNW arrays are designed to increase light absorption beyond the bandgap limitations by efficient light trapping and antireflection. Junction fabrication technique has made it possible to make very short carrier collection pathways that provide the opportunity to use lower-grade Si material and also reduce material quantity while maintaining high-quality photovoltaic performance. Additionally, peeling off and transfer of SiNW arrays or in situ growth of SiNWs directly on low-cost substrates can introduce another major cost reduction in photovoltaic industry. This chapter focuses on the recent progress and photovoltaic aspects of the state-of-the-art technologies for SiNW-based solar cell fabrication. Fabrication and performance of different types of photovoltaic devices based on SiNWs are summarized. The challenges face by the fabrication processes and drawbacks of these kinds of devices due to the use of SiNWs are also discussed. Several possible new techniques to face the problems of this emerging field are also included.

## 1 Introduction

In the last 45 years, demand of energy around the globe has been accelerated world energy consumption by more than double [1]. World's four-fifth consumption of energy comes from fossil fuel combustion, resulting in a 5.8% increase of $CO_2$ emissions. Currently, the renewable energy shares 19% of global energy consumption in 2012 and continued to grow strongly in 2013 [2]. Apart from other renewable

M. Dutta (✉) · N. Fukata
International Center for Materials Nanoarchitectonics (MANA), National Institute for Materials Science, Tsukuba, Japan
e-mail: sspmd.iacs@gamil.com

L. Thirugnanam
Department of Physics, Anna University, Chennai, India



S. Ikhmayies (ed.), *Advances in Silicon Solar Cells*,
https://doi.org/10.1007/978-3-319-69703-1_11

energy sources, conversion of solar energy has experienced a rapid growth in the last decade. By early 2014, at least 53 solar PV plants larger than 50 MW were operating in at least 13 countries. The world's 50 biggest plants reached cumulative capacity exceeding 5.1 GW by the end of 2013 [2]. Worldwide growth of photovoltaics, becoming a mainstream electricity source, has been fitting an exponential curve for more than two decades. By the end of 2014, cumulative photovoltaic capacity increased by 38,700 MW and reached at least 177 GW, sufficient to supply 1% of the world's total electricity consumption of currently 18,400 TWh [1, 2].

Silicon is currently the most dominant nontoxic commercial material for the production of solar cells due to its specific benefits over other materials. Si wafer-based solar cells accounted for about 90% of the total production in 2013. However, the vast majority of these commercial Si solar cells have been based on crystalline Si that can provide enough high efficiency for practical usage but not yet cost-competitive, chiefly due to the high cost of solar-grade Si wafers; half of the production cost (~ 1$ per watt) alone shares by Si wafers [3–5]. Highly efficient commercial Si solar cells require relatively large amounts of high-purity solar-grade Si to fully absorb incident sunlight, since Si has low absorption in the visible and near infrared region of the solar spectrum; these Si solar cells must also have long minority carrier diffusion lengths to ensure minimal carrier recombination and enhanced carrier collection [6–8]. In view of low-cost, abundance, and nontoxicity, use of polycrystalline and amorphous Si thin films is advantageous in the production of solar cells [9, 10]. However, these benefits are outweighed by their short carrier diffusion length and thus much lower efficiency than crystalline Si solar cells. Increasing effort has been devoted to address these problems in recent years by fabricating Si NW (SiNW) based solar cells. These SiNW solar cells exhibit a higher absorbance per unit thickness than commercial Si solar cells due to incident light trapping within the NW arrays and antireflection effect [7, 11]. SiNW radial p–n junction offers a shorter transportation length of photogenerated charge carriers [12]. As a consequence of this, it is possible to use lower-grade Si material with short minority carrier diffusion length for fabrication of these solar cells with sufficient conversion efficiency. Thus the use of lower grade and less amount of material opens a path to low-cost Si solar cell production.

## 2 Advantage of NW Geometry

In any solar cell, conversion of solar energy to electricity involves five fundamental physical steps: (1) absorption of photons, (2) exciton creation, (3) exciton separation into free electron and hole, (4) transport of free carrier to respective electrodes, and (5) collection of free carriers. Each step is subjected to losses. Efficiency of each of the steps controls the conversion efficiency of the cell. Instead of crystalline wafers or thin films, the use of NWs provides possibilities to minimize losses in each step at low cost. The potential cost benefits come primarily from the use of lower-grade Si material and reduction of amount of material reasonable to get moderate efficiencies.

## 2.1 Absorption and Exciton Formation

Absorption loss in planar Si solar cells arises mainly due to the reflection of solar light from the surface. As for an example, in the absence of any antireflection coating, planar Si solar cell losses more than 30% of incident light due to reflection from its surface [13]. This reflection at the interface arises from a difference in refractive index between Si and air medium [14]. As a common approach to reduce reflection, materials with refractive index between Si and air are used for antireflection coatings. Double and triple layers of coating are used instead of single-layer coating to enhance the antireflection property; however ideal condition needs continuous layer-graded refractive index. Yablonovitch et al. first derived the limit of a maximum light intensity enhancement $2n^2$ corresponding to a path length enhancement of $4n^2$ ($\sim$50 for silicon, depending on wavelength) compared with the incident beam for perfectly implemented randomized scheme, where n is the refractive index of the medium [15]. Garnet and Yang showed that the path length of incident solar radiation of SiNW array films in the AM1.5G spectrum increased by 73 times [16]. Several groups have reported the excellent antireflection and light-trapping properties of SiNW arrays [17–20]. Figure 1a shows the absorption and reflection properties of SiNW arrays fabricated on glass substrates [17, 18]. NW arrays showed low reflectance of less than 10% with a strong broadband optical absorption. This absorption by NW arrays was found to be higher than Si films of equivalent thickness.

Reflectance measurements of SiNW arrays on Si wafers fabricated by metal-catalyzed electroless etching technique showed suppression of light reflection over a wide spectral range compare to polished Si wafers (Fig. 1b). For SiNWs fabricated on single-crystal Si wafers, the reflectance is lower than 1.4% in the range of

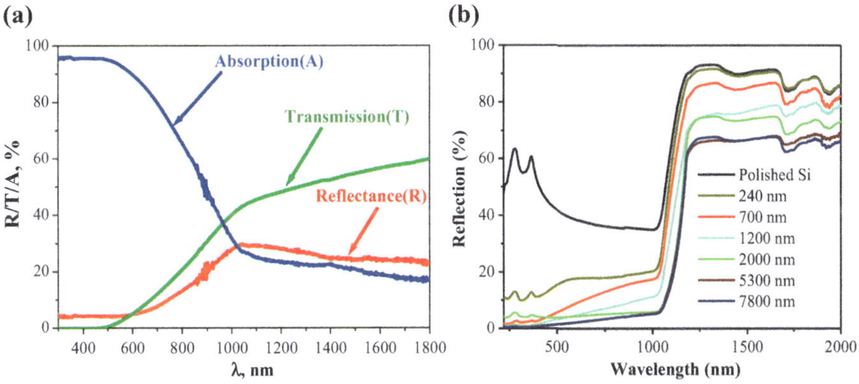


**Fig. 1** (**a**) Optical transmission (*T*), reflectance (*R*), and absorption ($A = 1\text{–}T\text{–}R$) of SiNWs prepared by etching of 2.7-μm-thick Si layers on glass (Reprinted with permission from Ref. [17]. Copyright (2009) American Chemical Society). (**b**) UV–vis reflection spectra of planar Si substrate and Si NW arrays with different lengths (Reproduced with permission from Ref. [20]. Copyright 2015, Elsevier)

300–600 nm. This antireflection as well as light-trapping property of SiNW arrays strongly depends on NW lengths, diameters, alignment, and spacings. Figure 1b shows the dependence of reflectance on NW array lengths [20]. Reflection decreases with the increase of NW length. In a more detailed experiment on VLS-grown Si wire arrays embedded in transparent polymer polydimethlysiloxane (PDMS), researchers from California Institute of Technology found that NWs obtained by quasi-periodic or random motifs led to lower overall optical absorption than well-aligned periodic wire array [21]. This group also has demonstrated a remarkable near-complete absorption of sunlight above the Si bandgap by using $SiN_x$ antireflection layer coated with Si wire arrays embedded in PDMS matrix structure filled with $Al_2O_3$ nanoparticle light scatterer. The absorption of the Si wire array in this structure exceeded the ray optics light-trapping limit for infrared wavelengths ($> 800$ nm) in planar structures. This remarkably low reflectance of the SiNW array is attributed to the advantage of incident light-trapping within the NW arrays. As a result collective light-scattering interactions among NWs that compelled the light to travel much longer distance by many turns than the NW array length and the ultrahigh surface areas of high-density SiNWs increase the absorption.

In Si core-shell NWs, over 100% enhancement of photocurrent can be obtained compare to only NWs without shell [22]. Simulation on the optical absorption of single core-shell NW revealed that by optimizing core diameter, shell thickness, and refractive index, this enhancement could be achieved. Conversion efficiency increases from 3.32% to 9.24% when periodically aligned Si nanopillar arrays fabricated on Si substrate, by a metal-catalyzed chemical etching process via the diameter-reduced polystyrene spheres as mask, were used as an antireflection layer of a planar p–n junction solar cell [23].

## *2.2 Charge Separation and Transportation*

A group of researchers from the University of California and Lawrence Berkeley National Laboratory has shown, using ab initio calculations, that axial charge separation in intrinsic SiNW of small diameter is possible by varying strain along the NW length [24]. In another report, the same researchers have shown theoretically that structural tapering of SiNW can also produce a strong electrostatic potential gradient by varying quantum confinement with diameter which can spatially separate the highest occupied and the lowest unoccupied states along the wire axis [25]. Thus designing solar cells by morphology control and separation of charge carriers by thermalization can open a new horizon for solar cells with minimum Auger recombination, a strong limiting factor for obtaining high efficiency. However these theoretical predictions are yet to be experimentally verified.

In radial p–n junction core-shell structures, excellent carrier separation and propagation take place via the built-in electric fields of radial p–n junctions [12, 26]. This short carrier collection in radial direction enables not only the effective collection of photogenerated charge carriers separated at the p–n junction via the electrodes but also the suppression of carrier recombination, because the carriers can be collected at the electrodes via a short transportation distance. This effect of radial junction in separation and transportation of charge carriers will be discussed in Sects. 4 and 5.

### *2.3 Carrier Collection and Cost*

Proper absorption of solar light in planar p–n junction solar cell demands increase in thickness of the base region with a carrier collection efficiency limited by minority carrier diffusion in the base region. This limitation restricts the use of low-cost and lower-grade Si materials having low minority carrier diffusion lengths, as absorbing base in planar p–n junction solar cells. In a theoretical work, Atwater and Lewis group [27] has shown that an array of radial p–n junction nanorods provides the solution to this problem by decoupling light absorption and carrier collection into orthogonal spatial directions. In this kind of device geometry, proper absorption of light occurs in long arrays in the direction of light with the benefit of short carrier collection in radial direction that allows effective collection of photogenerated charge carriers. These short collection lengths facilitate the use of materials with low minority carrier diffusion lengths maintaining high efficiency.

## 3 NW Synthesis Techniques

Since the first report on vapor–liquid–solid (VLS) mechanism of silicon wire growth by Wagner and Ellis [28] in 1964, significant progress have been made to develop facile and controlled methods for the fabrication of SiNWs in the last few decades. These several methods can be basically divided in two basic approaches: bottom-up and top-down. The bottom-up approach is an assembly process joining Si atom by atom for the growth of Si wires with diameters ranging from a few nanometer to several hundreds of nanometers, whereas in top-down approach dimensional reduction of bulk Si wafer fabricates SiNWs. Bottom-up approach includes methods like chemical vapor deposition (CVD) [28–43], molecular beam epitaxy [44, 45], reactive atmosphere annealing [46], laser ablation [47–49], and SiO evaporation technique [49–52] where VLS mechanism is a dominant growth mechanism, and apart from these some other techniques such as oxide-assisted growth (OAG) [53, 54], supercritical-fluid-based [55, 56], and solution-based growth [57] are also reported. On the other hand in top-down approach lithography [58, 59], reactive ion etching (RIE) [16, 60, 61] and metal-catalyzed electroless etching [62–70] are the commonly used techniques for the fabrication of SiNWs. Throughout details of these growth techniques would be beyond the scope of this chapter. Only widely used NW synthesis techniques for the fabrication of NW solar cells will be discussed in this chapter. Interested readers are referred to the review articles of Schmidt et al. [43] and Singh et al. [71] and the references therein.

## 3.1 Bottom-Up Fabrication

### 3.1.1 Chemical Vapor Deposition

Chemical vapor deposition (CVD) technique is particularly interesting not only because it is used to produce dense, high-aspect-ratio, and vertically aligned NW arrays but also because it is applicable to large-scale production in the semiconductor industry. Inside the CVD chamber, gaseous Si precursor, such as silane ($SiH_4$) or silicon tetrachloride ($SiCl_4$), is transported to the substrate surface for deposition. Depending on precursor type and temperature, the wire growth rate varies from nearly $10^{-2}$ to $10^3$ nm min$^{-1}$ [30, 32]. CVD has the advantage of tuning the wire diameter within a wide range from below 10 nm to several hundred micrometers and also can control the length of the wires by scheming the reaction time. CVD also offers the opportunity of controlling the doping profile inside NWs in both radial and axial direction by introducing extra precursor gas and changing the flow rate [33, 34, 39].

In VLS mechanism Si from vapor phase enters into the metal-Si alloy liquid droplet above the eutectic temperature. When this droplet becomes supersaturated, Si freezes out at the interface of droplet and Si (Fig. 2a). In lower temperature another growth mechanism comes into play called vapor–solid–solid (VSS) mechanism which deals with the metal catalyst particle in solid state instead of liquid state. In spite of all advantages of CVD method, one major drawback of this method is metal contamination.

Most of the Si wires grown by VLS process are made using Au as a catalyst. However, Au introduces two impurity energy levels within the Si bandgap: one acceptor level 0.54 eV below the conduction band and another donor level 0.29 eV above the valance band, which act as deep level traps that decrease carrier mobility, lifetime, and diffusion length [31, 40]. Several studies showed that Au diffuses along the outer surface during both growth and cooling [41, 42]. Au present on the tip and outer surface of Si wires causes a serious metal contamination problem in photovoltaic devices as it is difficult to remove due to its chemical inertness. As a result open-circuit voltages ($V_{oc}$) of solar cells fabricated using Au-catalyzed SiNW arrays and single NWs are found to be in very low range of 130–300 mV that outperforms the photovoltaic device performance. This drawback of Au catalyst enunciates search of alternative metal catalysts such as Al, Ag, Cu, Ni, Pt, etc. for the growth of Si wires [31, 39–41]. Wang et al. [31] has reported epitaxial growth of SiNWs on Si (111) substrate using Al as a catalyst as shown in Fig. 2(b). The growth mechanism was dominated by VSS growth as the growth temperatures (~430–490 °C) were much lower than the eutectic temperature (577 °C) of Al–Si binary phase diagram. In contrast to Au catalyst, Al contamination makes the NWs strongly p-type, but to avoid oxidation of Al catalyst, ultrahigh vacuum should be maintained throughout the process. NWs grown using Al as a catalyst showed tendency of tapering compared to the NWs grown using Au catalyst. Use of Ag as a catalyst gives the advantage of controlling the diameter from micrometer to

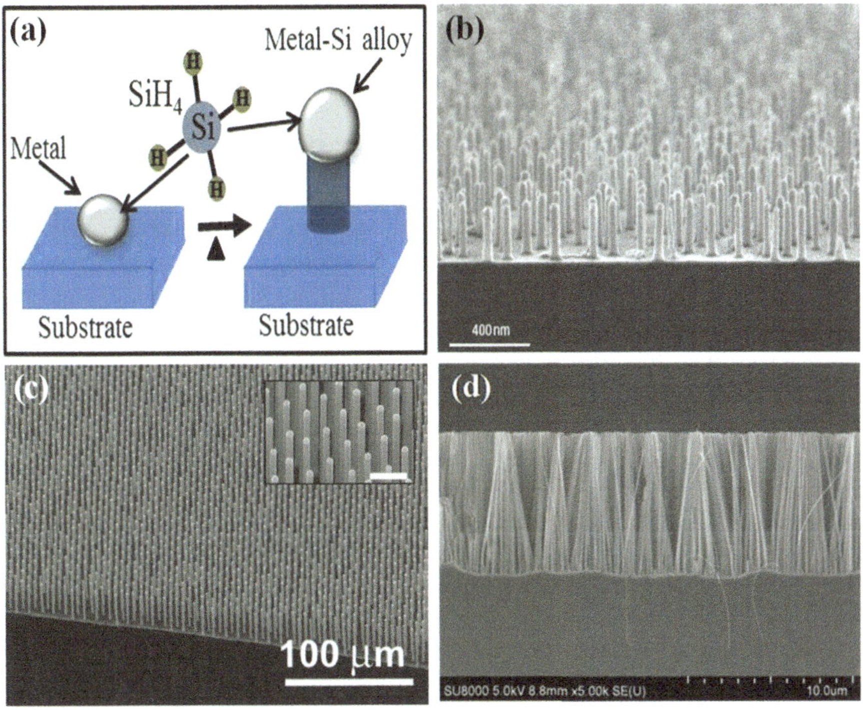


**Fig. 2** (**a**) Schematic of the VLS growth mechanism. (**b**) Tilted SEM cross-sectional image of Al-catalyzed SiNW arrays on Si (111) grown at 430 °C. Scale bar is 400 nm (Reproduced with permission from Ref. [31]. Copyright 2006, Nature Publishing Group). (**c**) Tilted SEM view of a Cu-catalyzed Si wire arrays (Reproduced with permission from Ref. [40]. Copyright 2007, American Institute of Physics). (**d**) Cross-sectional SEM image of SiNW arrays fabricated by Ag-catalyzed electroless etching technique. Scale bar is 10 μm

nanometer by controlling growth temperature with the opportunity of easy removal of superficial Ag using chemical methods [39]. Lower-temperature growth, than the eutectic temperature of Ag–Si system, results in the growth of NWs controlled by VSS mechanism rather than VLS. Cu-catalyzed high-temperature growth by using $H_2$:$SiH_4$ as a precursor gas mixed with $BCl_3$ (5% in $H_2$) resulted in p-type Si microwires (SiMWs) with diameter of 2 μm, length of ~100 μm, and an effective long electron minority carrier diffusion length of 10.5 ± 1 μm as copper is less detrimental to the minority carrier lifetime [36]. Four-point measurement confirmed the resistivity of the wires was 0.19 ± 0.02 Ω-cm that corresponds to an acceptor concentration of $(1.05 \pm 0.15) \times 10^{17}$ $cm^{-3}$. VLS growth technique using metal catalysts most of the time yields (111) oriented Si wires; however (100) orientation is rare.

## 3.2 Top-Down Fabrication

### 3.2.1 Metal-Catalyzed Electroless Etching (MCEE)

In top-down approach, metal-catalyzed electroless etching (MCEE) is a relatively new promising anisotropic etching technique that becomes most popular for its simple and inexpensive methods to fabricate wafer-scale SiNW arrays. The traditional top-down methods like electron beam lithography and RIE precisely can fabricate SiNW arrays with controlled diameter and spacing. However for these time-consuming and expensive methods, fabrication of SiNWs in large area still remains a challenge. The SiNWs produced by MCEE technique inherit the electrical characteristics of mother Si wafer.

In the early 1990s, behavior of ultrafine metal particles on Si wafer surface in HF—$H_2O_2$ wet chemical cleaning solutions was investigated [72, 73]. In 1997 preparation of porous Si layer by chemical etching in a HF, $HNO_3$, and $H_2O_2$ solution of Al-thin-film-covered c-Si was first reported [74]. It was found that in the presence of Al film, the incubation time necessary for the formation of porous Si was dramatically decreased. The first detailed insight of this recently popular chemical etching of Si surface in the presence of metal was reported in 2002 [62] by Li and Bohn. Light-emitting porous Si was produced by dipping Au, Pt, or Au/Pd thin-film ($< 10$ nm) coated (100) Si in HF and $H_2O_2$ solution. Porous Si with different morphologies and light-emitting properties was produced depending on doping profile of Si and the type of metal film. The presence of different metals showed different etching rates of Si surface. A reaction scheme was proposed to explain the metal-assisted etching process. Peng et al. [19, 63] in 2002 reported a wafer-scale production of SiNW array at room temperature by simply immersing Si wafers into HF–$AgNO_3$ solution. In addition to highly oriented SiNW arrays, Si nanostructures including porous Si and Si nanoholes also are produced by using an aqueous HF solution containing oxidizing agents, such as $Fe(NO_3)_3$ or $H_2O_2$ [64–66].

Different mechanisms are proposed to understand the reaction processes of dissolution of Si. These processes can be divided into three groups:

(I) Direct dissolution of Si in tetravalent state [62, 67, 68]

$$Si + 4h^+ + 4HF \rightarrow SiF_4 + 4H^+ \tag{1}$$

$$SiF_4 + 2HF \rightarrow H_2SiF_6 \tag{2}$$

(II) Direct dissolution of Si in divalent state [69, 70]

$$Si + 4HF_2^- \rightarrow SiF_6^{2-} + 2HF + H_2 \uparrow + 2e^- \tag{3}$$

(III) Si oxide formation and dissolution [64, 65]

$$Si + 2H_2O \rightarrow SiO_2 + 4H^+ + 4e^- \quad (4)$$

$$SiO_2 + 6HF \rightarrow H_2SiF_6 + 2H_2O \quad (5)$$

Through MCEE technique, fabrication of vertical SiNW arrays with controlled doping profile and orientation is possible. However, to control the position and NW diameter, template-assisted MCEE technique has been used. Highly ordered SiNW arrays with controlled diameter, length, and density have been achieved by combining MCEE technique and nanosphere lithography [66, 75]. The diameter of the NWs was determined by the diameter of the polystyrene spheres as mask. SiNW arrays with different structures, different cross-sectional shape, and perfect ordering were fabricated by using interference lithography followed by MCEE technique [76]. SiMW arrays were also prepared by combining photolithography with MCEE technique [77].

## 4 Junction Formation

After the synthesis of SiNWs, a junction must be established for quick electron-hole separation and their collection. Junctions can be fabricated either along the diameter (radial junction) or along the length (axial junction) of the NWs. Axial junction SiNWs can be prepared by etching Si wafers with planar p–n junctions formed earlier [23, 78]. Using the diameter-reduced polystyrene spheres as mask axial junction, Si NW arrays fabricated by MCEE technique showed an energy conversion efficiency up to 9.24% due to excellent antireflection property of the special periodical Si nanostructure [23].

Radial junctions are favorable compared to axial junctions for providing short collection paths on the order of NW diameter for the separated charge carriers, which reduce recombination rates of charge carriers and also allow the use of lower-grade Si with low minority carrier diffusion length. The advantages of radial junctions are realized by forming an oppositely doped Si shell layer around the Si core either by thermal diffusion or by depositing an extra oppositely doped thin Si layer by CVD deposition that can be single crystalline if grown epitaxially but more often is polycrystalline. These types of junctions are known as radial homojunctions. However, employing a diffusion process to fabricate homojunction usually results in undefined junction depths and sometimes even completely doped NWs [79–81]. By forming a completely doped n-type chemically etched slanting Si NW arrays on p-type Si (111) wafer, a conversion efficiency of 11.37% was achieved [79]. This high-conversion efficiency was attributed to the excellent antireflection property of 3–5-μm-long SiNW arrays and better electrical contacts due to the slantingly aligned structure. Diffusion process usually completely converts the whole Si NWs as the average diameter of chemically etched NWs is

~ 100 nm. As a result the junction forms within the wafer just immediately underneath of SiNW arrays that is similar to a planar p–n junction with Si NW arrays as surface texturing. By using NWs with larger diameter, radial p–n junction could be formed by dopant diffusion process in SiNWs [27]. So that dopant diffusion occurs at the surface of the NW without changing the core-doping type. Device modeling showed that short-circuit current density ($J_{sc}$) in the Si nanorod cell is essentially independent of the trap density, whereas in the planar geometry a high trap density leads to a very low $J_{sc}$, regardless of the trap density in the depletion region as illustrated in Fig. 3. As for example, for a 100-μm-thick Si solar cell, $J_{sc}$ in the radial p–n junction nanorod geometry slightly reduced from ~38 mA/cm$^2$ to ~35 mA/cm$^2$ for an increase in radius of nanorod from 100 nm to above 1 μm, whereas in the planar geometry, $J_{sc}$ dropped from 34 to 4 mA/cm$^2$ for a decrease in minority carrier diffusion length ($L_n$) from 100 μm to 100 nm related to the increase in density of recombination centers ($N_r$) from $10^{12}$ to $10^{18}$ cm$^{-3}$ in the quasi-neutral region of the shell layer. One thing should be remembered that in this simulation nanorod, radius is set equal to $L_n$ for optimal conditions. Ultimately these results showed that instead of high trap density in core and shell regions, $V_{oc}$ and the conversion efficiency of a radial p–n junction nanorod solar cells can remain high with the approximation that the depletion region trap density remains relatively low (Fig. 3). For example, a Si solar cell with $N_r$~$10^{14}$ cm$^{-3}$ in the depletion region and $N_r$~$7 \times 10^{16}$ cm$^{-3}$ in the shell and core regions (i.e., similar to $L_n = 100$ nm), the maximal efficiency of the radial p–n junction nanorod geometry was calculated to be 11%, compared to 1.5% in the planar geometry. If throughout the cell that same density of recombination centers was maintained, then simulation results showed that the efficiency of nanorod solar cell was double than that of planar geometry. Experimental investigations using low-grade Si material showed higher efficiency for radial p–n junction nanorod cell structures compared to planar cells [7]. Radial $n^+$–$p^+$ junction formed by diffusion method on 25-μm-thick pillar arrays with diameter 7.5 μm fabricated by deep reactive ion etching technique (DRIE) in $p^+$ crystalline Si (100) substrates with resistivity ~0.02 Ω cm (concentration ~ $5 \times 10^{18}$ cm$^{-3}$ and $L_n$ ~ 10 μm) showed nearly two fold increase in efficiency compared to planar cells.

Another type of radial junction can be formed on Si NWs by covering it with organic polymer or introducing a-Si layer between p and n layers. These radial heterojunction solar cells that combine an organic polymer with SiNWs not only require low-fabrication temperature due to the solution-based processes and low thermal stability of the polymers but also can reduce manufacturing costs by utilizing inexpensive and abundant organic polymers, in contrast to expensive p–n junctions consisting of only Si [17, 82]. Promising conversion efficiency up to 13.2% for such Si/organic hybrid solar cells has been demonstrated [83]. Introduction of a thin amorphous Si (a-Si) layer by CVD technique at the interface of p–n junction could effectively passivate the NW surface. Thus enhancement of photoconversion efficiency compared to radial p–n junction could be expected in radial p–i–n structures. Experimental realization of this enhancement of efficiency by introducing a thin a-Si layer has been observed several times [84–86].

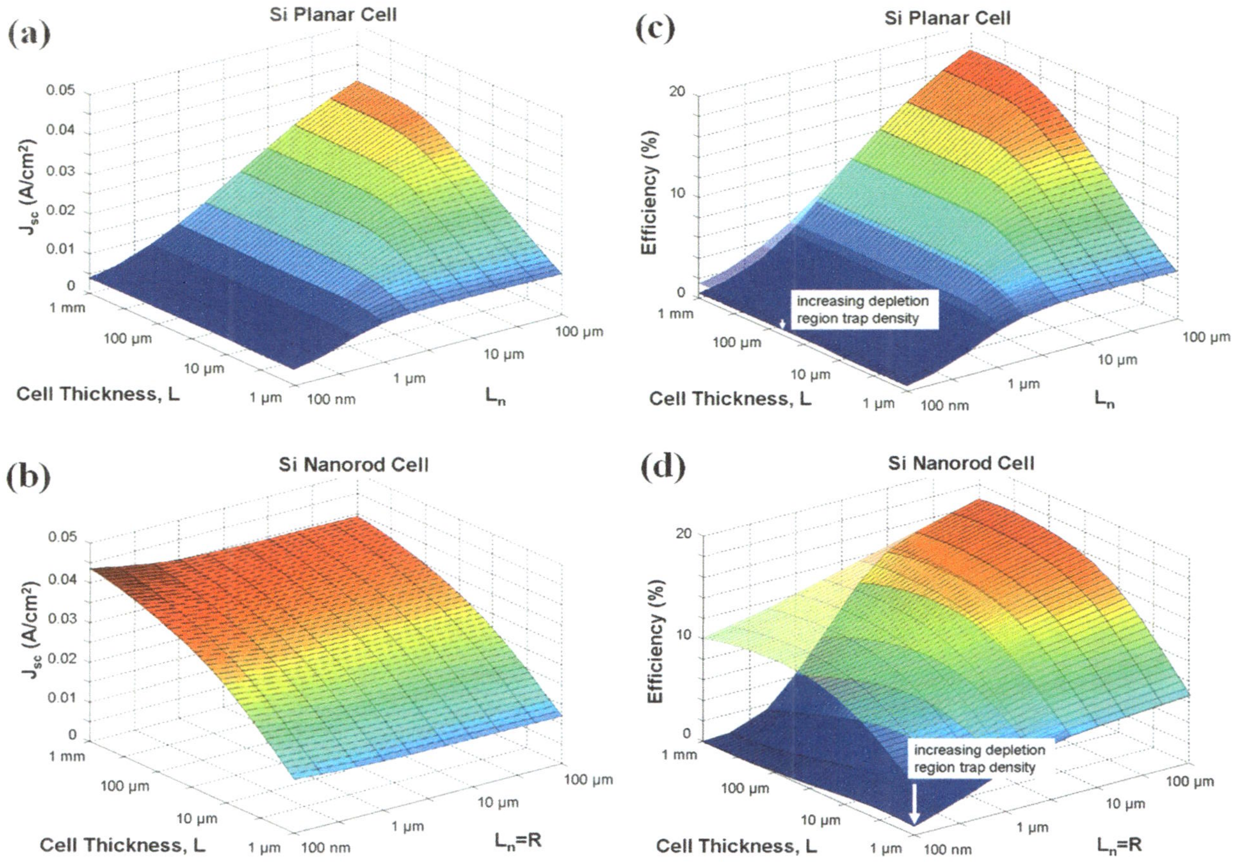


**Fig. 3** Short-circuit current density ($J_{sc}$) vs cell thickness $L$ and minority electron diffusion length $L_n$ for (**a**) a conventional planar p–n junction silicon cell and (**b**) a radial p–n junction nanorod silicon cell. Efficiency vs cell thickness $L$ and minority electron diffusion length $L_n$ for (**c**) a conventional planar p–n junction silicon cell and (**d**) a radial p–n junction nanorod silicon cell. In the radial p–n junction nanorod case, the cell radius $R$ is set equal to $L_n$ (Reproduced with permission from Ref. [27]. Copyright 2005, American Institute of Physics)

Some solar cell uses Schottky junction, a metal with a large work function in contact with SiNWs, to induce a barrier for carrier flow in only one direction similar to the p–n junction [87–89]. However, the barrier which restricts recombination of charge carriers at electrode is small in comparison with p–n junction. As a result saturation current density for Schottky junction is much larger than that of p–n junction. This is also a reason behind lower $V_{oc}$ of Schottky junction solar cells compared to p–n junction solar cells. Al–Si/Pt–Si Schottky junctions have been used to produce photovoltaic effect [87, 88]. Recently graphene has become the material of choice to produce Schottky junction with SiNWs that show relatively better performance than metal-Si Schottky junctions [90, 91].

# 5 Photovoltaics of Silicon NW Solar Cells

## *5.1 Single NW Photovoltaics*

Coaxial and axial single SiNW solar cells are emerging as potential candidate for nanoelectronic devices that require power as low as a few nanowatts.

### 5.1.1 Single NW Axial Junction Photovoltaics

Growth of axial single SiNW photovoltaic devices most commonly facilitated by VLS method. Growth of Si–Si axial junctions was achieved by changing the dopant gas precursors (e.g., $B_2H_6$ or $PH_3$) at the time of axial growth of NWs. Photovoltaic measurement on p–n junction single SiNW of length 25 μm under 1 sun AM 1.5 G illumination yields $V_{oc}$ and $I_{sc}$ values 0.12 V and 3.5 pA [92]. Improvement in photovoltaic properties was observed by moving from p–n to p–i–n junction as illustrated in Fig. 4. Introduction of i-segment = 2 μm and 4 μm in length improved $V_{oc}$ and $I_{sc}$ by 0.24 V, 14.0 pA and 0.29 V, 31.1 pA, respectively. The FF for the p–i–n NW device with $i = 4$ μm was 0.51 and yields a maximum power output of 4.6 pW per NW and conversion efficiency of 0.5%. These results showed that the introduction of i-segment reduces the leakage current and saturation current. To produce photovoltaic effect in a single SiNW without forming Si–Si p–n junction, electrical nanoscale joule heating [93] was selectively introduced beneath a single contact of Al [87]. The nature of induced junction was not clearly understood; however, the cause of rectifying behavior is partially attributed to the formation of Schottky barrier or p–n alloy or a combined effect of these two. Upon illumination this junction produced $V_{oc}$ of 0.19 V, $J_{sc}$ of 5 mA/cm$^2$, and FF of 0.4 that correspond to a conversion efficiency of 0.46%. Further improvement in photovoltaic effect was obtained by forming a Schottky junction between Pt and n-Si end of n-Si/n$^+$-Si NWs that were produced by a combination of colloidal lithography with DRIE [88]. Under solar light this Schottky junction showed a photovoltaic effect with $V_{oc}$

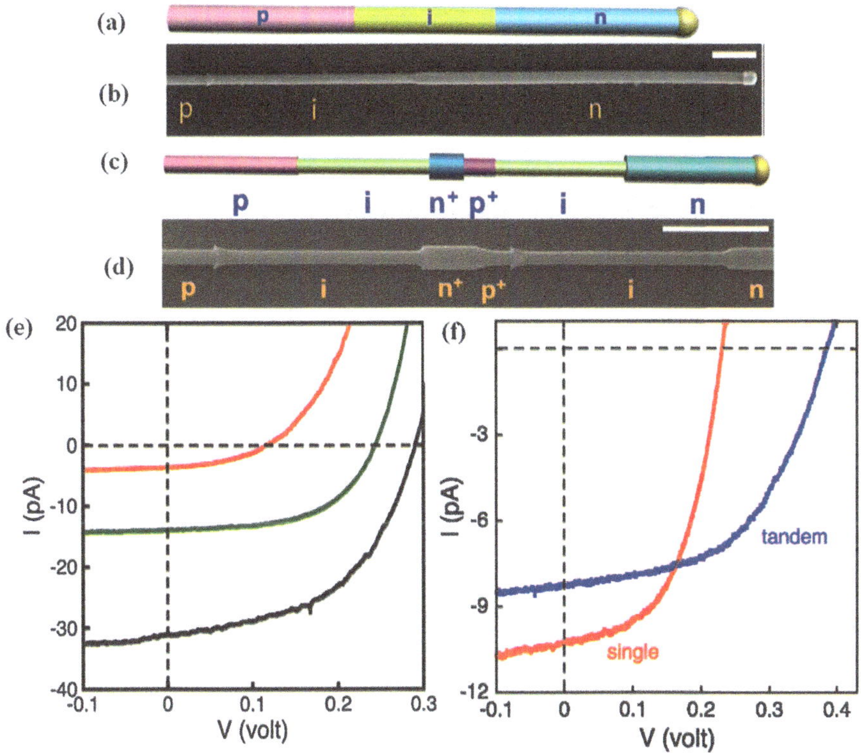


**Fig. 4** (**a**) Schematic and (**b**) SEM image of an axially modulated p–i–n SiNW. Growth is mediated with Au catalyst (gold-colored hemisphere). Scale bar is 1 μm. (**c**) Schematic and (**d**) SEM image of two p–i–n diodes integrated in series on a single NW (tandem axial SiNW). Scale bar is 1 μm. (**e**) *I*–*V* characteristics for single p–i–n SiNW with *i* region = 0 (red), 2 μm (green), and 4 μm (red), devices under AM 1.5G illumination with intensity 100 mW/cm$^2$. (**f**) *I*–*V* responses of single p–i–n SiNW photovoltaic device with *i* = 2 μm (red) and tandem SiNW (p–i–n$^+$–p$^+$–i–n) with *i* = 2 μm (blue) photovoltaic devices under AM 1.5G illumination (Reprinted with permission from Ref. [92]. Copyright right (2008) American Chemical Society)

of 0.30 V, $I_{sc}$ of 0.135 nA, and FF of 54% with an overall conversion efficiency of 1.81%.

The potential of introducing tandem cell structure on a single SiNW, by integrating two single-junction photovoltaic elements in series with a p–i–n$^+$–p$^+$–i–n axial modulation, has been also investigated (Fig. 4) [92]. Success of synthesis design was realized through SEM image of selectively etched NW by KOH due to the difference of etching rates p, n, and i segments of NW as illustrated in Fig. 4d. Photovoltaic measurements (Fig. 4f) on p–i (2 μm)–n single and p–i–n+–p+–i–n, *i* = 2 μm tandem SiNW yield $V_{oc}$ and $I_{sc}$ values of 0.23 V and 10.2 pA and 0.39 V and 8.2 pA, respectively. The tandem structure showed a 39% increase in power output compared to 2.3 pW for the single p–i–n NW cell with *i* = 2 μm. Treating the tunnel junction (heavily doped p+/n+) ideally, the $V_{oc}$ and output power of tandem cell should be doubled than that of a single cell, and the $I_{sc}$ should remain constant. However the smaller values of $V_{oc}$ compared to ideal values can be attributed to parasitic series resistances as a result of nonideal tunneling interface between the $n^+$ and $p^+$ segments.

### 5.1.2 Single NW Radial Junction Photovoltaics

For radial junction, single NW solar cell carrier separation takes place in the radial direction compared to the longer axial direction in axial junctions, with a carrier collection distance smaller or comparable to the minority carrier diffusion length [27]. Hence, photogenerated carriers can reach collecting electrodes without substantial bulk recombination. Figure 5 shows the schematic and SEM images of cross section of the coaxial p–i–n SiNW with a n-type Si shell, an intermediate i-Si layer, and a p-type core as well as the fabricated single NW device [86]. These p–i–n coaxial silicon NW solar cells under 1 sun illumination yield a power output of 50–200 pW per NW device and an apparent energy conversion efficiency of up to 3.4%. Thus these silicon NW devices individually or interconnected with each other could function as nanoscale power source for nanoelectronics in specific NW-based logic circuits as, for example, NW-based AND logic gate. Large surface areas of SiNWs enunciate huge surface recombination and become a determinantal

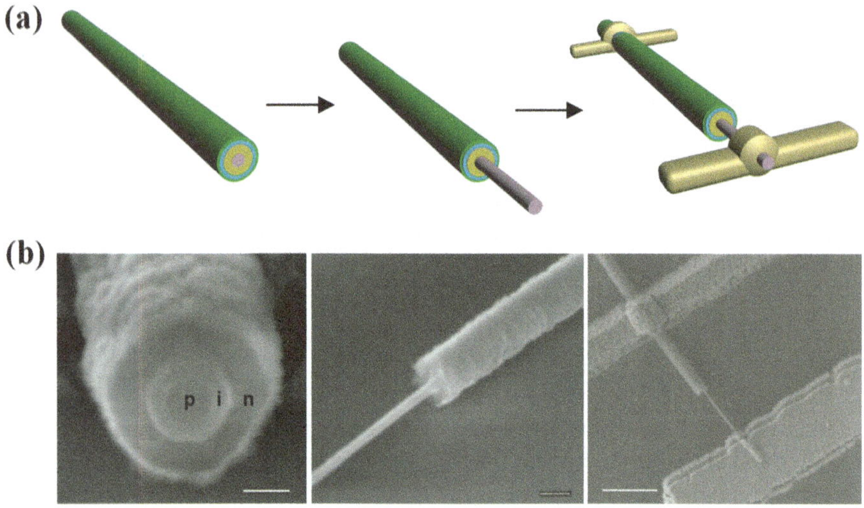


**Fig. 5** (**a**) Schematic of radial p–i–n SiNW photovoltaic device fabrication. Left, pink, yellow, cyan, and green layers correspond to the p-core, i-shell, n-shell, and PECVD-coated $SiO_2$, respectively. Middle, selective etching to expose the p-core. Right, metal contacts deposited on the p-core and n-shell. (**b**) SEM images corresponding to schematics in (**a**). Scale bars are 100 nm (left), 200 nm (middle), and 1.5 mm (right) (Reproduced with permission from Ref. [86]. Copyright 2007, Nature Publishing Group)

part of SiNW photovoltaic device performance. Appropriate surface passivation of single-wire radial p–n junction solar cells by a-Si:H and $SiN_x$:H coatings has increased the conversion efficiency up to 9% and $V_{oc}$ ~ 600 mV with a FF over 80%, the highest apparent efficiency to date for single NW device (Fig. 6) [94]. Si surfaces, coated with a-Si:H and $SiN_x$:H, have additional benefits for photovoltaic applications. a-Si:H forms a heterojunction to crystalline Si, enabling extremely high open-circuit voltages (up to 743 mV) for wafer-based Si solar cells [95], while a-$SiN_x$:H also functions as a versatile anti-reflective coating for crystalline Si [96]. Radial p–n junction was formed selectively by phosphorus diffusion at the upper part of NWs, and contacts made at the two ends of the NWs as shown in Fig. 6a, b. This device has both radial and axial parts. Since in the bottom segment of each wire no radial p–n junction was formed, in this portion minority carriers

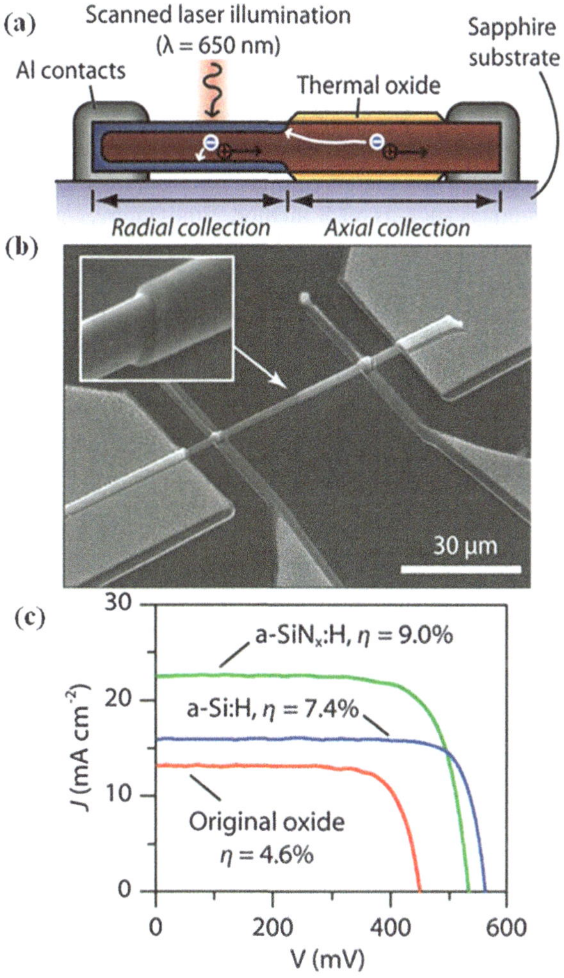


**Fig. 6** Single Si microwire photovoltaic device. (**a**) Schematic diagram and (**b**) SEM image. (**c**) Photovoltaic short-circuit current density and voltage characteristics of the device with different surface passivation under AM 1.5G, 100 mWcm$^{-2}$ illumination (Reproduced with permission from Ref. [94]. Copyright 2011, Royal Society of Chemistry)

must diffuse axially to reach the junction in order to be collected (Fig. 6a). Single NW device without any surface passivation showed conversion efficiency 4.6%, while in the presence of a-Si:H layer, this performance improved up to a conversion efficiency of 7.4%. Best conversion efficiency (9%) was obtained for NWs coated with a-$SiN_x$:H. Surface passivation by a-$SiN_x$:H and a-Si:H results in low surface recombination velocities and minority carrier diffusion lengths (<<70 cm/sec, 450–600 cm/sec for a-$SiN_x$:H, and > > 30 μm, 5–10 μm for a-Si:H). However, for unpassivated surface, recombination velocity and minority carrier diffusion length become >4 × $10^5$ cm/sec and <0.5 μm, respectively. Besides surface passivation, 40% enhancement in $J_{sc}$ and efficiency has been predicted in a-Si:H single-nanowire solar cells by engineering the cross section of the nanowire from circular into a front-opening crescent shape [97].

## *5.2 Silicon NW Array Photovoltaics*

Fabrication of SiNW array p–n and p–i–n junction solar cells has established the path for commercial production of low-cost Si solar cells in the near future. SiNW arrays used for these purposes have been fabricated using different fabrication techniques such as VLS method, lithography, ion etching, chemical etching techniques, etc. Among these, VLS method and chemical etching have been shown their potential for future large-scale SiNW array solar cell fabrication capability.

For the fabrication of homojunction (p–n), dopant diffusion technique has been used in large scale either adopting spin-on dopant or gas phase or chemical dopant diffusion process at high temperatures [98, 99]. Axial junction SiNW array solar cells are produced by etching pre-diffused planar p–n junction wafers [78, 100–102]. Efforts for the formation of radial p–n junction by dopant diffusion process most of the time result in completely doped opposite polarity NWs and junction formed somewhere inside the bulk material, especially for the SiNW arrays fabricated by chemical etching techniques as discussed before in Sect. 4 [19, 40, 80, 103]. Conversion efficiency of these kinds of solar cells, having vertically aligned selectively etched SiNW arrays of length ~ 4 μm and diameter in the range of 40–200 nm fabricated by MCEE technique on a p-type Si substrate, has reached 13.7% [104] where SiNW arrays converted n-type by phosphorus diffusion at 900 °C. Formation of radial junction needs dopant diffusion should be limited to the surface, so that polarity of inner core remains the same [7, 81, 105, 106]. Selective controlled dopant diffusion over only SiMW arrays (43–49 μm height, 2–3 μm in diameter with 7 μm pitch) grown by VLS method in the presence of $Al_2O_3$ scatterer and a-$SiN_x$:H surface passivation coating exhibited $V_{oc}$ ~ 500 mV, $J_{sc}$ ~ 24 mA/cm$^2$, and FF >65% with an overall conversion efficiency of 7.9% [107]. Contribution from the substrate holding the MW arrays in solar energy conversion was excluded as the scanning photocurrent microscopy measurements demonstrate an effective minority carrier diffusion length < 0.5 μm for electrons in the base region of the MWs.

To precisely control the junction depth for the formation of radial junction, the most elegant way is to deposit an oppositely charged Si layer (i.e., shell layer) around the NWs by CVD method [81, 84, 105, 108–116]. Optimizations of growth of polycrystalline p-Si shell layer over the SiNW arrays of length 7.8 μm fabricated by metal-catalyzed electroless etching of n-Si (111) wafer (resistivity of 1–50 Ω-cm) produce a conversion efficiency around 3.3% [26]. These p-Si shell layers were grown by cold-wall thermal CVD using $SiH_4$ and $B_2H_6$ as precursor and dopant gases, respectively. Cell performances are found to be strongly dependent on dopant gas fluxes and growth time. This low conversion efficiency instead of efficient light trapping by long NW arrays and short transportation path of photogenerated charge carriers was attributed to the high surface recombination enunciated by huge dangling bond-related surface defects present at the high surface area of the NWs [117]. Later critical NW length, necessary to obtain optimal cell performance with minimum carrier loss in radial p–n junction solar cell, was estimated to be ~ 700 nm [20]. In comparison with the planar cell, $V_{oc}$, $J_{sc}$, and FF of the corresponding solar cell were found to be increased from 0.38 V to 0.46 V, 14 mA/cm$^2$ to 24.4 mA/cm$^2$, and 0.34 to 0.63, respectively, with an overall increase in efficiency from 1.76% to 7% under the illumination of 100 mW/cm$^2$ AM1.5G solar simulator. Further passivation of the large number of surface defect states through oxidation by low-temperature UV ozone treatments offers the potential to enhance conversion efficiency from 7% to 8.4%. ESR measurements were used at room temperature to detect these defects and investigate the ozone annealing effect on defect passivation. In a further modification, ~1.5% enhancement of absolute efficiency was obtained by using two-step $H_2$ annealing process [118]. With the minimization of surface recombination, it is also necessary to minimize bulk recombination by synthesizing SiNW arrays on the thin Si wafers. As a low-cost replacement for thick wafer-based solar cells, thin wafer-based SiNW array solar cells have been investigated due to the low cost of raw materials and high efficiency resulting from reduced bulk recombination. With decreasing wafer thickness, all cell parameters were improved and that reflects reduction in bulk recombination loss [119]. In addition, the introduction of back surface field at the rear surface of these thinner solar cells further enhances the conversion efficiency by 36–45%. Under 100 mW/cm$^2$ illumination by an AM 1.5G solar simulator, these radial p–n junction cells with 700-nm-long SiNW array fabricated on 170-μm-thick Si (100) wafer (resistivity 27 Ω-cm) produce an average power conversion efficiency of 10.3% ($V_{oc}$, $J_{sc}$, and FF of 0.46 V, 32.1 mA/cm$^2$, and 0.7, respectively) and a maximum power conversion efficiency of 10.9% [119]. With these modifications front electrode prepared as buried contact fashion by self-aligned single-step lithography further could improve efficiency by ~ 7.82% by enhancing the collection efficiency of photogenerated charge carriers [120].

Surface recombination in SiNWs has also been minimized by inserting an intrinsic layer at the interface. Efficient radial a-Si/SiNW solar cells were fabricated by depositing hydrogenated intrinsic and p-type a-Si shell by PECVD on n-type SiNW arrays fabricated by MCEE technique [84]. A thin layer of Al-doped ZnO was coated on the top of a-Si layer by atomic layer deposition to define front electrode. Due to the low-temperature PECVD processing, such heterojunctions have many advantages such as better interface passivation by hydrogen and higher $V_{oc}$ compared to p–n homojunctions [85]. Under AM 1.5 illumination, this structure showed a conversion efficiency of 7.29% with $V_{oc}$, $J_{sc}$, and FF of 0.476 V, 27 mA/cm$^2$, and 0.562, respectively. Epitaxial growth of radial p–i–n junction on [103] oriented NWs of diameter ~ 300 nm and length up to ~ 10 μm by CVD exhibited single crystalline {110} faceted shell surfaces and raised the conversion efficiency up to 10% under AM 1.5 G illumination [109]. Vertical SiNW arrays acting as cores of radial junctions were fabricated using a combination of DRIE and inductively coupled plasma etching with passivation by Bosch etch processes to prepare epitaxy-ready radial shell growth. Introduction of very thin intrinsic poly-Si layer (50 nm) at the interface of radial p–n junction SiMW array (diameter of 3 ± 0.3 μm, spacing of 7 ± 0.2 μm, and length of 9 ± 1 μm) solar cells enhances the conversion efficiency from 7.2 to 9.3% (Fig. 7) [112]. SiMW arrays were fabricated on Si(100) wafer

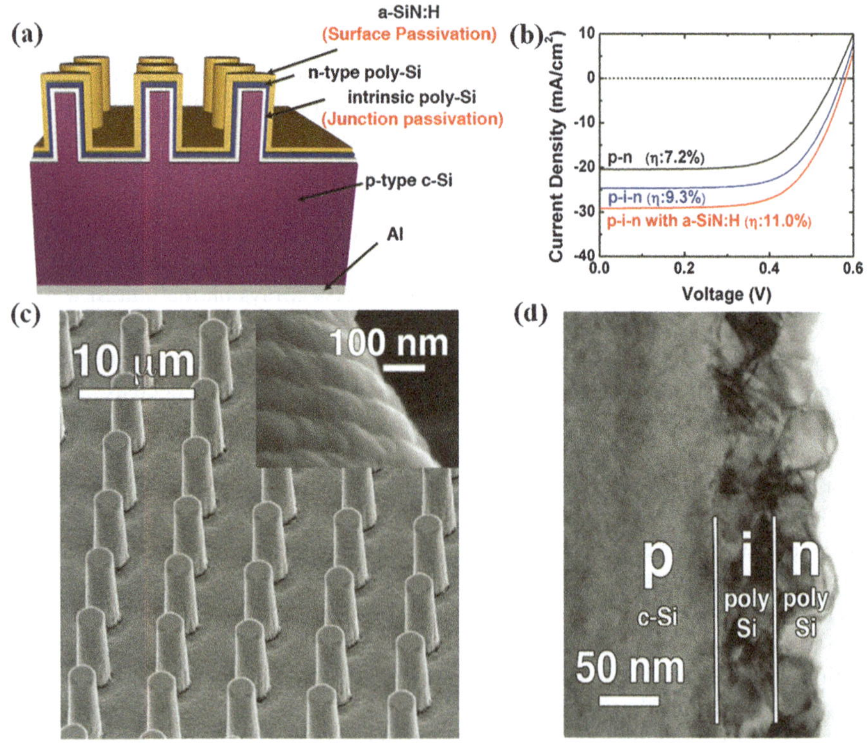


**Fig. 7** (**a**) Schematic of the radial p–i–n Si MW solar cells with a-SiN:H layer as an antireflection and surface passivation layer, (**b**) photovoltaic *J–V* characteristics of radial p–n and radial p–i–n Si MW solar cell with and without surface passivation. (**c**) SEM image of the Si microwire solar cells. Inset shows the uniformity of coating of the poly-Si layer and the a-SiN:H layer on the side walls. (**d**) TEM image of the p–i–n core-shell nanostructure (Reprinted with permission from Ref. [112]. Copyright (2011) American Chemical Society)

(0.1–0.9 Ω-cm) by DRIE, and a 50-nm-thick intrinsic poly-Si shell and a 40-nm-thick n-type poly-Si shell were deposited on the surface of the p-core SiMWs by CVD as shown in Fig. 7a, c, and d. Improvement in diode ideality factor from 2.45 ± 0.11 to 1.89 ± 0.14 and reduction in saturation current densities by more than one order were observed after insertion of intrinsic poly-Si layer between p-type SiMW core and n-type ploy-Si shell. These improvements clearly demonstrate the reduction in junction recombination by the passivation effect of the intrinsic layer. Further deposition of 80-nm-thick a-SiN:H layer by PECVD as a surface passivation and antireflection layer increases the mean $J_{sc}$ and the mean conversion efficiency by 25% and 19%, respectively. In situ growth of radial junction solar cells based on SiNWs arrays grown by VLS method was also achieved where growth of SiNWs and subsequent shell layers were done in the same CVD chamber by changing dopant gases and growth temperatures, showing PV responses [8, 114, 121, 122]. However the use of metal (e.g., Au) catalyst restricts the performance of these cells as these metal catalysts introduce trapped levels within the bandgap and huge carrier recombination occurred in these extra energy levels. Instead of this restriction growth by CVD technique has advantage to do deposition process at very low temperatures, even close to ambient, which allows temperature-sensitive substrates to be used [121, 123–125].

## 5.3 Silicon NW Hybrid Solar Cells

Hybrid cells based on SiNWs and organic semiconductors such as poly (3-hexylthiophene) P3HT [126–128], PEDOT:PSS [129–137], and others [138–141] have emerged as a potential cost-competitive photovoltaic structure by adopting low-temperature, scalable, and soluble processes of conjugated polymers to form heterojunction with silicon. The potential of this architecture is to combine the broadband optical absorption capability of the Si nanostructures with solution-based processes for simple and low-cost production that could make silicon-based solar cells a very viable option for next-generation photovoltaics. Core-shell geometry of these hybrid solar cells ensures efficient charge separation since the thickness of the polymer shells is comparable to the exciton diffusion length. Most of the conversion efficiency of the state-of-the-art hybrid solar cells is in the range of 10–13.6%. Different strategies have been used to achieve high efficiency: modifying the nanostructures [136, 142–145], modifying the interface between Si and polymers [146, 147], engineering the back contact [148, 149], incorporation of metal nanoparticles [150, 151], surface passivation [152], and optimization of front electrode [12, 83]. The hybrid solar cells with the short NW length of 200–500 nm have shown the best performance in terms of the highest power conversion efficiency and the largest $J_{sc}$ (Fig. 8a, b). Because long SiNW infiltration of polymers is difficult, so the coverage of polymers on the SiNWs is not complete, and also with the increase of NW length, the surface state-related defects increase which increases recombination. For SiNWs/PEDOT:PSS hybrid solar cells, employing

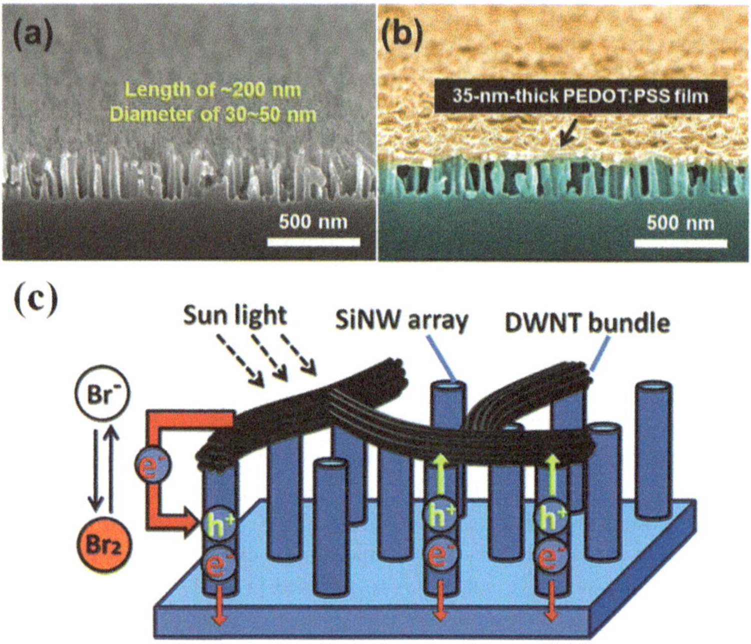


**Fig. 8** (**a**) The tilted cross-sectional SEM images of the 200-nm-long SiNWs fabricated by MCEE technique, (**b**) the corresponding morphology after coating with the PEDOT:PSS solution (Reproduced with permission from Ref. [83]. Copyright, 2006, Nature Publishing Group). (**c**) Schematic of hybrid solar cell composed of SiNW arrays, the DWNT film, and the redox electrolyte (Reproduced with permission from Ref. [164]. Copyright, (2009) American Chemical Society)

SiNW arrays fabricated by MCEE technique, maximum conversion efficiency 8.4–11.55% was obtained for NW lengths 200–570 nm [12, 146, 153]. These performance values were improved to 13.01% by using an intermediate 1,1-bis [(di-4-tolylamino)phenyl] cyclohexane (TAPC) layer to mitigate interface recombination [146]. A two-step surface treatment of SiNWs before polymer coating (a low-temperature ozone treatment followed by a HF etching process) also found to enhance the efficiency by 1.24% [137]. The performance of these SiNWs/PEDOT:PSS hybrid solar cells has been further improved by using a highly conductive, transparent, and stable gold (Au) mesh as front electrode, by a metal-transfer imprinting method [83]. The hybrid cells have shown a conversion efficiency of 13.2% with a $J_{sc}$ of 36.03 mA/cm$^2$, $V_{oc}$ of 539.2 mV, and FF of 67.8%. In a very recent work, power conversion efficiency up to 16.1% ($J_{sc}$ of 34 mA/cm$^2$, $V_{oc}$ of 607 mV, and FF of 78.3%) has been demonstrated by employing embedded metal front electrodes other than conventional electrode [154]. On the other hand, immersion of SiNW arrays in anisotropic tetramethylammonium hydroxide (TMAH) solution just for few minutes before PEDOT:PSS coating enhances the conversion efficiency by 69.5% [155]. In another modification step with SiNW/PEDOT:PSS, hybrid structure graphene oxide (GO) has further used to enhance light trapping [156]. These hybrid GO-blended cells showed large junction area compared to pure PEDOT:PSS structures. Again addition of graphene quantum dots (GQDs) in PEDOT:PSS has increased $J_{sc}$ and FF of hybrid solar cells from 32.11 to 36.26 mA/cm$^2$ and 62.85% to 63.87%, respectively [157]. These enhancements were attributed to the photon down-conversion process and the improved conductivity of PEDOT:PSS by appropriate addition of GQDs. Further, addition of Au nanoparticles (NPs) with PEDOT:PSS showed efficiency improvement by ~23% [151]. This improvement was attributed to electrical effects caused by Au NPs besides the traditional plasmonic effects. After addition of Au NPs, conductivity of the PEDOT:PSS layer was found to be slightly improved, and the built-in electric field of the device was also improved. Both of these suppressed recombination and enhanced charge separation and collection.

Fabrication of flexible SiNW solar cells was achieved through peeling off and transferring of SiNWs on low-cost and flexible substrates such as plastic sheets, metal foils, and glass slides [158–160]. Figure 9 shows such a SiNW array transfer and printing process where SiNW arrays formed by Ag-assisted chemical etching. SiNWs were peeled off from the substrate through the formation of horizontal crack on SiNWs by a two-step etching process and deposition of poly-(dimethylsiloxane) (PDMS) for providing mechanical support to freestanding NWs.

Recently, carbon/silicon (C/Si) heterojunction solar cells have attracted a great deal of interest. Among the carbon family, the discovery of carbon nanotubes (CNTs) and graphene has started revolution in the field of C/Si heterojunction solar cells [161–163]. Figure 8c illustrates hybrid solar cell formed by depositing a thin film of double-walled CNTs on SiNW array that showed a power conversion efficiency of 1.29% under AM 1.5G irradiation [164]. The CNT thin film also functioned as the transparent counter electrode. The cell performance could be tuned by varying the SiNW density. Core-shell heterojunction formed by coating carbon quantum dots on SiNW arrays, formed by MCEE technique, showed an excellent rectifying behavior with a rectification ratio of 103 at ±0.8 V in the dark and power conversion efficiency as high as 9.1% under AM 1.5G illumination [165]. The cause of this high efficiency was attributed to the optimized carrier transfer and collection capability as well as the improved optical absorption.

Emerging organic–inorganic lead halide perovskites could be promising as a new material for making hybrid solar cells with SiNWs. Theoretical study shows that solar cell that consists of a Si p–i–n nanowire array filled with $CH_3NH_3PbI_3$ could have an efficiency of 13.3% at a thin absorber thickness of 1.6 μm [166]. In this structure both the nanowires and perovskite absorb the incident light. As a result, the hybrid structure shows high absorption in a broad wavelength range of 300–800 nm. Here nanowires act as the channels for transporting photogenerated charge carriers.

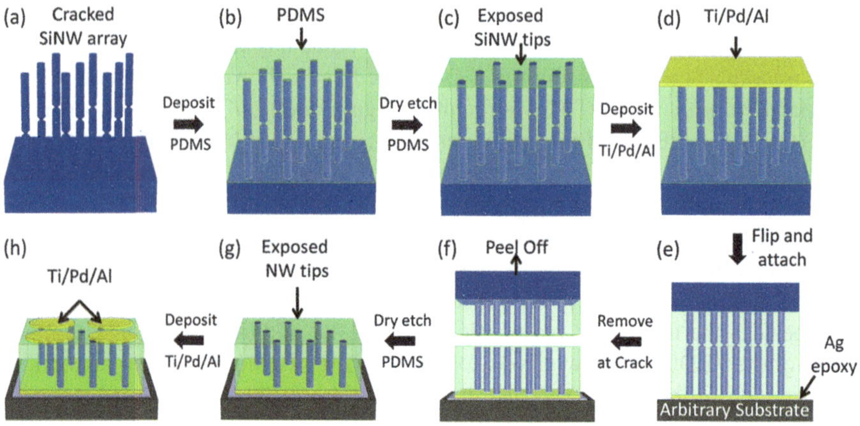


**Fig. 9** Schematic of the fabrication procedure for vertical SiNW array electronic devices on non-Si-based substrates. (**a**) Cracked vertical SiNW arrays are formed by inserting a water-soaking step between two consecutive Ag-assisted chemical etching steps. (**b**) SiNWs are filled with hexane-diluted PDMS for mechanical support and electrical insulation. (**c**) SiNW tips are exposed after dry etching of PDMS. (**d**) Metal contact is deposited over the exposed SiNW tips. (**e**) The metal contact side of the Si wafer is attached to an arbitrary substrate with a thin layer of silver epoxy in between. (**f**) The vertical SiNW array is separated from the donor Si wafer at the crack location by peeling off. (**g**) The other side of SiNW tips is exposed after dry etching of PDMS. (**h**) The other metal contact is deposited on the exposed SiNW tips over a shadow mask with 300 μm diameter holes to complete the vertical SiNW array device (Reprinted with permission from Ref. [158]. Copyright (2012) American Chemical Society)

## 5.4 *Non-radiative Energy Transfer-Based Silicon NW Solar Cells*

Recently, non-radiative energy transfer (NRET) has been proposed for hybrid nanostructures that combine absorbing components (e.g., quantum dots) with high-mobility semiconducting transport channels (e.g., Si). Use of colloidal semiconductor nanocrystals such as CdS, CdSe, and PbS quantum dots or layer of organic molecules in these hybrids provides the advantage of absorbing a wider range of the solar spectrum than Si layers are able to and has opened new possibilities in light harvesting exploiting the NRET process [167–176]. Use of NRET can overcome the limitations of low-charge transfer efficiency in charge transfer-based devices. Long-range dipole–dipole interaction involved in NRET process of these hybrids can exceed both radiative energy transfer and the short-range charge transfer process involved in charge transfer-based devices. Time-resolved photocurrent measurements showed that for excitation at the absorption peak of PbS QDs, the NRET efficiency to the adjacent SiNWs lies between 15 and 38%, and the photocurrent could be as large as three times than the photocurrent by direct light absorption in the SiNWs [171], whereas for a monolayer of colloidal

CdSe nanocrystals grafted on hydrogenated Si surface, NRET efficiency was estimated to be ~ 65% [168]. Efficient exciton harvesting in underlying radial p–n junction SiNW arrays has been achieved by energy transfer from nanocrystalline-Si (nc-Si) quantum dots with excellent carrier separation and propagation [119]. $J_{sc}$ of this nc-Si QD-modified SiNW solar cell increased by 18.5% with an overall maximum conversion efficiency improvement up to 12.9%. Analysis of time-dependent photoluminescence decay experiments revealed that NRET rate was ~ 1.6 times faster than the radiative energy transfer rate from the nc-Si QDs to the underlying Si layer. Similarly frequency down-conversion properties of inorganic QDs were utilized to enhance conversion efficiency of SiNW solar cells. Efficiency enhancement of 13% and 33% was observed from CdSe and CdS nanocrystals, respectively, integrated with radial p–n junction Si nanopillar solar cells with respect to reference cells without nanocrystals [177, 178].

## 5.5 *Silicon NW Schottky Junction*

As a low-cost alternative to conventional p–n junction Si solar cells, metal-Si Schottky junction solar cells were introduced. A metal with high work function was used to form Schottky barrier with n-Si (see Sects. 4 and 5.1.1). Formation of small barrier height cannot stop properly recombination of electrons at anode that results in large saturation current and low $V_{oc}$ typically ~ 0.3 V or less [87, 88, 90]. These drawbacks of metal-Si Schottky barrier can be removed by introducing a higher bandgap (> 1.1 eV) organic semiconductor with high mobility at the interface [179]. The choice of organic layer should be such that it can produce high offset between conduction band minimum of Si and LUMO of organic layer while maintaining low offset value between valence band maximum of Si and HOMO of organic layer. Advantage of radial Schottky junction has investigated by depositing 10 nm Pt film on a-SiNW arrays fabricated by RIE [180]. This core-shell Schottky barrier showed a 150% larger $J_{sc}$ compared to planar Schottky barrier. However $V_{oc}$ and FF of this core-shell structure became low due to the high surface area and the physical damage introduced on SiNW walls during RIE process. Recently, instead of metals graphene has been used to fabricate Schottky junction with SiNWs [181, 182]. Schottky barrier formed by a combination of graphene monolayer and SiNW arrays with an optimized length of NWs showed photovoltaic effect with conversion efficiency up to 2.15% by filling the interspace of SiNW array with graphene suspension [181]. These photovoltaic properties were further improved by chemical doping, and the Schottky junction between graphene/SiNWs arrays showed the energy conversion efficiencies of up to 2.86% under AM1.5 G illumination [182]. Control of the number of graphene layers and the doping level with appropriate surface passivation further improved the performance of these photovoltaic devices. By inserting electron-blocking layer such as thin layer of P3HT polymer between Si nanoarrays and graphene films, maximum power conversion efficiency as high as 8.71% has been obtained [91].

### 5.6 *Photoelectrochemical Solar Cells with Silicon NW Arrays as Photoanodes*

In SiNW array-based photoelectrochemical cells (PEC), NW array serves as photoelectrode immersed in liquid electrolyte with metal electrode as counter electrode. The PEC solar cells based on solely SiNW arrays showed inferior performance with conversion efficiency <5% and $J_{sc}$ value limited within 1–10 mA/cm$^2$ [183–186]. The presence of $SiN_x$ antireflective coating and scattering by transparent $Al_2O_3$ particles resulted in a peak external quantum efficiency of 89% and a broadband, near-unity internal quantum efficiency [187]. Investigation on the dependence of photovoltaic characteristics on the morphology of SiNWs showed that the best performance was obtained for NWs with length of 18 μm and diameter of 150 nm [188]. For thin NWs (diameter < 80 nm), un-depleted core region was so narrow that it could not transport charge carriers efficiently, and for NWs having length more than optimal value, surface states start to predominate the advantages of long SiNW arrays. This photovoltaic effect also found to depend on conductivity of SiNWs. Resistive NWs resulted in low photocurrent, while conductive NWs higher than an optimal value of doping level resulted in reduced space charge region and minority carrier diffusion length. Photovoltaic properties of these cells were also found to be dependent on pH of electrolyte solution. $V_{oc}$ was found to be decreased by ~ 44 mV per unit decrease in pH value [189]. This variation could be kept steady in the pH range of 3–9 by the introduction of a highly doped emitter on top of SiNWs. Metal nanoparticle (MNP) decorated SiNW arrays showed the enhancement in photoconversion efficiency of PEC cells [190]. Among the other MNPs, Pt NP leads to the highest performance. Optimization of Pt NP deposition leads to a conversion efficiency of 8.14% [191], and for carbon-passivated SiNWs, this conversion efficiency further enhanced to 10.86% [192]. In replacement of photo-corrosive and photo-oxidant conventional aqueous electrolyte $HBr/Br_2$, the use of ion liquid due to its superior thermal and chemical stability, high conductivity, non-corrosivity, and negligible vapor pressure resulted in conversion efficiency of 6% for Pt NP-decorated SiNWs prepared by a two-step chlorination/methylation procedure [193].

## 6 Future Outlook and Challenges

To make solar energy as a mainstream source of electricity, solar cells must become less expensive. To realize high-efficiency current state-of-the-art technology for the fabrication of planar Si cells imposes restrictions on the use of quality and quantity of the material. SiNWs in radial junction morphology have shown potential for cost-effective and high-performance solar cell production. SiNW-based solar cells open the possibility by relaxing the requirements impose by commercial solar cell fabrication. Advantages of using small amount of low-cost material and efficient light trapping could make SiNW-based solar cells with efficiency close to that of current planar cells. Despite the promising potential and rapid progress in last 10 years, many challenges remain to be addressed before the advantages of SiNW-based solar cells can be realized commercially. In particular these challenges include enhancement in surface and interface recombination with increase in surface area and roughness, increase of front contact resistance, difficulties in growth of SiNW arrays with uniform diameter, and pitch in wafer scale by low-cost methods especially on low-cost substrates. Surface or interface passivation by using conventional a-Si:H and a-$SiN_x$:H layer or inserting i-layer at interface has shown prospects in enhancing device performance by passivating surface and interface recombination centers [84, 94, 107, 194–196]. Alternatively surface and interface passivation of SiNW solar cells have been investigated by low-temperature ozone oxidation under UV radiation [20]. These passivation results urge for the development of more effective passivation methods in order to realize the benefits of huge surface and interface area of SiNWs over the planar morphology. For SiNW-based solar cells, another major challenge is the formation of uniform and continuous front contact. Due to the NW morphology, deposition of front contact becomes discontinuous, and as a result contact resistance increases. Some approaches have been taken to redeem this problem like specific etching of SiNWs on Si substrates so that the planar portion can be used to deposit continuous front contact or formation of polycrystalline p-Si matrix on n-SiNW arrays to provide continuous upper surface [20, 104]. In another approach mounting wax was used to fill the gaps between SiNW arrays, and then indium tin oxide was sputtered on the top to form front contact [107]. Slantingly aligned SiNW arrays also showed ability to form better front contacts [79]. Ability to transfer SiNW arrays by peeling off the NWs from the Si substrates or in situ growth of SiNWs on low-cost substrates lights the path to produce flexible solar cells using a small fraction of Si material needed for commercial solar cell production, thus removing a major cost hurdle of Si solar cell production [121, 158, 159]. For in situ growth of SiNWs and SiNW solar cells by VLS method, the presence of metal contamination at the top and side walls of NWs and at the interface needs major attention. Some work has been done in search of less inactive catalysts in this regard. In a recent work, conversion efficiency improvement from 2.11% to 6.18% was obtained by treating PECVD grown SiNWs by an optimal concentration of hydrochloric acid to reduce the indium contamination, used as a catalyst to grow SiNWs [197]. In addition to these, further advancement in light trapping in SiNW arrays has been recently proposed by fabricating decagonal and elliptical SiNWs arrays [198, 199]. SiNW and nanopillar arrays when used as an antireflection layer for conventional solar cells enhanced the efficiency up to 22% without using conventional antireflection layers [200]. To accomplish this, a thin $Al_2O_3$ layer was deposited on the nanostructure surface to achieve excellent chemical and electrical passivation. This potential of high rate of cost reduction and increment of efficiency for SiNW solar cells justify further future research and development.

## References


1. 2014 Key World Energy Statistics (IEA, 2014, p 6). http://www.iea.org/publications/freepublications/ publication/keyworld 2014.pdf. Accessed 1 June 2015
2. Renewables 2014 Global Status Report (REN21, 2014, p. 21). http://www.ren21.net/Portals/0/documents/Resources/GSR/2014/GSR2014_KeyFindings_low%20res.pdf. Accessed 1 June 2015
3. M. Green, Prog. Photovolt. **17**, 183 (2009)
4. E.A. Alsema, Prog. Photovolt. **25**, 17 (2000)
5. T. Song, S.-T. Lee, B. Sunn, Nano Energy **1**, 654 (2012)
6. S. Xu, S.Y. Huang, I. Levchenko, H.P. Zhou, D.Y. Wei, S.Q. Xiao, L.X. Xu, W.S. Yan, K. Ostrikov, Adv. Energy Mater. **1**, 373 (2011)
7. H.P. Yoon, Y.A. Yuwen, C.E. Kendrick, G.D. Barber, N.J. Podraza, J.M. Redwing, T.E. Mallouk, C.R. Wronski, T.S. Mayer, Appl. Phys. Lett. **96**, 213503 (2010)
8. O. Gunawan, S. Guha, Sol. Energy Mater. Sol. Cells **93**, 1388 (2009)
9. R. Guo, H. Huang, P. Chang, L. Lu, X. Chen, X. Yang, Z. Fan, B. Zhu, D. Li, Nano Energy **8**, 141 (2014)
10. X. Pi, L. Zhang, D. Yang, J. Phys. Chem. C **116**, 21240 (2012)
11. Y. Li, Q. Chen, D. He, J. Li, Nano Energy **7**, 10 (2014)
12. K. Sato, M. Dutta, N. Fukata, Nanoscale **6**, 6092 (2014)
13. M.A. Green, *Solar Cells: Operating Principles, Technology and System Applications* (Prentice-Hall, Englewood Cliffs, 1982), 274 pp
14. E. Hecht, *Optics* (Pearson, San Francisco, 2002), p. 698
15. E. Yablonovitch, G. Cody, IEEE Trans. Electron Devices **29**(2), 292 (1982)
16. E. Garnett, P. Yang, Nano Lett. **10**, 1082 (2010)
17. V. Sivakov, G. Andr, A. Gawlik, A. Berger, J. Plentz, F. Falk, S.H. Christiansen, Nano Lett. **9**, 1549 (2009)
18. L. Tsakalakos, J. Balch, J. Fronheiser, B.A. Korevaar, O. Sulima, J. Rand, A.D. Kumar, U. Rapol, J. Nanophotonics **1**, 013552 (2007)
19. K.Q. Peng, Y. Xu, Y. Wu, Y.J. Yan, S.T. Lee, J. Zhu, Small **1**, 1062 (2005)
20. M. Dutta, N. Fukata, Nano Energy **11**, 219 (2015)
21. M.D. Kelzenberg, S.W. Boettcher, J.A. Petykiewicz, D.B. Turner-Evans, M.C. Putnam, E.L. Warren, J.M. Spurgeon, R.M. Briggs, N.S. Lewis, H.A. Atwater, Nat. Mater. **9**, 239 (2010)
22. W.F. Liu, J.I. Oh, W.Z. Shen, IEEE Electron Device Lett. **32**, 45 (2011)
23. X. Li, J. Li, T. Chen, B. Tay, J. Wang, H. Yu, Nanoscale Res. Lett. **5**, 1721 (2010)
24. Z. Wu, J.B. Neaton, J.C. Grossman, Nano Lett. **9**, 2418 (2009)
25. Z. Wu, J.B. Neaton, J.C. Grossman, Phys. Rev. Lett. **100**, 246804 (2008)
26. M. Dutta, N. Fukata, J. Nanosci. Nanotechnol. **15**, 4339 (2015)
27. B.M. Kayes, H.A. Atwater, N.S. Lewis, J. Appl. Phys. **97**, 114302 (2005)
28. R.S. Wagner, W.C. Ellis, Appl. Phys. Lett. **4**, 89 (1964)
29. Y. Cui, L.J. Lauhon, M.S. Gudiksen, J.F. Wang, C.M. Lieber, Appl. Phys. Lett. **78**, 2214 (2001)
30. V.A. Nebol'sin, A.A. Shchetinin, A.A. Dolgachev, V.V. Korneeva, Inorg. Mater. **41**, 1256 (2005)
31. Y.W. Wang, V. Schmidt, S. Senz, U. Gosele, Nat. Nanotechnol. **1**, 186 (2006)
32. S. Kodambaka, J. Tersoff, M.C. Reuter, F.M. Ross, Phys. Rev. Lett. **96**, 096105 (2006)
33. H. Schmid, M.T. Bjork, J. Knoch, H. Riel, W. Riess, J. Appl. Phys. **103**, 024304 (2008)
34. C. Mouchet, L. Latu-Romain, C. Cayron, E. Rouviere, C. Celle, J.P. Simonato, Nanotechnology **19**, 335603 (2008)
35. B.A. Wacaser, M.C. Reuter, M.M. Khayyat, C.Y. Wen, R. Haight, S. Guha, F.M. Ross, Nano Lett. **9**, 3296 (2009)

36. M.C. Putnam, D.B. Turner-Evans, M.D. Kelzenberg, S.W. Boettcher, N.S. Lewis, H.A. Atwater, Appl. Phys. Lett. **95**, 163116 (2009)
37. V.T. Renard, M. Jublot, P. Gergaud, P. Cherns, D. Rouchon, A. Chabli, V. Jousseaume, Nat. Nanotechnol. **4**, 654 (2009)
38. Y. Wu, Y. Cui, L. Huynh, C.J. Barrelet, D.C. Bell, C.M. Lieber, Nano Lett. **4**, 433 (2004)
39. M. Dutta, L. Thirugnanam, K. Sato, N. Fukata, Mater. Express **3**, 85 (2013)
40. B.M. Kayes, M.A. Filler, M.C. Putnam, M.D. Kelzenberg, N.S. Lewis, H.A. Atwater, Appl. Phys. Lett. **91**, 103110/1–3 (2007)
41. J.B. Hannon, S. Kodambaka, F.M. Ross, R.M. Tromp, Nature London **440**, 69 (2006)
42. G.S. Doerk, N. Ferralis, C. Carraro, R. Maboudian, J. Mater. Chem. **18**, 5376 (2008)
43. V. Schmidt, J.V. Witteman, S. Senz, U. Gosele, Adv. Mater. **21**, 2681 (2009)
44. P.D. Kanungo, N. Zakharov, J. Bauer, O. Breitenstein, P. Werner, U. Goesele, Appl. Phys. Lett. **92**, 263107 (2008)
45. L. Schubert, P. Werner, N.D. Zakharov, G. Gerth, F.M. Kolb, L. Long, U. Gösele, T.Y. Tan, Appl. Phys. Lett. **84**, 4968 (2004)
46. E.S. Greiner, J.A. Gutowski, W.C. Ellis, J. Appl. Phys. **32**, 2489 (1961)
47. Y.F. Zhang, Y.F. Tang, N. Wang, C.S. Lee, I. Bello, S.T. Lee, Appl. Phys. Lett. **72**, 1835 (1998)
48. Y.-H. Yang, S.-J. Wu, H.S. Chiu, P.-I. Lin, Y.-T. Chen, J. Phys. Chem. B **108**, 846 (2004)
49. Y.F. Zhang, Y.H. Tang, H.Y. Peng, N. Wang, C.S. Lee, I. Bello, S.T. Lee, Appl. Phys. Lett. **75**, 1842 (1999)
50. F.M. Kolb, H. Hofmeister, R. Scholz, M. Zacharias, U. Go¨sele, D.D. Ma, S.T. Lee, J. Electrochem. Soc. **151**, G472 (2004)
51. N. Wang, Y.H. Tang, Y.F. Zhang, D.P. Yu, C.S. Lee, I. Bello, S.T. Lee, Chem. Phys. Lett. **283**, 368 (1998)
52. N. Wang, Y.H. Tang, Y.F. Zhang, C.S. Lee, I. Bello, S.T. Lee, Chem. Phys. Lett. **299**, 237 (1999)
53. S.T. Lee, N. Wang, Y.F. Zhang, Y.H. Tang, Mater. Res. Bull. **24**, 36 (1999)
54. W.S. Shi, H.Y. Peng, Y.F. Zheng, N. Wang, N.G. Shang, Z.W. Pan, C.S. Lee, S.T. Lee, Adv. Mater. **12**, 1343 (2000)
55. T. Hanrath, B.A. Korgel, Adv. Mater. **15**, 437 (2003)
56. X.M. Lu, T. Hanrath, K.P. Johnston, B.A. Korgel, Nano Lett. **3**, 93 (2003)
57. T. Heitsch, D.D. Fanfair, H.Y. Tuan, B.A. Korgel, J. Am. Chem. Soc. **130**, 5436 (2008)
58. R. Juhasz, N. Elfstrom, J. Linnros, Nano Lett. **5**, 275 (2005)
59. S.F. Hu, W.C. Weng, Y.M. Wan, Solid State Commun. **130**, 111 (2004)
60. C.M. Hsu, S.T. Connor, M.X. Tang, Y. Cui, Appl. Phys. Lett. **93**, 133109 (2008)
61. H.D. Tong, S. Chen, W.G. van der Wiel, E.T. Carlen, A. van den Berg, Nano Lett. **9**, 1015 (2009)
62. X. Li, P.W. Bohn, Appl. Phys. Lett. **77**, 2572 (2000)
63. K.Q. Peng, Y.J. Yan, S.P. Gao, J. Zhu, Adv. Mater. **14**, 1164 (2002)
64. K.Q. Peng, Y. Wu, H. Fang, X.Y. Zhong, Y. Xu, J. Zhu, Angew. Chem. Int. Ed. **44**, 2737 (2005)
65. K.Q. Peng, J.J. Hu, Y.J. Yan, Y. Wu, H. Fang, Y. Xu, S.T. Lee, J. Zhu, Adv. Funct. Mater. **16**, 387 (2006)
66. K.Q. Peng, M.L. Zhang, A.J. Lu, N.B. Wong, R.Q. Zhang, S.T. Lee, Appl. Phys. Lett. **90**, 163123 (2007)
67. S. Chattopadhyay, X.L. Li, P.W. Bohn, J. Appl. Phys. **91**, 6134 (2002)
68. C.Y. Chen, C.S. Wu, C.J. Chou, T.J. Yen, Adv. Mater. **20**, 3811 (2008)
69. K.Q. Peng, H. Fang, J.J. Hu, Y. Wu, J. Zhu, Y.J. Yan, S. Lee, Chem–Eur J **12**, 7942 (2006)
70. C. Chartier, S. Bastide, C. Levy-Clement, Electrochim. Acta **53**, 5509 (2008)
71. N. Singh, K.D. Buddharaju, S.K. Manhas, A. Agarwal, S.C. Rustagi, G.C. Lo, N. Balasubramanian, D.-L. Kwong, IEEE Trans. Electron Devices **55**, 3107 (2008)
72. T. Ohmi, in *Proceedings of 8th Workshop Ultra Clean Technology*, Tokyo, 1990, p 5

73. H. Morinaga, T. Futatsuki, T. Ohmi, E. Fuchita, M. Oda, C. Hayashi, J. Electrochem. Soc. **142**, 966 (1995)
74. D. Dimova Malinovska, M. Sendova Vassileva, N. Tzenov, M. Kamenova, Thin Solid Films **297**, 9 (1997)
75. Z.P. Huang, H. Fang, J. Zhu, Adv. Mater. **19**, 744 (2007)
76. W.K. Choi, T.H. Liew, M.K. Dawood, Nano Lett. **8**, 3799 (2008)
77. X. Wang, K.Q. Peng, X.L. Wu, S.T. Lee, Electrochem. Commun. **12**, 509 (2010)
78. S.M. Wong, H.Y. Yu, Y.L. Li, J.S. Li, F. Wang, M.F. Yang, N. Singh, P.G.Q. Lo, D.L. Kwong, IEEE Trans. Electron Devices **58**, 3224 (2011)
79. H. Fang, X. Li, S. Song, Y. Xu, J. Zhu, Nanotechnology **19**, 255703 (2008)
80. B.-R. Huang, Y.-K. Yang, T.-C. Lin, W.-L. Yang, Sol. Energy Mater. Sol. Cells **98**, 357 (2012)
81. X. Wang, K.L. Pey, C.H. Yip, E.A. Fitzgerald, D.A. Antoniadis, J. Appl. Phys. **108**, 124303 (2010)
82. M. Wright, A. Uddin, Sol. Energy Mater. Sol. Cells **107**, 87 (2012)
83. K.–.T. Park, H.–.J. Kim, M.–.J. Park, J.–.H. Jeong, J. Lee, D.–.G. Choi, J.–.H. Lee, J.–.H. Choi, Sci. Rep. **5**, 12093 (2015)
84. G. Jia, M. Steglich, I. Sill, F. Falk, Sol. Energy Mater. Sol. Cells **96**, 226 (2012)
85. T. Mishima, M. Taguchi, H. Sakata, E. Maruyama, Sol. Energy Mater. Sol. Cells **95**, 18 (2011)
86. B. Tian, X. Zheng, T.J. Kempa, Y. Fang, N. Yu, G. Yu, J. Huang, C.M. Lieber, Nature **449**, 885 (2007)
87. M.D. Kelzenberg, D.B. Turner-Evans, B.M. Kayes, M.A. Filler, M.C. Putnam, N.S. Lewis, H.A. Atwater, Nano Lett. **8**, 710 (2008)
88. Y. Qu, L. Liao, R. Cheng, Y. Wang, Y.-C. Lin, Y. Huang, X. Duan, Nano Lett. **10**, 1941 (2010)
89. C.H. Ruan, Y.J. Lin, J. Appl. Phys. **114**, 143710 (2013)
90. J.J. Zeng, Y.J. Lin, Appl. Phys. A Mater. Sci. Process **116**, 581 (2014)
91. X.Z. Zhang, C. Xie, J.S. Jie, X. Zhang, Y. Wu, W.J. Zhang, J. Mater. Chem. A **1**, 6593 (2013)
92. T.J. Kempa, B. Tian, D.R. Kim, J. Hu, X. Zheng, C.M. Lieber, Nano Lett. **8**, 3456 (2008)
93. I. Park, Z. Li, A.P. Pisano, R.S. Williams, Nano Lett. **7**, 3106 (2007)
94. M.D. Kelzenberg, D.B. Turner-Evans, M.C. Putnam, S.W. Boettcher, R.M. Briggs, J.Y. Baek, N.S. Lewis, H.A. Atwater, Energ. Environ. Sci. **4**, 866 (2011)
95. E. Maruyama, A. Terakawa, M. Taguchi, Y. Yoshimine, D. Ide, T. Baba, M. Shima, H. Sakata, M. Tanaka Sanyo's challenges to the development of high-efficiency HIT solar cells and the expansion of HIT business. Conference Record of the IEEE 4th World Conference on Photovoltaic Energy Conversion (WCPEC), Waikoloa, HI, May 2006
96. P. Doshi, G.E. Jellison, A. Rohatgi, Appl. Optics **36**, 7826 (1997)
97. Z. Yang, G. Cao, A. Shang, D.Y. Lei, C. Zhang, P. Gao, J. Ye, X. Li, Nanoscale Res. Lett. **11**, 233 (2016)
98. M.–.F. Ng, S.W. Tong, Nano Lett. **12**, 6133 (2012)
99. G. Dong, F. Liu, J. Liu, H. Zhang, M. Zhu, Nanoscale Res. Lett. **8**, 544 (2013)
100. D. Kanematsu, S. Yata, A. Terakawa, M. Tanaka, M. Konagai, Jpn. J. Appl. Phys. **54**, 08KA09 (2015)
101. S.M. Wong, H.Y. Yu, Y.L. Li, J.S. Li, F. Wang, M.F. Yang, N. Singh, P.G.Q. Lo, D.L. Kwong Boosting short circuit current with rationally designed periodic Si nanopillar surface texturing for thin film solar cell. Paper presented at IEEE International Electron Devices Meeting (IEDM), San Francisco, CA, 6–8 December 2010
102. R.A. Puglisi, C. Garozzo, C. Bongiorno, S. Di Franco, M. Italia, G. Mannino, S. Scalese, A.L. Magna, Sol. Energy Mater. Sol. Cells **132**, 118 (2015)
103. F. Hui, L. Xudong, S. Shuang, X. Ying, Z. Jing, Nanotechnology **19**, 255703 (2008)
104. D. Kumar, S.K. Srivastava, P.K. Singh, M. Husain, V. Kumar, Sol. Energy Mater. Sol. Cells **95**, 215 (2011)

105. J.-S. Hwang, M.-C. Kao, J.-M. Shiu, C.-N. Fan, S.-C. Ye, W.-S. Yu, H.-M. Lin, T.-Y. Lin, S. Chattopadhyay, L.-C. Chen, K.-H. Chen, J. Phys. Chem. C **115**, 21981 (2011)
106. S.-H. Baek, H.S. Jang, J.H. Kim, Curr. Appl. Phys. **11**, S30–S33 (2011)
107. M.C. Putnam, S.W. Boettcher, M.D. Kelzenberg, D.B. Turner-Evans, J.M. Spurgeon, E.L. Warren, R.M. Briggs, N.S. Lewis, H.A. Atwater, Energ. Environ. Sci. **3**, 1037 (2010)
108. E.C. Garnett, P. Yang, J. Am. Chem. Soc. **130**, 9224 (2008)
109. J. Yoo, S.A. Dayeh, W. Tang, S.T. Picraux, Appl. Phys. Lett. **102**, 093113 (2013)
110. G. Jia, B. Eisenhawer, J. Dellith, F. Falk, A. Thøgersen, A. Ulyashin, J. Phys. Chem. C **117**, 1091 (2013)
111. C.E. Kendrick, H.P. Yoon, Y.A. Yuwen, G.D. Barber, H. Shen, T.E. Mallouk, E.C. Dickey, T.S. Mayer, J.M. Redwing, Appl. Phys. Lett. **97**, 143108 (2010)
112. D.R. Kim, C.H. Lee, P.M. Rao, I.S. Cho, X. Zheng, Nano Lett. **11**, 2704 (2011)
113. O. Gunawan, S. Guha, Sol. Energy. Mat. Sol. **93**, 1388 (2009)
114. L. Yu, B.O. Donnell, M. Foldyna, P.R.i. Cabarrocas, Nanotechnology **23**, 194011 (2012)
115. D. Kohen, C. Morin, P. Faucherand, A. Brioude, S. Perraud, Phys. Status Solidi (a) **211**, 1143 (2014)
116. M.–.D. Ko, T. Rim, K. Kim, M. Meyyappan, C.–.K. Baek, Sci. Rep. **5**, 11646 (2015)
117. Y. Li, M. Li, R. Li, P. Fu, L. Chu, D. Song, Appl. Phys. Lett. **106**, 091908 (2015)
118. W. Jevasuwan, K.C. Pradel, T. Subramani, J. Chen, T. Takei, K. Nakajima, Y. Sugimoto, N. Fukata, Jpn. J. Appl. Phys. **56**, 04CP01 (2017)
119. M. Dutta, L. Thirugnanam, P.V. Trinh, N. Fukata, ACS Nano **9**, 6891 (2015)
120. F. Khan, S.–.H. Baek, J.H. Kim, Sol. Energy **137**, 122–128 (2016)
121. L. Tsakalakos, J. Balch, J. Fronheiser, B.A. Korevaar, O. Sulima, J. Rand, Appl. Phys. Lett. **91**, 233117 (2007)
122. B. Eisenhawer, I. Sill, F. Falk, Phys. Status Solidi (a) **210**, 695 (2013)
123. G. Jia, I. Höger, A. Gawlik, J. Dellith, L.R. Bailey, A. Ulyashin, F. Falk, Phys. Status Solidi (a) **210**, 728 (2013)
124. X. Xie, X. Zeng, P. Yang, H. Li, J. Li, X. Zhang, Q. Wang, Phys. Status Solidi (a) **210**, 341 (2013)
125. J. Cho, B. O'Donnell, L. Yu, K.–.H. Kim, I. Ngo, P.R. i Cabarrocas, Prog. Photovolt. **21**, 77 (2013)
126. W. Wang, X. Li, L. Wen, Y. Zhao, H. Duan, B. Zhou, T. Shi, X. Zeng, N. Li, Y. Wang, Nanoscale Res. Lett. **9**, 238 (2014)
127. S.B. Dkhila, R. Ebdellib, W. Dachraouia, H. Faltakhc, R. Bourguigac, J. Davenasb, Synth. Met. **192**, 74 (2014)
128. S.–.H. Tsai, H.–.C. Chang, H.–.H. Wang, S.–.Y. Chen, C.–.A. Lin, S.–.A. Chen, Y.–.L. Chueh, J.H. He, ACS Nano **5**, 9501 (2011)
129. Z. Liang, M. Su, H. Wang, Y. Gong, F. Xie, L. Gong, H. Meng, P. Liu, H. Chen, W. Xie, J. Chen, ACS Appl. Mater. Interfaces **7**, 5830 (2015)
130. X. Gong, Y. Jiang, M. Li, H. Liu, H. Ma, RSC Adv. **5**, 10310 (2015)
131. W. Lu, Q. Chen, B. Wang, L. Chen, Appl. Phys. Lett. **100**, 023112 (2012)
132. W.–.R. Wei, M.–.L. Tsai, S.–.T. Ho, S.–.H. Tai, C.–.R. Ho, S.–.H. Tsai, C.–.W. Liu, R.–.J. Chung, J.–.H. He, Nano Lett. **13**, 3658 (2013)
133. X. Shen, B. Sun, D. Liu, S.-T. Lee, J. Am. Chem. Soc. **133**, 19408 (2011)
134. L. He, D. Lai, H. Wang, C. Jiang, Rusli, Small **8**, 1664 (2012)
135. P.R. Pudasaini, F. Ruiz-Zepeda, M. Sharma, D. Elam, A. Ponce, A.A. Ayon, ACS Appl. Mater. Interfaces **5**, 9620 (2013)
136. H. Jeong, H. Song, Y. Pak, K. Kwon, K. Jo, H. Lee, G.Y. Jung, Adv. Mater. **26**, 3445 (2014)
137. J. Wang, H. Wang, A.B. Prakoso, A.S. Togonal, L. Hong, C. Jiang, Rusli, Nanoscale **7**, 4559 (2015)
138. L. He, C. Jiang, D.L. Rusli, H. Wang, Appl. Phys. Lett. **99**, 021104 (2011)
139. H. Karaagac, Phys. Status Solidi (a) **211**, 2503 (2014)

140. K. Golap, A. Sudip, A. Hare Ram, A. Rakesh, S. Tetsuo, S. Maheshwar, K. Wakita, U. Masayoshi, J. Phys. D Appl. Phys. **42**, 115104 (2009)
141. L. Wen, X. Liu, N. Yang, J. Zhai, C. Huang, Y. Li, L. Jiang, Appl. Phys. Lett. **97**, 253111 (2010)
142. S.-C. Shiu, J.-J. Chao, S.-C. Hung, C.-L. Yeh, C.-F. Lin, Chem. Mater. **22**, 3108 (2010)
143. S.A. Moiz, A. Muhammad Nahhas, H.–.D. Um, S.–.W. Jee, H.K. Cho, S.–.W. Kim, J.–.H. Lee, Nanotechnology **23**, 145401 (2012)
144. L. Hong, X. Wang, H. Zheng, L. He, H. Wang, H. Yu, Rusli, Appl. Phys. Lett. **104**, 053104 (2014)
145. J. He, P. Gao, M. Liao, X. Yang, Z. Ying, S. Zhou, J. Ye, Y. Cui, ACS Nano **9**, 6522 (2015)
146. P. Yu, C.–.Y. Tsai, J.–.K. Chang, C.–.C. Lai, P.–.H. Chen, Y.–.C. Lai, P.–.T. Tsai, M.–.C. Li, H.–.T. Pan, Y.–.Y. Huang, C.-I. Wu, Y.–.L. Chueh, S.–.W. Chen, C.–.H. Du, S.–.F. Horng, H.–.F. Meng, ACS Nano **7**, 10780 (2013)
147. J.P. Thomas, K.T. Leung, Adv. Funct. Mater. **24**, 4978 (2014)
148. S. Jeong, E.C. Garnett, S. Wang, Z. Yu, S. Fan, M.L. Brongersma, M.D. McGehee, Y. Cui, Nano Lett. **12**, 2971 (2012)
149. Y. Zhang, F. Zu, S.–.T. Lee, L. Liao, N. Zhao, B. Sun, Adv. Energy Mater. **4**, 1300923 (2014)
150. K. Liu, S. Qu, X. Zhang, F. Tan, Z. Wang, Nanoscale Res. Lett. **8**, 88 (2013)
151. X. Shena, Z. Xia, L. Chen, S. Lia, J. Zhao, Electrochim. Acta **222**, 1387–1392 (2016)
152. F. Zhang, D. Liu, Y. Zhang, H. Wei, T. Song, B. Sun, ACS Appl. Mater. Interfaces **5**, 4678 (2013)
153. H.–.J. Syu, S.–.C. Shiu, C.–.F. Lin, Sol. Energy Mater. Sol. Cells **98**, 267 (2012)
154. H.–.D. Um, D. Choi, A. Choi, J.H. Seo, K. Seo, ACS Nano **11**, 6218–6224 (2017)
155. X. Shen, B. Ma, L. Chen, J. Zhao, RSC Adv. **7**, 688–693 (2017)
156. K. Uma, T. Subramani, H.–.J. Syu, T.–.C. Lin, C.–.F. Lin, J. Appl. Phys. **117**, 105102 (2015)
157. M.-L. Tsai, W.–.R. Wei, L. Tang, H.–.C. Chang, S.–.H. Tai, P.–.K. Yang, S.P. Lau, L.–.J. Chen, J.–.H. He, ACS Nano **10**, 815–821 (2016)
158. J.M. Weisse, C.H. Lee, D.R. Kim, X. Zheng, Nano Lett. **12**, 3339 (2012)
159. M. Sharma, P.R. Pudasaini, F. Ruiz-Zepeda, D. Elam, A.A. Ayon, ACS Appl. Mater. Interfaces **6**, 4356 (2014)
160. J.Y. Kwon, D.H. Lee, M. Chitambar, S. Maldonado, A. Tuteja, A. Boukai, Nano Lett. **12**, 5143 (2012)
161. F. Wang, D. Kozawa, Y. Miyauchi, K. Hiraoka, S. Mouri, Y. Ohno, K. Matsuda, Nat. Commun. **6**, 6305 (2015)
162. Y. Song, X. Li, C. Mackin, X. Zhang, W. Fang, T. Palacios, H. Zhu, J. Kong, Nano Lett. **15**, 2104 (2015)
163. X. Li, Z. Lv, H. Zhu, Adv. Mater. **27**, 6549 (2015)
164. Q. Shu, J. Wei, K. Wang, H. Zhu, Z. Li, Y. Jia, X. Gui, N. Guo, X. Li, C. Ma, D. Wu, Nano Lett. **9**, 4338 (2009)
165. C. Xie, B. Nie, L. Zeng, F.–.X. Liang, M.–.Z. Wang, L. Luo, M. Feng, Y. Yu, C.-Y. Wu, Y. Wu, S.–.H. Yu, ACS Nano **8**, 4015 (2014)
166. X. Yan, C. Zhang, J. Wang, X. Zhang, X. Ren, Nanoscale Res. Lett. **12**, 1–6 (2017)
167. S. Lu, A. Madhukar, Nano Lett. **7**, 3443–3451 (2007)
168. H.M. Nguyen, O. Seitz, D. Aureau, A. Sra, N. Nijem, Y.N. Gartstein, Y.J. Chabal, A.V. Malko, Appl. Phys. Lett. **98**, 161904 (2011)
169. S. Chanyawadee, R.T. Harley, M. Henini, D.V. Talapin, P.G. Lagoudakis, Phys. Rev. Lett. **102**, 077402 (2009)
170. G. Allan, C. Delerue, Phys. Rev. B: Condens. Matter Mater. Phys. **75**, 195311 (2007)
171. S. Lu, Z. Lingley, T. Asano, D. Harris, T. Barwicz, S. Guha, A. Madhukar, Nano Lett. **9**, 4548 (2009)
172. H.M. Nguyen, O. Seitz, W. Peng, Y.N. Gartstein, Y.J. Chabal, A.V. Malko, ACS Nano **6**, 5574 (2012)

173. O. Seitz, L. Caillard, H.M. Nguyen, C. Chiles, Y.J. Chabal, A.V. Malko, Appl. Phys. Lett. **100**, 021902 (2012)
174. M. Achermann, M.A. Petruska, S. Kos, D.L. Smith, D.D. Koleske, V.I. Klimov, Nature **429**, 642 (2004)
175. S. Blumstengel, S. Sadofev, C. Xu, J. Puls, F. Henneberger, Phys. Rev. Lett. **97**, 237401 (2006)
176. G. Heliotis, G. Itskos, R. Murray, M.D. Dawson, I.M. Watson, D.D.C. Bradley, Adv. Mater. **18**, 334 (2006)
177. B. Güzeltürk, E. Mutlugün, X. Wang, K.L. Pey, H.V. Demir, Appl. Phys. Lett. **97**, 093111 (2010)
178. H.–.C. Chen, C.-C. Lin, H.-W. Han, Y.-L. Tsai, C.–.H. Chang, H.–.W. Wang, M.–.A. Tsai, H.-C. Kuo, P. Yu, Opt. Express **19**, A1141 (2011)
179. S. Avasthi, S. Lee, Y.-L. Loo, J.C. Sturm, Adv. Mater. **23**, 5762 (2011)
180. Z. Jia, X. Yueqing, W. Qi, C. Yi Amorphous silicon core-shell nanowire Schottky solar cells. Paper published in 35th IEEE Photovoltaic Specialists Conference (PVSC), Honolulu, HI, 20–25 June 2010
181. C. Xie, P. Lv, B. Nie, J. Jie, X. Zhang, Z. Wang, P. Jiang, Z. Hu, L. Luo, Z. Zhu, L. Wang, C. Wu, Appl. Phys. Lett. **99**, 133113 (2011)
182. G.F. Fan, H.W. Zhu, K.L. Wang, J.Q. Wei, X.M. Li, Q.K. Shu, N. Guo, D.H. Wu, ACS Appl. Mater. Interfaces **3**, 721 (2011)
183. J.R. Maiolo, B.M. Kayes, M.A. Filler, M.C. Putnam, M.D. Kelzenberg, H.A. Atwater, N.S. Lewis, J. Am. Chem. Soc. **129**, 12346 (2007)
184. J.M. Spurgeon, S.W. Boettcher, M.D. Kelzenberg, B.S. Brunschwig, H.A. Atwater, N.S. Lewis, Adv. Mater. **22**, 3277 (2010)
185. S.W. Boettcher, J.M. Spurgeon, M.C. Putnam, E.L. Warren, D.B. Turner-Evans, M.D. Kelzenberg, J.R. Maiolo, H.A. Atwater, N.S. Lewis, Science **327**, 185 (2010)
186. K.Q. Peng, X. Wang, S.-T. Lee, Appl. Phys. Lett. **92**, 163103 (2008)
187. J.M. Spurgeon, R.M. Briggs, N.S. Lewis, H.A. Atwater, Nat. Mater. **9**, 239 (2010)
188. G. Yuan, H. Zhao, X. Liu, Z.S. Hasanali, Y. Zou, A. Levine, D. Wang, Angew. Chem. **121**, 9860 (2009)
189. E.L. Warren, S.W. Boettcher, M.G. Walter, H.A. Atwater, N.S. Lewis, J. Phys. Chem. C **115**, 594 (2010)
190. S.W. Boettcher, E.L. Warren, M.C. Putnam, E.A. Santori, D. Turner-Evans, M.D. Kelzenberg, M.G. Walter, J.R. McKone, B.S. Brunschwig, H.A. Atwater, N.S. Lewis, J. Am. Chem. Soc. **133**, 1216 (2011)
191. K.-Q. Peng, X. Wang, X.-L. Wu, S.-T. Lee, Nano Lett. **9**, 3704 (2009)
192. X. Wang, K.-Q. Peng, X.-J. Pan, X. Chen, Y. Yang, L. Li, X.-M. Meng, W.-J. Zhang, S.-T. Lee, Angew. Chem. Int. Ed. **50**, 9861 (2011)
193. X. Shen, B. Sun, F. Yan, J. Zhao, F. Zhang, S. Wang, X. Zhu, S. Lee, ACS Nano **4**, 5869 (2010)
194. P. Yang, X. Zeng, X. Xie, X. Zhang, H. Li, Z. Wang, RSC Adv. **3**, 24971 (2013)
195. K.T. Li, X.Q. Wang, P.F. Lu, J.N. Ding, N.Y. Yuan, Nanoscale Res. Lett. **8**, 396 (2013)
196. Z. Zuo, K. Zhu, G. Cui, W. Huang, J. Qu, Y. Shi, Y. Liu, G. Ji, Sol. Energy Mater. Sol. Cells **125**, 248 (2014)
197. X. Zhang, X. Zeng, S. Zhang, F. Liu, Mater. Sci. Semicond. Process. **41**, 457–461 (2016)
198. H. Mohamed, O.H. Mohamed Farhat, F.F. Areed Nihal, A.O. Salah Sabry, Opt. Eng. **53**, 117105 (2014)
199. Y. Wu, Z. Xia, Z. Liang, J. Zhou, H. Jiao, H. Cao, X. Qin, Opt. Express **22**, A1292 (2014)
200. H. Savin, P. Repo, G. von Gastrow, P. Ortega, E. Calle, M. Garín, R. Alcubilla, Nat. Nanotechnol. **10**, 624–629 (2015)

# Index


**A**
Absorption coefficient, 2
Absorption loss, 285, 301
Acid etching, 208, 210, 211
Active layer
  light absorption, 54
  SC, 54
  solar cell based on gradient particulate structure, 101–102
Active layer thickness, 121
Alkaline etching, 208–211
Al–Si/Pt–Si Schottky junctions, 310
Aluminum (Al) gratings, 119
Amorphous and microcrystalline Si PV cells
  a-SiGe alloy, 164
  contact electrodes, 163
  microcrystalline (μc-Si) solar cells, 163
  PECVD process, 162, 163
  spectral responses, 164
  vacuum-based deposition methods, 162
Amourphous Si (a-Si), 162, 163, 171, 251–261
Angular functions, 61
Annealing, 303, 315
Antireflection, 284, 300–302, 307, 316, 317, 323
Antireflection coating (ARC) layer
  active layer thickness, 121
  grating period and duty cycle, 121–122
  grating thickness, 120
  grating width, 113
  refractive index, 112–113
  thickness, 113
Au-Si system, 278
Axial junction, 274–275

**B**
Back reflector (BR), 110, 111, 114, 119–121, 125, 126, 129, 132–134
Balance of system (BOS), 270
Band defects, 258–259
Bandgap mechanisms, 252
Beer's law, 283
Biodegradable thin-film Si PV cells
  biological compatibility, 172, 173
  biomedical surgery, diagnosis and therapy, 171
  dissolvable energy devices, 171
  materials degradation, 171, 172
Black silicon, 236–237
Bottom-up and top-down approaches, 277
Bottom-up approach, 303
Bottom-up fabrication, 277
Branz's model, 256

**C**
Charge carriers, 300, 302, 303, 307, 310, 315, 319, 322
Chemical etching method, 207, 208
Chemical mist deposition (CMD), 145, 148–157
Chemical vapor deposition (CVD), 303
$CO_2$ reduction, 230, 261
Complex refractive index, 54–56, 62, 66, 74, 80, 92
Conduction band (CB), 262
Conductive DBR thin films
  current–voltage characteristics, 18–21
  device performance, 15



S. Ikhmayies (ed.), *Advances in Silicon Solar Cells*,
https://doi.org/10.1007/978-3-319-69703-1

Conductive DBR thin films (*cont.*)
on glass substrates, 12–14
metal substrates, 14, 15
modified heterojunction silicon wafer solar cell, 16
peak reflectance wavelength, 9, 11
photo-generation current density, 18, 19
rear interface reflectance, 16
reflectance, absorptance and transmittance, 16–18
wavelength-dependent complex refractive indices, 9
Conductive thin films, 9, 10
Contact formation, 282
Crystalline silicon (c-Si), 54–60, 63, 69, 70, 80–90, 92–94, 96, 98, 99, 103, 230
and amorphous silicon, 9
based photovoltaics, 1
$SiO_2$/Si films, 21
Cuprous oxide, 245

D
Dangling bonds, 181, 182, 187, 255
hydrogen atoms, 39
hydrogen-passivated silicon surface, 29
localization, 43
quantum confinement effect, 31
$Si_{28}H_{32}$ quantum dot, 38
Decorated silicon
cuprous oxide, 245
gold nanofilms, 242–244
graphene, 246–247
molybdenum sulfide, 244
platinum nanoparticles, 238–242
Deep reactive ion etching technique (DRIE), 308, 310, 316, 317
Density of states (DOS), 38–40
Device-scale nanowire arrays, 283
Dielectric diffraction grating, *see* Distributed Bragg reflector (DBR)
Diffusion length, 54
Dimethyl sulfoxide (DMSO), 222
Distributed Bragg reflector (DBR), 112–113
ARC (*see* Antireflection coating (ARC) layer)
calculated peak reflectance, 5, 6
calculated reflectance and absorbance, 7
center wavelength and number of pairs, 114
conductive schemes, 8
degree of absorption, 110
description, 110
Euler's formula, 7
extinction coefficients, 5
FDTD method, 112
gap–midgap ratio, 4, 5
grating period and thickness and duty cycle, 114, 115
light absorption, 111
light-trapping mechanism, 110
metal nanoparticles/nanogratings, 110
metal reflectors and photonic crystals, 111
one-dimensional photonic crystal, 4
optical reflectance, target wavelength, 4
performance analysis, 116, 122–125
periodic stacks, 3
photonic bandgaps, 3
red and infrared wavelength regions, 111
refractive index, 6
structure, 112
surface plasmon resonance, 110
surface reflections, 111
Taylor series expansion, 6
Drude model, 141
Duty cycle, 114, 115, 121, 122

E
Electrical properties, 288–291
Electroless etching, 301, 303, 305, 315
Electron spin resonance (ESR) signal, 181
Electron-beam lithography, 279
External quantum efficiency (EQE), 143, 164, 283
Extinction coefficient, 5–7, 10, 14, 18

F
Fabry-Perot resonance, 285
Finite-difference time-domain (FDTD) method, 120, 125
Flexible and stretchable cells, 167
Floating bond model, 256
Fresnel formulae, 62
Fujishima-Honda experiment, 230

G
GaAs nanowire, 274, 276
Gas-solid interface, 278
Generalized gradient approximation (GGA), 36
Geometrical junction, 275
Gold nanofilms, 242–244
Graphene, 246–247

Grating, *see* Dielectric diffraction grating
Grating period, 114, 115, 121, 122
Grating thickness, 114, 115, 120

**H**

$H_2$ generation, 240, 243
Heterojunction, 212, 218–222, 224, 225, 308, 313, 317, 319
Heterojunction semiconductor, 251
Heterojunction silicon wafer solar cell, 9, 15–18, 20, 21
Heterojunction solar cells, 140
HF vapor etching, 180, 181
High efficiency, 206, 212, 224, 225
Highest occupied molecular orbitals (HOMO), 29
Homogeneous plane-parallel silicon plate, 54, 71, 102
Homo-junction/hetero-junction, 271
Hybrid silicon solar cell, *see* Organic–inorganic hybrid silicon solar cell
Hybrid solar cell (HSC) fabrication
  bandgap metal-oxide NCs, 188
  fullerene derivatives, 188
  light trapping, exciton dissociation and carrier transportation, 188
  nonthermal plasma, 188
  particle surface elements, 191, 192
  polymer semiconductors, 192–194
  Si NC/PTB7 mass ratio, 194–196
  Si NC-based, 188–191
Hydrogen collision model, 256
Hydrogen gas, 259

**I**

Impurity-related model, 254
Indium tin oxide (ITO), 138
Integral absorption coefficient, 70, 71, 87, 89–91, 93–101, 103

**J**

Junction formation, 280–281
J-V curves, 191, 193, 196–198

**L**

Lambertian light trapping scheme, 285
Legendre functions, 62
Light absorption, 129, 130, 283
  analysis of, 72
  c-Si, 54
  efficiency of, 54, 87
  enhancement of, 103
  monolayers, 73
  number of monolayers $N_{ml}$, 92
  SC performance, 54
Light trapping, 213–217, 225, 231, 237, 300, 315, 319, 322, 323conductive materials, 8
  conductive thin films, 9, 10
  crystalline silicon-based photovoltaics, 1 (*see also* Distributed Bragg reflector (DBR))
  doped semiconductor films, 8
  FreeSnell, commercial software, 8
  front and rear surfaces, solar cell, 2
  long- and short-wavelength photons, 2
  periodic gratings, 3
  photo-generation, 2, 3
  process parameters, 9
  single-layer/double-layer antireflective coating, 2 (*see also* Thin-film silicon solar cells)
  UV-VIS-IR spectrophotometer, 8
  wavelength-dependent absorption coefficient, 2
  wavelength-dependent gap–midgap ratio, 8
  *See also* Conductive DBR thin films
Localized surface plasmon resonances (LSPRs), 118
Lowest unoccupied molecular orbitals (LUMO), 29

**M**

MCEE technique, 307, 315
MeOH/EG cosolvent
  carrier transport parameters, 142
  conductivity, 142
  cosolvent addition, 142
  device performance, 144
  EG concentration, 142
  electron recombination suppression, 144
  EQE spectra, 143
  *J*-V curves, 144
  mean squared error, 141
  MeOH/EG concentration ratio, 142
  performance parameters, 145
  photovoltaic performance, 145
  SE analysis, 143
  TL oscillator and Drude model, 141
Metal nanograting
  light absorption and field analyses, 133, 134

Metal nanograting (*cont.*)
RCWA, 132
silver nanoparticles, 132
Metal nanostructures, 110, 117, 118, 125, 133
Metal-assisted chemical etching (MACE) process, 208, 213–216
Metal-catalyzed electroless etching (MCEE), 231, 306, 307
Metallic cathode, 230
Micro- and nanoholes, 233–235
Micro-crystalline (μc-Si) solar cells, 162–164
Microvoid-induced dangling bonds, 256–257
Mie-type resonances, 285
Molybdenum sulfide, 244
Monocrystalline Si (c-Si), 171
Monocrystalline silicon surface, 207
Mott–Schottky measurements, 235
Multi-junction nanowire solar cells, 276
Multijunction solar cells, 259, 276
Multiple exciton generation (MEG) process, 177

**N**
Nanobelt (NB) arrays, 236
Nanobelts, 236
Nanoelectronics, 292
Nanoscale Si dots
band gaps, 34
based LED, 35
Brillouin zone, 34
excitonic luminescence, silicon nanocrystals, 34
indirect semiconductors, 33
QCSE, 34
recombination pavements, 33
visible photoluminescence, 34
Nanosphere lithography, 279
Nanowire fabrication, 277–280
Nanowires (NWs), 231
Near-infrared photons, 22
Non-radiative energy transfer (NRET), 320, 321
Nonthermal plasma, 178, 188, 198
NW geometry
absorption and exciton formation, 301, 302
carrier collection and cost, 303
charge separation and transportation, 302, 303
fundamental physical steps, 300
NW synthesis techniques, 304–307
bottom-up fabrication
Al-catalyzed SiNW arrays, 305
CVD technique, 304, 305
VLS growth mechanism, 305
fabrication of SiNWs, 303
junction formation, 307–310
top-ddown fabrication
MCEE, 306, 307
VLS mechanism, 303

**O**
Optical absorption spectra, 29, 31, 44, 45, 48
Optical anisotropy
photovoltaic performance, 140
SE, 139
Optical reflectance, 4, 7, 10, 12–14, 16, 18, 22
Organic–inorganic hybrid silicon solar cell, 206, 218–225
fabrication process, 207
description, 225
electrodes, 223
heterojunction, 219–222
substrate cleaning, 218, 219
heterojunction, 206
nontoxic Earth-abundant resources, 206
passivation technologies (*see* Passivation technologies)
PEDOT
PSS, 206
photon-electric conversion devices, 206
silicon nanocone, 216, 217
silicon nanowire, 213–216
silicon pyramid, 211, 212
ultrathin silicon wafer, 207–210, 212
Oxide-assisted growth (OAG), 303

**P**
Parasitic absorption
optical reflectance, 7
targeted wavelength regions, 5
Particle surface elements, 191, 192
Passivation methods, 181
Passivation technologies
ALD, 224
$HNO_3$-oxidized device, 224
organic–inorganic hybrid silicon solar cells, 224
silicon/PEDOT\:PSS devices, 224
silicon/PEDOT\:PSS interface, 224
Percus–Yevick approximation, 77, 78
Performance analysis
absorption and quantum efficiency, 116
absorption peaks, 122

absorption spectra, 131
designed silicon solar cells, 124
electric field profile, 124
front and backside metallic gratings, 132
grating (Si/$SiO_2$) bilayers, 116
J–V curves, 116
light absorption behavior, 122, 123
Photo-active electrodes, 230
Photoconversion efficiency (PCE), 177
Photocurrent density, 232, 246–249, 259
Photocurrent density–potential curve, 249
Photoelectrochemical cells (PECs), 230
Photoelectrodes, 259
Photo-generation current density, 18, 19
Photolithographic patterning, 282
Photoluminescence (PL), 184
measurements, 31–33
porous silicon, 27
at room temperature, 27
in silicon nanostructures, 28
Photovoltaic (PV) electricity, 269
Photovoltaic (PV) technologies
cell performance, 162
flexible, stretchable and/or biodegradable formats, 162
microelectronic processing methods., 162
Si-based semiconductor devices, 161 (*see also* Single-crystalline Si PV cells)
wearable/implantable biomedical electronic and photonic systems, 162
*See also* Amorphous and microcrystalline Si PV cells
Photovoltaic devices, 206, 212, 215, 216, 304, 310, 311, 321
Planar p–n junction solar cell, 302, 303
Planar solar cells, 292
Plasma-enhanced chemical vapor deposition (PECVD), 138, 162
Plasmonic modes, 118, 125, 126, 135
Platinum nanoparticles, 238–242
Poly (3,4-ethylenedioxythiophene), 139–140
poly(styrene sulfonate) (PEDOT), 137–157
CMD, 148–157
c-Si semiconductor technology, 138
depth profile, 145–148
heterojunction solar cells, 140
hybrid optoelectronic devices, 138
ITO, 138
MeOH/EG cosolvent, 140–145
molecular structure, 138, 139
n-Si solar cells, 137
optical anisotropy (*see* Optical anisotropy)
TCO and PE-CVD, 137
transparent hole-conducting polymer, 138
Polycrystalline Si (poly-Si), 171–173
Polydimethlysiloxane (PDMS), 302, 319, 320
Polymer, 302, 308, 317, 318, 321
Polymer semiconductors, 192–194
Porous nanostructures, 231

**Q**
Quantum confinement
and surface passivation, 31
effects, 33
optical spectra profiles, 44
PLs, 27
Si quantum dots, 28
terminated and non-terminated Si quantum dots, 29
Quantum-confined Stark effect (QCSE), 34
Quasicrystalline approximation (QCA), 55, 72, 77, 81, 83, 86, 91, 94
formalism of, 73
small- and wavelength-sized particles, 55, 102
TMM, 72

**R**
Radial distribution function (RDF), 75
Radial junction, 272–274
Radial junction structure, 273
Radial p–i–n Si MW solar cells, 316
Radial p–i–n SiNW photovoltaic device fabrication, 312
Radial p–n junction, 300, 302, 303, 308, 309, 313–316, 321
Reactive ion etching (RIE), 279, 303
Recombination centers, 237
Reflection
amplitude coherent transmission, 74
amplitude transmission, 62
coefficients, 65, 67, 78–80, 83, 91–93
coherent transmission, 60, 66, 73
energy total transmission, 65, 66
incoherent transmission, 61
QCA, 72
Refractive index, 112
ARC, 112–113

Refractive index (*cont.*)
DBR unit blocks, 4
defined, 4
extinction coefficient, 6, 10
target peak reflectance wavelength, 3
Rigorous coupled-wave analysis (RCWA) method, 125, 132

S
Second World War, 230
Semiconductor nanowire, 270
Semiconductor nanowire geometry, 285
Shockley-Queisser efficiency, 291
Shockley-Queisser limit, 276, 292
Shockley-Read-Hall (SRH) recombination, 290
Si manufacturing technology, 271
Si nanowire solar cells, 300–307, 310–314
challenges, 323
nontoxic commercial material, 300
NRET, 320, 321
NW geometry (*see* NW geometry)
NW synthesis techniques (*see* NW synthesis techniques)
PEC, 322
photovoltaics, 300
polycrystalline and amorphous, 300
polycrystalline p-Si matrix, 323
quality and quantity of material, 322
renewable energy sources, 299
silicon NW hybrid solar cells, 317–319
silicon NW Schottky junction, 321
single NW photovoltaics
axial junction, 310–312
radial junction, 312–314
SiNW and nanopillar arrays, 323
SiNW array p–n and p–i–n junction, 314–316
SiNW radial p–n junction, 300
Si NC/PTB7 mass ratio, 194–196
Si NC-based HSCs, 188–191
Si photocathodes, 231, 240, 261
Si wafer, 231, 245, 247
Si/SiMH heterojunction, 237
Silicon (Si) quantum dots, 31–33
DFT simulations, 29
direct-gap semiconductor nanostructures, 27
doping/defects, 31
electronic structure, 28, 29
HOMO and LUMO, 29
multiple excitons, 28
nanocrystal size distributions, 30
nanomaterials, 26
nano-miniaturization, 27 (*see also* Nanoscale Si dots)
optical active devices, 27
optoelectronic devices, 26
oxide-passivated nanocrystals, 27
PL (*see* Photoluminescence (PL))
plane-wave pseudopotential framework, 30
quantum confinement effect, 28, 31
radiative efficiency, Si–PL, 30
semiempirical modifications, 30
sizeable scale synthesis, 26
solid-state semiconducting nanostructures, 26
structure, 35
theoretical and computational modeling, 28 (*see also* Unpassivated and passivated Si quantum dots)
Silicon carbide (SiC), 248–251
anodic photocurrent, 251
photoanode, 251
photocatalytic behavior, 250
structures, 249
Silicon membrane, 207, 225
Silicon nanocone, 216, 217
Silicon nanocrystals (Si NCs) synthesis, 179, 180
crystallinity, 196, 197
ESR signal, 181
HF vapor, 180
HF vapor etching, 181
inorganic semiconductor films, 178
material optical properties, 177
MEG process, 177
optical and electrical properties, 178
oxidization, 181
quantum confinement effect, 177
Si–OH- and Si–$H_x$-related peaks, 182
Si–O–Si stretching mode, 180
thermal annealing, 182
Silicon nanoparticles (Si NPs) synthesis
aqueous solution, 178
crystallinity of, 182–184
electrical properties, 185–188
experimental setup, 179
optical properties, 184, 186
plasma operating conditions, 178
size of, 182, 183
VHF nonthermal plasma, 178
Silicon nanowire solar cells
advantages, 283
axial junction, 274, 275
contact formation, 282
device structure, 271, 272

electronic and photonic devices, 270
geometrical junction, 275
geometry, 271
hetero-junction, 276
homo-junction, 276
J-V curves, 275
optical and electrical behavior, 273
optical properties, 283–288
optoelectronic device, 270
photovoltaic technologies, 269
power conversion efficiency, 271
PV industry, 269
radial junction, 272–274
RIE, 279
Silicon NW Schottky junction, 321
Silicon particles
individual submicron spherical, 54
monodisperse spherical, 80
monolayer of, 63, 64, 68
plane-parallel monolayers of, 54
scattering and absorption, 55
single monolayer of, 69
single spherical crystalline, 55, 102
submicron crystalline, 72, 90
Silicon pyramid, 211–213
Silicon wafer, 207–211, 216, 218–221, 224, 225
Single scattering approximation (SSA)
homogeneous plane-parallel plate, 62
monolayer of silicon particles, 63, 64
monolayers, 55
multilayer and equivalent plane-parallel plate, 65–71
QCA, 82
reflection coefficients, 60–62
spectral absorption and transmission, 60–62
Single-crystalline Si PV cells
controlled spalling process, 166
CVD, 166
material and mechanical design methods, 167
mechanical, 168–170
on-grid and off-grid power plant systems, 164
ribbon PV cells, 166
SOI substrates, 166
SOI wafer substrates, 164
Staebler-Wronski effect, 164
substrate equirements, 167, 168
Si–Si bond, 253, 255
Solar cell structures
light absorption, 129, 130
TE and TM, 127, 129
Solar-to-hydrogen (STH), 231, 232
Solid-liquid interface, 278, 279
Spectral absorption coefficient, 68–70, 81, 85, 87, 88, 93, 94, 96, 100, 103
Staebler-Wronski effect, 164, 253–257
Substrate junction, 275–276
Surface passivation, 290, 291, 313, 314, 316, 317, 321
devices dependences, 200
J–V curves, solar cells fabrication, 197, 198 (*see also* Passivation methods)
Si NC/PTB7 hybrid solar cell performance, 199
TFTs parameters, 200
Surface plasmon polaritons (SPPs), 118
Surface recombination, 312, 314, 315
Surface saturation, 50
Surface states
energy gap, 38
HOMO and LUMO, 43
localization, 38
Surface wettability, 218, 219, 224

**T**

Tauc-Lorentz (TL) oscillator, 141
Thick c-Si wafers, 166
Thin film transistors (TFTs), 185–188
Thin-film silicon solar cells, 110
advantages, 117
blazed dual-grating periodicity structures, 125
DBR (*see* Distributed Bragg reflector (DBR))
geometrical parameters, grating, 126
grating parameters, 130–131
hybrid light-trapping structure, 126
light absorption enhancement, 117
metal nanostructures, 117
metallic and dielectric nanostructures, 118
nonconventional energy sources, 109
one-dimensional photonic crystal, 110
PBC and PML, 126
performance analysis, 131–132
photonic and plasmonic mode gratings, 125
plasmon coupling, 117
PMS, 118
RCWA and FDTD, 125
reduced cell thickness, 117
silicon technology, 109
SiOx antireflection coating layer, 118 (*see also* Solar cell structures)
SPPs and LSPRs, 118

Thin-film silicon solar cells (*cont.*)
  TM polarization, 118, 125
  ultrathin, 119
  ultrathin aluminum (Al) gratings, 119
  ultrathin design parameters, 119
Ti/Au mesh film, 216
Top-down techniques, 279
Transfer matrix method (TMM), 55, 65, 71, 91
Transparent conductive oxide (TCO), 8, 137, 138
Transverse electric (TE) polarization, 127, 129
Transverse magnetic (TM) polarization, 118, 125, 127, 129

**U**
Ultrathin silicon wafer, 207–210, 212
Unpassivated and passivated Si quantum dots
  bonding character, 41–44
  electronic properties, 38–41
  GGA, 36
  Hellmann–Feynman force, 36
  optical properties, 44–49
  pseudopotential ab initio simulations, 36
  size effect, 36–38
Urbach tail, 257

**V**
Vapor–liquid–solid (VLS) method, 213, 233, 239, 271, 277, 302–305, 310, 314, 317, 323
Vapor–solid–solid (VSS), 279, 304, 305
Vertical SiNW array electronic devices, 320
Very high frequency (VHF) nonthermal plasma, 178

**W**
Water splitting, 259–261
Wavelength-dependent absorption, 272
Wet chemical etching, 280

**Z**
Zonyl FSH addition, 220

The manufacturer's authorised representative in the EU is Springer Nature Customer Service Centre GmbH, Europaplatz 3, 69115 Heidelberg, Germany. If you have any concerns regarding our products, please contact ProductSafety@springernature.com

Printed and bound by CPI Group (UK) Ltd, Croydon, CR0 4YY
30/09/2026
02225229-0013